# AAPS Introductions in the Pharmaceutical Sciences

## Volume 5

**Series Editor**

Claudio Salomon, National University of Rosario
Rosario, Argentina

**Founding Editor**

Robin Zavod, Chicago College of Pharmacy
Midwestern University
Downers Grove, IL, USA

The *AAPS Introductions in the Pharmaceutical Sciences* book series is designed to support pharmaceutical scientists at the point of knowledge transition. Springer and the American Association of Pharmaceutical Scientists (AAPS) have partnered again to produce a second series that juxtaposes the *AAPS Advances in the Pharmaceutical Sciences* series. Whether shifting between positions, business models, research project objectives, or at a crossroad in professional development, scientists need to retool to meet the needs of the new scientific challenges ahead of them. These educational pivot points require the learner to develop new vocabulary in order to effectively communicate across disciplines, appreciate historical evolution within the knowledge area with the aim of appreciating the current limitations and potential for growth, learn new skills and evaluation metrics so that project planning and subsequent evolution are evidence-based, as well as to simply "dust the rust off" content learned in previous educational or employment settings, or utilized during former scientific explorations. The *Introductions* book series will meet these needs and serve as a quick and easy-to-digest resource for contemporary science.

Ioannis Nikolakakis • Ioannis Partheniadis
Editors

# Amorphous Solid Dispersions

## A Comprehensive Guide to Theory, Manufacturing, Processing, Regulatory Status and Clinical Translation

Springer

aaps®

*Editors*
Ioannis Nikolakakis
Laboratory of Pharmaceutical Technology, School of Pharmacy, Faculty of Health Sciences
Aristotle University of Thessaloniki
Thessaloniki, Greece

Ioannis Partheniadis
Department of Pharmacy, Faculty of Health and Medical Sciences
University of Copenhagen
Copenhagen, Denmark

ISSN 2522-834X ISSN 2522-8358 (electronic)
AAPS Introductions in the Pharmaceutical Sciences
ISBN 978-3-032-24882-4 ISBN 978-3-032-24883-1 (eBook)
https://doi.org/10.1007/978-3-032-24883-1



This Springer imprint is published by the registered company Springer Nature Switzerland AG
The registered company address is: Gewerbestrasse 11, 6330 Cham, Switzerland

*To Sofia and Manolis, for your love,*
*understanding and support.*
*To my family, friends, and all those who*
*inspired this journey.*

# Preface

Many newly discovered molecules, as well as molecules in the development pipelines of pharmaceutical companies, comprise drugs intended for the treatment of severe, potentially life-threatening chronic diseases such as cancer, hepatitis C, and cystic fibrosis. Unfortunately, although of high pharmacological value, these molecules are of high molecular weight and high lipophilicity, making them poorly water-soluble and less permeable through the gastric epithelium. Since dissolution in the gastric fluids is a prerequisite for absorption into systemic circulation, poor solubility poses a main problem for bioavailability. To tackle this problem, the academic community and industries have intensified their research activities and have come up with enabling technologies that combine established engineering methods with the use of suitable generally safe auxiliaries.

These strategies aim to convert crystalline drugs into stable amorphous forms, thereby increasing the drug's apparent solubility, in order to maintain supersaturation for a long enough time and improve absorption. Most of these engineering methods, such as spray drying and hot-melt extrusion, are continuous processes, thus suitable for modern high-production-yield manufacturing of drug products. Complementing these methods with downstream processing has led to the formulation of the amorphous solid polymeric dispersions (ASDs) and co-amorphous systems (CAMS) into standard oral dosage forms, easy to handle and administer, with effective bioavailability.

Industrial and academic research have made great advances in this field, and many amorphous drug products have reached the US and European markets. Furthermore, there are numerous literature resources now available to researchers to teach and provide guidance for ASD and CAMS development of poorly water-soluble drugs.

The book is organized into five parts:

I. Introduction and Fundamentals—Chaps. 1, 2, and 3
II. Manufacturing and Engineering Technologies—Chaps. 4, 5, and 6
III. Emerging and Alternative ASD Platforms—Chaps. 7 and 8
IV. Downstream Processing and Patient-Centric Design—Chaps. 9 and 10
V. Biopharmaceutical Evaluation and Translation—Chaps. 11, 12, and 13

The book begins with an introductory chapter on Amorphous Solid Dispersions. A structural classification of the ASDs is presented first, based on the physical state of the components and the number of phases present, followed by polymer stabilization mechanisms. It continues with thermodynamic and kinetic considerations that govern the physical stability and performance of ASDs. Emphasis is given to glass transition and secondary relaxation temperatures as global and early stability indices. The importance of the interplay between the drug, the polymer, and ASD processing parameters, such as the temperature in thermal methods and the choice of solvent in solvent evaporation processes, is stressed. The introductory chapter ends with a section of the historical evolution and current market landscape of ASD-based products.

Chapter 2 delves into nucleation and crystallization processes and relates relaxation to crystallization kinetics. Local molecular mobility plays a key role in physical stabilization. Subsequently, attention has been directed to the studies investigating the stability of drug-polymer binary mixtures. Stabilization using mesoporous materials is subsequently discussed, as its mechanism is completely different from that of polymers. This understanding could contribute to the improvement of ASD technology and may enable the design of innovative ASDs with high drug loading.

Chapter 3 provides a structured and mechanistic perspective on rational polymer selection for the successful development of ASDs, with a particular focus on antifungal drugs as model compounds. This chapter particularly emphasizes that polymers are not inert excipients but active design elements whose informed selection is pivotal for translating ASD technologies into robust and clinically reliable drug delivery systems.

Chapter 4 deals with the processing of ASDs using hot-melt extrusion (HME). This is becoming a popular ASD processing method. The combination of thermal energy with mechanical shearing has been established as one of the more effective continuous manufacturing methods for the preparation of ASDs. The mechanisms of amorphization and the benefits of the method are described, as well as the risks when thermolabile drugs are processed and how to predict safe processing.

Chapter 5 focuses on solvent evaporation approaches, particularly spray drying and freeze drying, as key techniques in the preparation of ASDs. Recent developments in polymer technologies, nozzle design, and process analytical tools have greatly increased the robustness and application of solvent-based ASD methods, despite issues with stability, residual solvent removal, and large-scale manufacture. The challenges of using solvent evaporation methods are discussed, and any new developments under consideration will be elucidated to conclude the chapter.

Chapter 6 deals with mechanochemical approaches for the preparation of amorphous solid dispersions. Mechanochemistry, i.e., the induction of physical or chemical reactions by mechanical processing, has emerged as a versatile, environment-friendly route for ASD preparation, offering a solvent-free alternative to conventional methods. By supplying mechanical energy through milling, extrusion, or acoustic mixing, crystalline materials are transformed into disordered states through defect accumulation and lattice disruption. The chapter reviews the theoretical principles of mechanochemistry and tribochemistry, describing how the mode of stress application, energy input, and reactor environment influence amorphization. Special attention is given to drug–polymer systems and the role of intermolecular interactions in stabilizing mechanically generated amorphous phases. Key process parameters, such as milling intensity, temperature control, and humidity, are discussed in relation to product stability and reproducibility. Furthermore, emerging technologies like reactive extrusion and resonant acoustic mixing (RAM) are evaluated as scalable tools for green and sustainable pharmaceutical manufacturing.

Chapter 7 presents in detail various aspects of co-amorphous drug delivery systems (CAMS) and evolving perspectives. The information in this chapter takes CAMS from early fundamental research work to recent advancements (2021–2025). The authors propose that recently developed CAMS formulated into solid dosage forms have the potential to become a new dosage form that overcomes dissolution problems of poorly soluble drugs. Their expectation is strongly based on three critical quality attributes that the CAMS products satisfy: (i) excellent stability, (ii) great dissolution improvement, (iii) downstream processability and, in particular, compactibility and tableting feasibility.

Chapter 8 presents formulation strategies for developing high-drug-loading amorphous solid dispersions, including the use of a combination of polymers, coprecipitation methods, chemical and thermal crosslinking, the spray drying-granulation hybrid approach, electrospinning, atomic layer coating (ALC), and protein-based approaches.

Chapter 9 is based on extensive, carefully screened scientific literature, presenting first trends in the preparation methods of ASDs in the last 2.5 decades with statistical bar and pie charts. This is followed by an extensive presentation and discussion of the overall processability and tableting feasibility of ASDs, with a comparison of the different ASD preparation methods. The influence of polymer type/grade and composition of the formulation on tableting and ASD's chemical stability is also examined. The role of compression pressure, dwell time, drug load, and the crystallinity of tableting excipients on the physical stability of processed ASDs and the stability of the ASDs in the final dosage forms is considered. At the end of the chapter, ASD products in the pharmaceutical market collected from EMA publicly available reports and from data available from the FDA in the last 2.5 decades are presented and discussed.

Chapter 10 focuses on the improvement of pediatric drug delivery through ASDs. Key considerations for pediatric use are the selection of excipients with established safety in children, adapting manufacturing techniques like spray drying and hot-melt extrusion to produce age-appropriate formats, and implementing robust stability strategies to preserve the amorphous state. Clinical successes, such as ASD-based formulations for cystic fibrosis, HIV, fungal infections, and immunosuppression, clearly demonstrate their practical value in pediatric care, not only enhancing drug performance but also improving adherence through more acceptable dosage forms.

Chapter 11 critically summarizes published model-independent and model-dependent methodologies for the assessment of in vivo performance of ASDs in adults, summarized from in vitro data. Based on published information to date, a better understanding of the luminal fate of ASDs, especially in the fed state, and the harnessing of the uncertainty of global regulatory acceptance of non-compendial in vitro testing methods would further increase the reliability of the in vitro assessment of luminal performance of orally administered ASDs.

Chapter 12 describes comprehensively the regulatory guidelines that are related to ASD-based products. In particular, the text delves into and analyzes the guidelines on stability, impurities, specifications, the enhanced quality approach, immediate-release solid oral dosage forms, and the SmPC of ASD-based medicinal products. An example of the pharmaceutical development and regulatory authorization of a medicinal product throughout its lifecycle is provided.

Chapter 13 explores the future perspectives of ASDs, highlighting a transition from empirical "trial-and-error" methods to a systematic, data-driven paradigm. Artificial Intelligence (AI) and Machine Learning (ML) are presented as foundational tools for predicting drug-polymer miscibility, solubility enhancement, and long-term physical stability. The chapter discusses the shift toward continuous manufacturing (CM) via hot-melt extrusion, integrated with Process Analytical Technology (PAT) for real-time monitoring and closed-loop control. Furthermore, the point-of-care manufacture of personalized ASDs is examined through the lens of 3D printing technologies, such as fused deposition modeling (FDM), which allow for customized dosage strengths and complex geometries for controlled drug release. Ultimately, the integration of material science and digital twin technologies is positioned to overcome current hurdles in clinical translation and regulatory alignment, rendering previously "unformulatable" compounds accessible by 2030.

Thessaloniki, Greece — Ioannis Nikolakakis
Copenhagen, Denmark — Ioannis Partheniadis

# Acknowledgments

We sincerely thank our friends and colleagues who have contributed to the publication of this book covering a very important, rapidly growing topic on the formulation of poorly water-soluble drugs.

# Contents

## Part III Emerging and Alternative ASD Platforms

# Editors and Contributors

## About the Editors

**Ioannis Nikolakakis** is a Professor at the School of Pharmacy, Faculty of Health Sciences at the Department of Pharmacy, Aristotle University of Thessaloniki. His research interests span various areas related to powder processing, solid-liquid interactions, hot-melt extrusion, pellets, tableting, and biomaterials. He has been the coordinator of research programs and industrial projects for the development of pellets and tablets for oral absorption and sustained release, the formulation of hygroscopic drugs and the formulation of poorly soluble drugs using self-emulsifying systems and amorphous drug systems for absorption improvement. He serves on the editorial boards of the journals *Drug Development and Research* and of the journal *Processes*.

**Ioannis Partheniadis** is a Postdoctoral Fellow at the Department of Pharmacy, Faculty of Health and Medical Sciences, University of Copenhagen. His work on pharmaceutical technology comprises novel thermal and spectroscopic characterization methods, conventional and novel formulation methods including tableting, hot-melt extrusion, needleless electrospinning as well as the development of amorphous and co-amorphous systems. He has extensive collaboration with industry for the development of solid and semi-solid drug formulations.

## Contributors

**Hisham Al-Obaidi** The School of Pharmacy, University of Reading, Reading, UK

**Panagiotis Barmpalexis** Laboratory of Pharmaceutical Technology, School of Pharmacy, Faculty of Health Sciences, Aristotle University of Thessaloniki, Thessaloniki, Greece

Natural Products Research Centre of Excellence-AUTH (NatPro-AUTH), Center for Interdisciplinary Research and Innovation (CIRI-AUTH), Thessaloniki, Greece

**Konstantina Chachlioutaki** Laboratory of Pharmaceutical Technology, School of Pharmacy, Faculty of Health Sciences, Aristotle University of Thessaloniki, Thessaloniki, Greece

Center for Interdisciplinary Research and Innovation (CIRI-AUTH), Thessaloniki, Greece

**Edwin Chun Yin Chong** The School of Pharmacy, University of Reading, Reading, UK

**Dennis Douroumis** School of Life and Health Sciences, Department of Health Sciences, University of Nicosia -Athens, Athens, Greece

**Jelena Djuris** Department of Pharmaceutical Technology, Faculty of Pharmacy, University of Belgrade, Belgrade, Serbia

**Dimitrios G. Fatouros** Laboratory of Pharmaceutical Technology, School of Pharmacy, Faculty of Health Sciences, Aristotle University of Thessaloniki, Thessaloniki, Greece

Center for Interdisciplinary Research and Innovation (CIRI-AUTH), Thessaloniki, Greece

**Marita Gandaridou** Laboratory of Pharmaceutical Technology, School of Pharmacy, Faculty of Health Sciences, Aristotle University of Thessaloniki, Thessaloniki, Greece

**Margarita Maria Gkantanidou** Laboratory of Pharmaceutical Technology, School of Pharmacy, Faculty of Health Sciences, Aristotle University of Thessaloniki, Thessaloniki, Greece

**Svetlana Ibric** Department of Pharmaceutical Technology, Faculty of Pharmacy, University of Belgrade, Belgrade, Serbia

**Kyriakos Kachrimanis** Laboratory of Pharmaceutical Technology, School of Pharmacy, Faculty of Health Sciences, Aristotle University of Thessaloniki, Thessaloniki, Greece

Center for Interdisciplinary Research & Innovation, Aristotle University of Thessaloniki, Thessaloniki, Greece

**Afroditi Kapourani** Laboratory of Pharmaceutical Technology, School of Pharmacy, Faculty of Health Sciences, Aristotle University of Thessaloniki, Thessaloniki, Greece

**Kohsaku Kawakami** Research Center for Macromolecules and Biomaterials, National Institute for Materials Science, Tsukuba, Ibaraki, Japan

Graduate School of Science and Technology, University of Tsukuba, Tsukuba, Ibaraki, Japan

**Siva S. Kolipaka** Centre for Research Innovation (CRI), University of Greenwich, Chatham, UK

**Melpomeni Liogka** Laboratory of Pharmaceutical Technology, School of Pharmacy, Faculty of Health Sciences, Aristotle University of Thessaloniki, Thessaloniki, Greece

Center for Interdisciplinary Research and Innovation (CIRI-AUTH), Thessaloniki, Greece

**Maria Malamatari** Institute of Pharmaceutical Science, Faculty of Life Sciences & Medicine, Franklin-Wilkins Building, King's College London, London, UK

Institute for Globally Distributed Open Research and Education (IGDORE), London, UK

**Inês C. B. Martins** Department of Pharmacy, Faculty of Health and Medical Sciences, University of Copenhagen, Copenhagen, Denmark

**Djordje Medarevic** Department of Pharmaceutical Technology, Faculty of Pharmacy, University of Belgrade, Belgrade, Serbia

**Ioannis Nikolakakis** Laboratory of Pharmaceutical Technology, School of Pharmacy, Faculty of Health Sciences, Aristotle University of Thessaloniki, Thessaloniki, Greece

**Seth P. Nowaseb** The School of Pharmacy, University of Reading, Reading, UK

School of Pharmacy, University of Namibia, Windhoek, Namibia

**Ioannis Partheniadis** Department of Pharmacy, Faculty of Health and Medical Sciences, University of Copenhagen, Copenhagen, Denmark

**Aleksandar Pavic** Institute of Molecular Genetics and Genetic Engineering, University of Belgrade, Belgrade, Serbia

**Thomas Rades** Department of Pharmacy, Faculty of Health and Medical Sciences, University of Copenhagen, Copenhagen, Denmark

**Christos Reppas** Department of Pharmacy, School of Health Sciences, National and Kapodistrian University of Athens, Zografou, Greece

**Maria Vertzoni** Department of Pharmacy, School of Health Sciences, National and Kapodistrian University of Athens, Zografou, Greece

# Part I
# Introduction and Fundamentals

# Chapter 1
# Introduction to Amorphous Solid Dispersions

**Ioannis Partheniadis and Ioannis Nikolakakis**

**Abstract** The modern pharmaceutical drug discovery and development landscape is increasingly dominated by APIs characterized as new chemical entities (NCEs) with high molecular weights and lipophilicity. Consequently, 75–90% of these compounds possess poor water solubility. Often classified as BCS Class II or IV, these APIs face significant therapeutic hurdles due to dissolution rate-limited absorption and inconsistent pharmacokinetic profiles. While traditional strategies such as salt formation and particle size reduction are widely utilized, they are often limited by chemical structure or fail to improve inherent equilibrium solubility. Amorphous Solid Dispersions (ASDs) have emerged as a robust platform to overcome these challenges by utilizing the "amorphous advantage." By disrupting the highly ordered crystalline lattice, ASDs eliminate the lattice energy barrier, providing higher enthalpy and free energy to drive significantly increased apparent aqueous solubility. This elevated chemical potential facilitates the "spring effect," generating rapid supersaturation to enhance oral bioavailability. This chapter explores the thermodynamic and kinetic principles behind ASDs, their role in modern drug development, and the challenges of maintaining physical stability against recrystallization.

**Keywords** Amorphous solid dispersions · Bioavailability enhancement · Supersaturation (spring and parachute) · Glass transition temperature · Rule of Five

I. Partheniadis (✉)
Department of Pharmacy, Faculty of Health and Medical Sciences, University of Copenhagen, Copenhagen, Denmark
e-mail: ioannis.partheniadis@sund.ku.dk

I. Nikolakakis
Laboratory of Pharmaceutical Technology, School of Pharmacy, Faculty of Health Sciences, Aristotle University of Thessaloniki, Thessaloniki, Greece
e-mail: yannikos@pharm.auth.gr

 
I. Nikolakakis, I. Partheniadis (eds.), *Amorphous Solid Dispersions*, AAPS Introductions in the Pharmaceutical Sciences 5,
https://doi.org/10.1007/978-3-032-24883-1_1

## 1.1 The Solubility Challenge in Modern Drug Development

In recent decades, the pharmaceutical landscape has undergone a paradigm shift driven by high-throughput screening and combinatorial chemistry. While efficient for identifying high-affinity binders, these methodologies have skewed the molecular properties of new chemical entities (NCEs) toward higher molecular weights and increased lipophilicity [1–3]. Consequently, a substantial majority of the modern developmental pipeline consists of poorly water-soluble compounds. Current data suggest that approximately 40% of marketed drugs and 75–90% of NCEs exhibit low aqueous solubility [1, 4, 5]. Under the Biopharmaceutics Classification System (BCS), these molecules are typically categorized as Class II (low solubility, high permeability) or Class IV (low solubility, low permeability) [1, 6].

For these hydrophobic substances, the primary constraint on therapeutic efficacy is dissolution rate-limited absorption. Inadequate dissolution within the gastrointestinal tract prevents the drug from reaching the systemic circulation at concentrations sufficient for a pharmacological effect [6–8]. To achieve therapeutic plasma levels, clinicians often resort to higher doses, which increases the potential for dose-dependent toxicity. For example, the volume of aqueous fluid required to dissolve a therapeutic dose of danazol at gastric pH exceeds the physiological capacity of the human gastrointestinal tract [9]. Furthermore, these compounds often demonstrate significant food effects, where oral bioavailability fluctuations due to the patient's nutritional state may lead to inconsistent pharmacokinetic profiles [10, 11].

Traditional formulation strategies often prove insufficient for addressing the solubility requirements of contemporary NCEs. Salt formation, though a standard approach for ionizable weak acids and bases, is fundamentally limited by the drug's pKa and chemical structure. Estimates indicate that only 20–30% of NCEs possess the chemical properties necessary for stable salt formation, leaving a significant portion of the pipeline requiring alternative solubilization technologies [12]. Physical modifications, such as micronization and nanocrystal formation, are applied to increase the specific surface area of solid drug particles and accelerate the dissolution rate, as described by the Nernst-Brunner equation [13, 14]. However, while surface area reduction can improve dissolution, it does not alter the drug's inherent equilibrium solubility. For highly hydrophobic compounds where equilibrium solubility is the limiting factor, micronization often fails to provide a meaningful increase in bioavailability [1, 8]. Lipid-based delivery systems, including self-emulsifying drug delivery systems (SEDDS) and liposomes, offer an alternative by facilitating micellar solubilization [5, 15]. However, these systems are frequently associated with limited drug solubility in the oil phase, high inter-subject variability, manufacturing complexity, and stability issues related to organic solvent requirements [16]. These limitations have prompted an industry-wide transition toward amorphous solid dispersions (ASDs) as a more robust platform for enhancing the delivery of BCS Class II and IV compounds [1].

The "amorphous advantage" is derived from the disordered molecular arrangement of the amorphous state relative to the highly ordered crystalline lattice [6]. In the crystalline state, significant energy is required to disrupt the lattice structure during dissolution [6, 17]. Because the amorphous state lacks long-range order, this lattice energy barrier is absent. Amorphous materials, therefore, possess higher enthalpy, entropy, and free energy than their crystalline counterparts. This elevated chemical potential serves as the thermodynamic driving force for achieving a significantly higher apparent aqueous solubility [18]. This increased solubility enables the formation of a supersaturated solution, in which the dissolved drug concentration exceeds the thermodynamic equilibrium solubility of the crystalline form [17, 19]. This metastable equilibrium is often referred to as "amorphous solubility" (not a true thermodynamic term) [20]. Theoretically, the magnitude of this solubility increase is related to the difference in Gibbs free energy between the amorphous and crystalline phases. However, due to this high-energy state, amorphous forms are thermodynamically unstable and prone to devitrification (recrystallization) with time into the more stable, lower-energy crystalline form [6].

The kinetic benefit of the amorphous forms is evidenced by an increased dissolution rate, governed by the Noyes-Whitney equation, which establishes that the dissolution rate is directly proportional to saturation solubility ($C_s$) [21, 22]. The higher $C_s$ of the amorphous form thus result in a more rapid generation of supersaturation, a phenomenon frequently termed the "spring" effect [23, 24]. When formulated as an ASD, the drug is typically dispersed within a hydrophilic polymeric matrix that improves wettability and increases the effective surface area for dissolution [25]. For an ASD to be clinically effective, the formulation must also provide a "parachute" effect, wherein the polymer kinetically inhibits drug precipitation, maintaining the supersaturated state for a duration sufficient to allow for systemic absorption [23, 24].

## 1.2 Structural Classification and Thermodynamic Basis of Amorphous Solid Dispersions

The physical state of a pharmaceutical solid exists on a continuum ranging from a highly ordered crystalline lattice to a disordered amorphous state. The fundamental distinction between these states resides in the nature of their molecular periodicity. A crystalline solid is defined by three-dimensional long-range order, maintaining translational, orientational, and conformational symmetry over domains typically exceeding $10^3$ molecules [6]. In contrast, an amorphous solid lacks long-range translational symmetry. However, these materials are not entirely stochastic; they retain short-range order involving nearest-neighbor interactions over molecular distances, which are too small to exhibit Bragg diffraction [26]. Consequently, the amorphous state is formally described as a "frozen" supercooled liquid that possesses the mechanical properties of a solid but the disordered molecular distribution of a liquid.

These structural differences result in divergent physicochemical properties. Analytically, the absence of long-range periodicity in amorphous solids results in a characteristic broad "halo" in X-ray powder diffraction (XRPD) patterns and the absence of a defined melting point ($T_m$) during thermal analysis [26]. Thermodynamically, the amorphous state is a non-equilibrium phase characterized by higher internal energy, enthalpy, entropy, and specific volume relative to the corresponding crystalline form [27]. Because amorphous molecules are not packed efficiently, the intermolecular attractive forces are weaker than those within a constrained crystal lattice.

These attributes dictate the performance of Amorphous Solid Dispersions (ASDs). While the lattice energy of a crystalline drug acts as a barrier to dissolution, the disordered structure of the amorphous state significantly lowers the energy required for molecules to enter the solvent phase [6, 25]. This creates a thermodynamic driving force that enhances apparent aqueous solubility and accelerates dissolution rates. However, because the amorphous state is metastable, it possesses a persistent propensity to revert to the lower-energy crystalline state. Therefore, ASDs require the incorporation of polymeric carriers to provide kinetic and thermodynamic stabilization [6].

### *1.2.1 Classification of Solid Dispersions*

A "solid dispersion" refers to a system where one or more active pharmaceutical ingredients (APIs) are dispersed within an inert carrier matrix [28]. These systems are classified based on the physical state of the components and the number of phases present.

(i) *Solid Solutions vs. Glass Solutions.* Historically, a "solid solution" referred to a crystalline carrier where solute molecules were incorporated into the crystal lattice either substitutionally or interstitially [29]. In modern ASD literature, however, the term is often used synonymously with a glass solution. A glass solution is a homogenous, single-phase system where the drug is molecularly dispersed within an amorphous polymer matrix. This state is characterized by a single, composition-dependent glass transition temperature ($T_g$).

(ii) *Glass Solutions vs. Glass Suspensions.* The distinction between these two systems depends on component miscibility. A glass solution is a molecularly dispersed, single-phase system, which is generally preferred for maximizing bioavailability [30]. Conversely, if the drug concentration exceeds its miscibility limit in the polymer, or if phase separation occurs during storage, a glass suspension (or amorphous solid suspension) is formed. This is a two-phase system where the drug exists as amorphous clusters or drug-rich domains within the polymer. Such systems typically exhibit two distinct $T_g$ values and represent a higher risk for subsequent recrystallization.

### *1.2.2 Mechanisms of Polymer Stabilization*

The polymeric matrix is the primary determinant of ASD stability. It functions via two main mechanisms:

(i) *Thermodynamic Stabilization*. The polymer forms a miscible mixture with the drug, effectively lowering the chemical potential of the API and reducing the thermodynamic driving force for crystallization [31].
(ii) *Kinetic Stabilization*. Through the "anti-plasticizing" effect, polymers with high $T_g$ values increase the overall $T_g$ of the dispersion [6]. By increasing the system's viscosity and reducing free volume, the polymer restricts the molecular mobility of the drug, thereby delaying nucleation and crystal growth [32]. Furthermore, specific intermolecular interactions, such as hydrogen bonding or ionic forces, immobilize drug molecules and disrupt self-association.

Finally, the polymer influences the "spring and parachute" mechanism during dissolution [17, 33]. The high-energy amorphous drug provides the "spring" (rapid supersaturation), while the polymer acts as the "parachute," kinetically inhibiting precipitation and maintaining high intraluminal concentrations of drug in the ASDs, thus facilitating absorption.

## 1.3 Thermodynamic and Kinetic Fundamentals of Amorphous Solid Dispersions

The physical stability and performance of ASDs are governed by the interplay between thermodynamic driving forces and kinetic barriers. The glass transition temperature ($T_g$) represents the critical boundary in the phase behavior of these systems, marking the second-order phase transition between a rigid, glassy state and a mobile, supercooled liquid state [34, 35].

### *1.3.1 Thermodynamic Considerations and Solubility*

From a thermodynamic perspective, the amorphous phase is a non-equilibrium phase possessing higher Gibbs free energy, enthalpy, and entropy relative to the crystalline. This elevated chemical potential provides the driving force for enhanced apparent solubility, which is often reported at magnitudes 10 to 1600 times higher than the crystalline equilibrium solubility [18, 36]. Because the amorphous form lacks a periodic lattice, the energy required to disrupt intermolecular forces during dissolution (lattice energy) is absent, allowing drug molecules to enter the solution more readily.

However, this high-energy state renders ASDs thermodynamically unstable. The driving force for recrystallization is defined by the difference in chemical potential ($\Delta\mu$) between the disordered and ordered phases. This potential increases as the temperature decreases below the equilibrium melting point ($T_m$), a phenomenon dictated by the degree of supercooling [37]. Consequently, the system possesses a persistent thermodynamic drive to revert to the lower-energy crystalline form to achieve equilibrium.

### 1.3.2 Kinetic Stability and Molecular Mobility

While thermodynamics determines the propensity to crystallize, kinetics determine the rate at which this occurs. The $T_g$ serves as the primary indicator of molecular mobility, which is a prerequisite for the diffusion and surface integration required for crystal nucleation and growth.

(i) *Glassy State Stability*. Below the $T_g$, global molecular motions (known as $\alpha$-relaxations) are significantly restricted by high matrix viscosity. A common industrial heuristic, the "$T_g$–50 K rule," suggests that storing amorphous materials at least 50 K below their $T_g$ reduces molecular mobility to negligible levels over practical timescales [38].
(ii) *Secondary Relaxations*. Despite being below the $T_g$, instability can still manifest through local, non-cooperative motions termed $\beta$-relaxations. These secondary relaxations require lower activation energy and can drive localized reorganization and eventual crystallization even within a rigid glass [39].
(iii) *Fragility (m)*. The kinetic stability is also influenced by the "fragility" of the system. This parameter describes the rate at which molecular mobility changes as the temperature approaches $T_g$ [40]. "Fragile" glass-formers exhibit rapid changes in mobility with minor temperature fluctuations, potentially increasing the risk of recrystallization compared to "strong" glass-formers.

### 1.3.3 Manufacturing and Formulation Implications

$T_g$ also dictates the operational parameters for thermal processing, such as hot melt extrusion (HME). Processing temperatures are typically maintained 20–30 °C above the $T_g$ of the polymer to ensure sufficient melt flow and mixing [41, 42]. This creates a design conflict: while high $T_g$ polymers are preferred for long-term kinetic stabilization (the anti-plasticizing effect), the required high processing temperatures may induce thermal degradation of the active pharmaceutical ingredient (API).

Furthermore, the $T_g$ of a miscible, single-phase ASD can be predicted using the Gordon–Taylor equation (Eq. 1.1) [43]:

$$T_{\mathrm{g,mix}} = \frac{w_1 T_{\mathrm{g},1} + K w_2 T_{\mathrm{g},2}}{w_1 + K w_2} \tag{1.1}$$

Positive deviations from this predicted value often indicate strong intermolecular interactions, such as hydrogen bonding, which further immobilize the drug [44]. Conversely, environmental factors like moisture act as potent plasticizers, significantly lowering the system $T_g$ and increasing the free volume available for drug reorientation and phase separation [45].

### 1.3.4 The "Spring and Parachute" Mechanism

The translation of these principles into in vivo performance is best described by the kinetic generation and maintenance of supersaturation:

(i) *The "Spring."* Upon exposure to dissolution media, the high-energy amorphous solid dissolves rapidly. This rapid influx of drug molecules, driven by the lack of a lattice energy barrier, generates a state of supersaturation where the concentration exceeds the crystalline equilibrium solubility [20, 46].

(ii) *The "Parachute."* Because this supersaturated state is metastable, the drug is prone to rapid precipitation [23]. The polymeric carrier acts as a "parachute" by sterically hindering drug-drug aggregation, increasing the viscosity of the boundary layer, and forming specific interactions that inhibit nucleation. This maintains a "supersaturation window" for a duration sufficient to allow for systemic absorption.

## 1.4 Formulation and Manufacturing Workflow

The development of successful ASDs necessitates the establishment of a robust design space, i.e., a multidimensional combination of formulation variables and process parameters that ensures physical stability, chemical purity, and biological efficacy. The primary objective within this framework is to facilitate the transition of a crystalline drug into a high-energy amorphous state while simultaneously mitigating the thermodynamic propensity for recrystallization or amorphous-amorphous phase separation (AAPS) [47]. This workflow integrates in-silico thermodynamic predictions with manufacturing feasibility studies, often employing "frontloading" techniques to screen polymeric carriers using minimal quantities of the active pharmaceutical ingredient (API). The selection of both the manufacturing technology and the polymer matrix is governed by a decision-based hierarchy centered on three primary criteria: solubility of drug/polymer in organic solvent when solvent evaporation is used, drug-polymer miscibility, and the thermal stability profile of the API when thermal methods are applied for the preparation of ASDs.

Solubility in volatile organic solvents is the decisive factor when evaluating solvent-based manufacturing technologies, such as spray drying or fluid-bed processing. For these methodologies to be industrially viable, the API and the polymeric carrier must share a common solvent system that allows for sufficient concentration, typically exceeding 5 mg/mL for initial screening and significantly higher for commercial scale-up [48]. The choice of solvent is critical; it must not only ensure a homogeneous distribution of components but also possess favorable properties regarding toxicity, viscosity, and atomization characteristics. If an API exhibits insufficient solubility in pharmaceutically acceptable, ICH-compliant solvents, solvent-based processing may be precluded, necessitating a transition toward fusion-based manufacturing routes [49].

The second critical criterion is the miscibility between the drug and the polymer, which serves as the fundamental driver for the long-term physical stability of the ASD. Evaluating miscibility involves assessing the strength of molecular interactions, such as hydrogen bonding or ionic forces, which reduce the chemical potential of the drug and provide a thermodynamic barrier to phase separation [47, 50]. In the early formulation phase, miscibility is often predicted using the Flory-Huggins interaction parameter ($\chi$) or Hansen solubility parameters, where similar cohesive energy densities indicate potential compatibility [51, 52]. Advanced computational tools, such as COSMO-RS, are increasingly utilized to calculate activity coefficients, where lower values correlate with stronger drug-polymer interactions and a reduced driving force for nucleation [53]. Experimentally, these predictions are verified through techniques such as polarized light microscopy (PLM) or differential scanning calorimetry (DSC), where the presence of a single, composition-dependent $T_g$ confirms a homogeneous, single-phase system [48].

Thermal stability serves as the primary constraint for fusion-based manufacturing processes, most notably HME [54]. This technology requires the API to withstand elevated temperatures and intense mechanical shear without undergoing chemical degradation. APIs with melting points exceeding 200 °C or those characterized as thermolabile are generally unsuitable for traditional HME unless significant melting point depression can be achieved through formulation additives [55, 56]. For such thermally sensitive compounds, specialized technologies like KinetiSol® may be employed, utilizing frictional energy to achieve fusion within extremely short residence times to mitigate degradation risks [57]. Beyond manufacturing feasibility, the thermal profile of the final formulation dictates its shelf-life stability. ASDs are typically engineered to maintain a $T_g$ significantly higher than the intended storage temperature to restrict the molecular mobility required for crystal growth.

Ultimately, defining the processing window requires a deep understanding of the drug-polymer-process interplay. In solvent-based methods, the design space is constrained by evaporation rates and diffusion coefficients, where rapid drying is essential to "trap" the drug molecularly before phase separation can occur. In fusion-based methods, the space is defined by specific mechanical energy, barrel temperature, and residence time. By integrating theoretical miscibility models with an understanding of how material properties respond to thermal and mechanical stress, formulators can establish a design space that ensures that the ASD maintains its supersaturation advantage and physical integrity throughout its lifecycle.

## 1.5 Historical Evolution and Current Market Landscape

The transition of ASDs from a theoretical concept to an industrial standard reflects a broader necessity to address the increasing lipophilicity of the global drug pipeline. The conceptual framework for these systems was established in 1961 by Sekiguchi and Obi, who demonstrated that dispersing a poorly soluble drug within a physiologically inert, water-soluble carrier could significantly enhance its dissolution rate [58]. This foundational work initiated progression through several developmental generations. Early formulations focused on crystalline carriers such as urea or sugars, which primarily improved drug wettability but offered limited stabilization for the amorphous state. Subsequent generations introduced synthetic and semi-synthetic polymers, such as povidone and cellulose derivatives, which serve as stabilizing matrices. Contemporary third-generation ASDs have further refined this approach by incorporating surfactants or self-emulsifying agents [59]. These additives facilitate higher drug loading and prevent the rapid precipitation of the drug upon its release into the aqueous gastrointestinal environment, thereby maintaining the necessary state of supersaturation for absorption.

Commercialization of these systems accelerated in the 1990s as manufacturing technologies matured and regulatory pathways became more established. A seminal milestone was the 1992 approval of Sporanox® (itraconazole), which utilized a fluid-bed bead layering process to coat sugar spheres with a drug-polymer solution [1]. This was followed by the transformative 2007 reformulation of Kaletra® (lopinavir/ritonavir) [60]. Originally marketed as a soft-gelatin capsule that required refrigeration to maintain stability, the transition to an ASD tablet manufactured via hot-melt extrusion eliminated the need for cold-chain distribution and significantly reduced the dosing frequency for patients [61]. These early successes proved that amorphous materials could be stabilized and manufactured at scale, paving the way for the current era of supersaturating drug delivery systems.

The contemporary market landscape is characterized by an increasing reliance on ASD technology to rescue molecules that reside in the "Beyond Rule of 5" chemical space [62, 63]. Between 2012 and 2023, the US Food and Drug Administration (FDA) approved 48 drug products containing 36 unique amorphous active pharmaceutical ingredients [64]. Data analysis indicates that approximately 61% of these amorphous drugs violate at least one Lipinski criterion, specifically regarding molecular weight and lipophilicity, necessitating advanced solid-state interventions to ensure adequate oral bioavailability. This trend is particularly evident in the antiviral and antineoplastic sectors, which together account for more than half of the approved ASD market. These therapeutic classes often involve high-melting-point compounds with significant lattice energies, making the "amorphous advantage" a critical requirement for clinical efficacy.

Current market dynamics also reflect a shift toward high-drug-load formulations and patient-centric dosage forms [64]. Because many modern amorphous drugs require high therapeutic doses, research has intensified into maintaining stability at high drug-to-polymer ratios to minimize the physical size of the final dosage form. While the oral tablet remains the predominant format, there has been a notable diversification into granules and pellets, such as those used for Kalydeco® and Orkambi®, to facilitate administration in pediatric and geriatric populations [65]. Furthermore, while spray drying and hot-melt extrusion remain the primary industrial processing methods, the recent approval of formulations utilizing electrospraying indicates a maturing field that continues to adopt specialized engineering solutions to meet the unique physicochemical challenges of next-generation chemical entities [64].

## 1.6 Summary and Perspective

Amorphous solid dispersions (ASDs) are an essential formulation strategy for addressing the low aqueous solubility of modern drug candidates. By removing the crystalline lattice and creating a disordered molecular state, ASDs bypass the lattice energy barrier that typically limits dissolution. This results in a significant increase in the apparent solubility and chemical potential of the drug, which drives faster dissolution and enables the generation of supersaturated solutions in the gastrointestinal tract.

However, the high-energy nature of the amorphous state makes it inherently unstable. Without stabilization, drug molecules naturally tend to reorganize into a lower-energy crystalline form, which eliminates the solubility advantages. The success of an ASD, therefore, depends on the selection of a polymer matrix that can provide both kinetic and thermodynamic stabilization. As discussed, the polymer achieves this by increasing the glass transition temperature ($T_g$) of the system (thereby reducing molecular mobility) and by forming specific molecular interactions that prevent drug-drug aggregation.

During the dissolution process, the performance of the ASD is defined by the "spring and parachute" mechanism. The amorphous drug provides the "spring" by rapidly creating a state of supersaturation, while the polymer acts as the "parachute" by maintaining that concentration long enough for systemic absorption to occur. Failure to maintain this state leads to drug precipitation and reduced bioavailability.

From a broader perspective, the success of ASDs has fundamentally challenged traditional concepts of "drug-likeness." For years, Lipinski's Rule of Five served as a rigid boundary, suggesting that molecules with high molecular weights ($M_W > 500$) and high lipophilicity ($\log P > 5$) were poor candidates for oral delivery. However, the rise of ASD technology has enabled the clinical success of several molecules that reside in the "Beyond Rule of 5" (bRo5) space. These technologies prove that high $M_W$ and extreme lipophilicity are no longer insurmountable barriers; rather, they are engineering challenges that can be managed through advanced amorphous solid-state interventions.

As the pharmaceutical industry continues to develop molecules with high molecular weights and lipophilicity, the role of ASDs will become increasingly prominent. While current manufacturing standards like spray drying and hot-melt extrusion are well-established, new engineering approaches and a deeper understanding of polymer science are required to manage complex formulations. The following chapters will explore these concepts in detail.

## References


1. Jermain SV, Brough C, Williams RO. Amorphous solid dispersions and nanocrystal technologies for poorly water-soluble drug delivery – an update. Int J Pharm. 2018;535(1):379–92.
2. Karagianni A, Kachrimanis K, Nikolakakis I. Co-amorphous solid dispersions for solubility and absorption improvement of drugs: composition, preparation, characterization and formulations for oral delivery. Pharmaceutics. 2018;10(3):50.
3. Tao N, Yan Z, Wang X, Wang Y, Ji L, Qiu L, et al. A facile way to enhance the therapeutic efficacy of hydrophobic drugs via amorphous solid dispersions. AAPS J. 2025;27(2):63.
4. Loftsson T, Brewster ME. Pharmaceutical applications of cyclodextrins: basic science and product development. J Pharm Pharmacol. 2010;62(11):1607–21.
5. Nikolakakis I, Partheniadis I. Self-emulsifying granules and pellets: composition and formation mechanisms for instant or controlled release. Pharmaceutics. 2017;9(4)
6. Baghel S, Cathcart H, O'Reilly NJ. Polymeric amorphous solid dispersions: a review of amorphization, crystallization, stabilization, solid-state characterization, and aqueous solubilization of biopharmaceutical classification system class II drugs. J Pharm Sci. 2016;105(9):2527–44.
7. Dressman JB, Vertzoni M, Goumas K, Reppas C. Estimating drug solubility in the gastrointestinal tract. Adv Drug Deliv Rev. 2007;59(7):591–602.
8. Zhao YH, Abraham MH, Le J, Hersey A, Luscombe CN, Beck G, et al. Rate-limited steps of human oral absorption and QSAR studies. Pharm Res. 2002;19(10):1446–57.
9. Childs SL, Kandi P, Lingireddy SR. Formulation of a danazol cocrystal with controlled supersaturation plays an essential role in improving bioavailability. Mol Pharm. 2013;10(8):3112–27.
10. Courtney R, Wexler D, Radwanski E, Lim J, Laughlin M. Effect of food on the relative bioavailability of two oral formulations of posaconazole in healthy adults. Br J Clin Pharmacol. 2004;57(2):218–22.
11. Thombre AG, Herbig SM, Alderman JA. Improved ziprasidone formulations with enhanced bioavailability in the fasted state and a reduced food effect. Pharm Res. 2011;28(12):3159–70.
12. Serajuddin AT, Pudipeddi M. Salt-selection strategies. In: Handbook of pharmaceutical salts: properties, selection, and use, vol. 2. Zurich/Weinheim/New York: Vperlag Helvetica Chemica Acta/Wiley-VCH; 2002.
13. Dokoumetzidis A, Macheras P. A century of dissolution research: from Noyes and Whitney to the biopharmaceutics classification system. Int J Pharm. 2006;321(1):1–11.
14. Möschwitzer JP. Drug nanocrystals in the commercial pharmaceutical development process. Int J Pharm. 2013;453(1):142–56.
15. Hauss DJ. Oral lipid-based formulations. Adv Drug Deliv Rev. 2007;59(7):667–76.
16. Zheng W, Jain A, Papoutsakis D, Dannenfelser RM, Panicucci R, Garad S. Selection of oral bioavailability enhancing formulations during drug discovery. Drug Dev Ind Pharm. 2012;38(2):235–47.

17. Newman A, Knipp G, Zografi G. Assessing the performance of amorphous solid dispersions. J Pharm Sci. 2012;101(4):1355–77.
18. Hancock BC, Parks M. What is the true solubility advantage for amorphous pharmaceuticals? Pharm Res. 2000;17(4):397–404.
19. Pandi P, Bulusu R, Kommineni N, Khan W, Singh M. Amorphous solid dispersions: an update for preparation, characterization, mechanism on bioavailability, stability, regulatory considerations and marketed products. Int J Pharm. 2020;586:119560.
20. Taylor LS, Zhang GGZ. Physical chemistry of supersaturated solutions and implications for oral absorption. Adv Drug Deliv Rev. 2016;101:122–42.
21. Brough C, Williams RO 3rd. Amorphous solid dispersions and nano-crystal technologies for poorly water-soluble drug delivery. Int J Pharm. 2013;453(1):157–66.
22. Gibaldi M, Feldman S. Establishment of sink conditions in dissolution rate determinations. Theoretical considerations and application to nondisintegrating dosage forms. J Pharm Sci. 1967;56(10):1238–42.
23. Brouwers J, Brewster ME, Augustijns P. Supersaturating drug delivery systems: the answer to solubility-limited oral bioavailability? J Pharm Sci. 2009;98(8):2549–72.
24. Sun DD, Lee PI. Evolution of supersaturation of amorphous pharmaceuticals: the effect of rate of supersaturation generation. Mol Pharm. 2013;10(11):4330–46.
25. Van den Mooter G. The use of amorphous solid dispersions: a formulation strategy to overcome poor solubility and dissolution rate. Drug Discov Today Technol. 2012;9(2):e79–85.
26. Klug HP, Alexander LE. X-ray diffraction procedures: for polycrystalline and amorphous materials. New York: Wiley; 1974.
27. Stillinger FH. Supercooled liquids, glass transitions, and the Kauzmann paradox. J Chem Phys. 1988;88(12):7818–25.
28. Chiou WL, Riegelman S. Pharmaceutical applications of solid dispersion systems. J Pharm Sci. 1971;60(9):1281–302.
29. Moore MD, Wildfong PL. Aqueous solubility enhancement through engineering of binary solid composites: pharmaceutical applications. J Pharm Innov. 2009;4(1):36–49.
30. Laitinen R, Priemel PA, Surwase S, Graeser K, Strachan CJ, Grohganz H, et al. Theoretical considerations in developing amorphous solid dispersions. In: Amorphous solid dispersions: theory and practice. New York: Springer; 2014. p. 35–90.
31. Marsac PJ, Shamblin SL, Taylor LS. Theoretical and practical approaches for prediction of drug-polymer miscibility and solubility. Pharm Res. 2006;23(10):2417–26.
32. Kaushal AM, Gupta P, Bansal AK. Amorphous drug delivery systems: molecular aspects, design, and performance. Crit Rev Ther Drug Carrier Syst. 2004;21(3):133–93.
33. Ashwathy P, Anto AT, Sudheesh MS. A mechanistic review on the dissolution phase behavior and supersaturation stabilization of amorphous solid dispersions. Drug Dev Ind Pharm. 2021;47(1):1–11.
34. Bengtzelius U, Gotze W, Sjolander A. Dynamics of supercooled liquids and the glass transition. J Phys C Solid State Phys. 1984;17(33):5915–34.
35. Debenedetti PG, Stillinger FH. Supercooled liquids and the glass transition. Nature. 2001;410(6825):259–67.
36. Graeser KA, Patterson JE, Zeitler JA, Rades T. The role of configurational entropy in amorphous systems. Pharmaceutics. 2010;2(2):224–44.
37. Kou X, Zhou L. Stability of amorphous solid dispersion. In: Amorphous solid dispersions: theory and practice. New York: Springer; 2014. p. 515–44.
38. Hancock BC, Shamblin SL, Zografi G. Molecular mobility of amorphous pharmaceutical solids below their glass transition temperatures. Pharm Res. 1995;12(6):799–806.
39. Bhattacharya S, Suryanarayanan R. Local mobility in amorphous pharmaceuticals--characterization and implications on stability. J Pharm Sci. 2009;98(9):2935–53.
40. Donth E-J. The glass transition: relaxation dynamics in liquids and disordered materials. Heidelberg: Springer; 2001.

41. Brown C, DiNunzio J, Eglesia M, Forster S, Lamm M, Lowinger M, et al. Hot-melt extrusion for solid dispersions: composition and design considerations. In: Amorphous solid dispersions: theory and practice. New York: Springer; 2014. p. 197–230.
42. Chokshi RJ, Sandhu HK, Iyer RM, Shah NH, Malick AW, Zia H. Characterization of physico-mechanical properties of indomethacin and polymers to assess their suitability for hot-melt extrusion process as a means to manufacture solid dispersion/solution. J Pharm Sci. 2005;94(11):2463–74.
43. Gordon M, Taylor JS. Ideal copolymers and the second-order transitions of synthetic rubbers. i. non-crystalline copolymers. J Appl Chem. 1952;2(9):493–500.
44. Nair R, Nyamweya N, Gönen S, Martínez-Miranda LJ, Hoag SW. Influence of various drugs on the glass transition temperature of poly(vinylpyrrolidone): a thermodynamic and spectroscopic investigation. Int J Pharm. 2001;225(1–2):83–96.
45. Hancock BC, Zografi G. The relationship between the glass transition temperature and the water content of amorphous pharmaceutical solids. Pharm Res. 1994;11(4):471–7.
46. Kawakami K. Supersaturation and crystallization: non-equilibrium dynamics of amorphous solid dispersions for oral drug delivery. Expert Opin Drug Deliv. 2017;14(6):735–43.
47. Rumondor AC, Stanford LA, Taylor LS. Effects of polymer type and storage relative humidity on the kinetics of felodipine crystallization from amorphous solid dispersions. Pharm Res. 2009;26(12):2599–606.
48. Anane-Adjei AB, Jacobs E, Nash SC, Askin S, Soundararajan R, Kyobula M, et al. Amorphous solid dispersions: utilization and challenges in preclinical drug development within AstraZeneca. Int J Pharm. 2022;614:121387.
49. Huang S, Williams RO 3rd. Effects of the preparation process on the properties of amorphous solid dispersions. AAPS PharmSciTech. 2018;19(5):1971–84.
50. Tian Y, Booth J, Meehan E, Jones DS, Li S, Andrews GP. Construction of drug-polymer thermodynamic phase diagrams using Flory-Huggins interaction theory: identifying the relevance of temperature and drug weight fraction to phase separation within solid dispersions. Mol Pharm. 2013;10(1):236–48.
51. Hancock BC, York P, Rowe RC. The use of solubility parameters in pharmaceutical dosage form design. Int J Pharm. 1997;148(1):1–21.
52. Zhao Y, Inbar P, Chokshi HP, Malick AW, Choi DS. Prediction of the thermal phase diagram of amorphous solid dispersions by Flory-Huggins theory. J Pharm Sci. 2011;100(8):3196–207.
53. Zeneli E, Bohets H, Mebenga FN, Holm R, Tistaert C, Kuentz M. Guiding excipient selection for physically stable amorphous solid dispersions: a combined in-vitro in-silico approach. Eur J Pharm Sci. 2025;212:107152.
54. Mendonsa N, Almutairy B, Kallakunta VR, Sarabu S, Thipsay P, Bandari S, et al. Manufacturing strategies to develop amorphous solid dispersions: an overview. J Drug Deliv Sci Technol. 2020;55:101459.
55. Haser A, Huang S, Listro T, White D, Zhang F. An approach for chemical stability during melt extrusion of a drug substance with a high melting point. Int J Pharm. 2017;524(1):55–64.
56. LaFountaine JS, McGinity JW, Williams RO 3rd. Challenges and strategies in thermal processing of amorphous solid dispersions: a review. AAPS PharmSciTech. 2016;17(1):43–55.
57. Ellenberger DJ, Miller DA, Williams RO 3rd. Expanding the application and formulation space of amorphous solid dispersions with KinetiSol®: a review. AAPS PharmSciTech. 2018;19(5):1933–56.
58. Sekiguchi K, Obi N. Studies on absorption of eutectic mixture. I. A comparison of the behavior of eutectic mixture of sulfathiazole and that of ordinary sulfathiazole in man. Chem Pharm Bull. 1961;9(11):866–72.
59. Kaushik R, Budhwar V, Kaushik D. An overview on recent patents and technologies on solid dispersion. Recent Pat Drug Deliv Formul. 2020;14(1):63–74.
60. Tambe S, Jain D, Meruva SK, Rongala G, Juluri A, Nihalani G, et al. Recent advances in amorphous solid dispersions: preformulation, formulation strategies, technological advancements and characterization. Pharmaceutics. 2022;14(10):2203.

61. Klein CE, Chiu YL, Awni W, Zhu T, Heuser RS, Doan T, et al. The tablet formulation of lopinavir/ritonavir provides similar bioavailability to the soft-gelatin capsule formulation with less pharmacokinetic variability and diminished food effect. J Acquir Immune Defic Syndr. 2007;44(4):401–10.
62. Lipinski CA. Drug-like properties and the causes of poor solubility and poor permeability. J Pharmacol Toxicol Methods. 2000;44(1):235–49.
63. Stegemann S, Moreton C, Svanbäck S, Box K, Motte G, Paudel A. Trends in oral small-molecule drug discovery and product development based on product launches before and after the Rule of Five. Drug Discov Today. 2023;28(2):103344.
64. Moseson DE, Tran TB, Karunakaran B, Ambardekar R, Hiew TN. Trends in amorphous solid dispersion drug products approved by the U.S. Food and Drug Administration between 2012 and 2023. Int J Pharm X. 2024;7:100259.
65. Meruva S, Singaraju AB, Vinjamuri BP, Ternik R, Stagner WC. Current state of minitablet product design: a review. J Pharm Sci. 2024;113(5):1123–54.

# Chapter 2
# Physical Stability of Amorphous Drugs and Amorphous Solid Dispersions

**Kohsaku Kawakami**

**Abstract** The use of the amorphous state is one of the most promising strategies for overcoming the challenges associated with drugs with low aqueous solubility. However, the development of amorphous products often encounters unexpected issues originating from their non-equilibrium nature. The most drastic structural transformation in amorphous solids is crystallization, which involves nucleation and crystal growth. Direct experimental detection of crystallization is impossible until the crystals have grown to a sufficiently large size. Thus, progress in nucleation theory is crucial for achieving precise control of crystallization. Relaxation is a structural change to attain a more energetically favorable state and is relevant to the physical and chemical stability of amorphous drugs. Because $\alpha$ relaxation represents global molecular mobility, it is related to crystallization kinetics, unless nucleation is the limiting step. Local molecular mobility, which may be detected as secondary relaxations, sometimes plays a key role in physical stabilization. In this chapter, physical stability of amorphous drugs is discussed with focus on their crystallization and relaxation behaviors. Subsequently, attention has been directed to studies investigating the stability of drug-polymer binary mixtures. Stabilization using mesoporous materials is subsequently discussed, as its mechanism is completely different from that of polymers. This understanding could contribute to the improvement of ASD technology and may enable the design of innovative ASDs with high drug loading.

**Keywords** Amorphous solid dispersion · Nucleation · Crystal growth · Crystallization · Fragility · Relaxation

---

K. Kawakami (✉)
Research Center for Macromolecules and Biomaterials, National Institute for Materials Science, Tsukuba, Ibaraki, Japan

Graduate School of Science and Technology, University of Tsukuba, Tsukuba, Ibaraki, Japan
e-mail: kawakami.kohsaku@nims.go.jp

 
I. Nikolakakis, I. Partheniadis (eds.), *Amorphous Solid Dispersions*, AAPS Introductions in the Pharmaceutical Sciences 5,
https://doi.org/10.1007/978-3-032-24883-1_2

## 2.1 Introduction

Owing to a paradigm shift in drug discovery research, the proportion of poorly soluble drugs increased rapidly in the 1990s. Earlier discovery efforts relied on a phenotypic approach, that is, a classical trial-and-error methodology where candidate compounds were screened against cells, tissues, or whole organisms. Various physicochemical and biological factors affecting the efficacy of candidates, including solubility, protein binding, and metabolism, were automatically evaluated in this process. However, current discovery research primarily depends on a target-based approach, where candidate compounds are tested against specific targets, including receptors and enzymes, based on hypotheses of action mechanisms [1]. Although this procedure provides a clear strategy for molecular design, it often allows compounds with physicochemical and/or biological problems, such as low aqueous solubility, to progress through screening. Moreover, low solubility compounds may show false-positive results owing to nonspecific interactions caused by, e.g., crushed compounds in aqueous media.

Modification of molecular structure is the first strategy to overcome low solubility, followed by selection of optional solid forms, including the use of salts and cocrystals [2]. Classical formulation approaches such as micronization and the addition of surfactants may also be used at this stage. Unless these strategies are effective, formulation-based approaches become necessary. Among these, amorphous solid dispersions (ASD) is one of the most effective formulations for enhancing the oral absorption of poorly soluble drugs [2–5]. Amorphous solids can offer supersaturation during dissolution because of their high-energy state. In comparison with other formulation technologies that increase equilibrium solubility, supersaturation-based formulations may have the advantage of enhancing transmembrane permeation of drugs by increasing drug activity [6].

However, the main disadvantages of ASD technology include the need for specialized manufacturing facilities, difficulty in predicting oral absorption in humans, and concerns about storage stability. In particular, the assessment of long-term storage stability is time-consuming, because application of accelerated stability studies used for conventional crystalline formulations requires caution [7]. Although many ASDs are already commercially available, most drugs developed using ASD technology possess unnecessarily high physical stability [8]. A deeper understanding of the crystallization behavior of amorphous drugs is essential to further expand the applicability of ASD technology.

Figure 2.1 presents the enthalpy–temperature relationships of the liquid, glass, and crystal phases. A liquid generally crystallizes when cooled down below its melting temperature ($T_m$), but it can be supercooled without crystallization by applying a sufficiently fast cooling rate to reach the glass transition temperature ($T_g$), where the viscosity of the material is ca. $10^{12}$ Pa·s, regardless of compound. Global molecular mobility significantly decreases below $T_g$; however, the mobility just below $T_g$ is still sufficient for crystallization [9]. The molecular mobility is almost frozen at approximately $T_g$–50 °C, known as the Kauzmann temperature ($T_k$). The amorphous

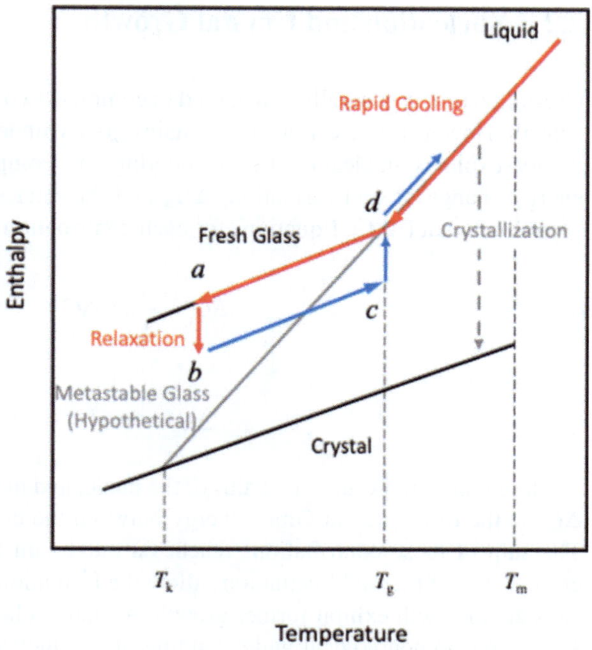


**Fig. 2.1** Enthalpy-temperature relationship of liquid, glass, and crystal. The red arrows show the process of transformation from liquid to glass upon cooling, followed by relaxation under isothermal condition. The blue arrows represent the trace when the relaxed glass is heated in DSC. $T_m$ melting temperature, $T_g$ glass transition temperature, $T_k$ Kauzmann temperature. See text for further details

state is generally considered physically stable if it is stored below $T_k$. However, local molecular mobility remains still active in this temperature range and may contribute to nucleation. This chapter discusses the molecular dynamics of amorphous drugs below $T_g$.

ASDs contain polymeric excipients mainly for two important reasons. The first involves assisting in the dissolution of amorphous drugs. Various hydrophilic polymers are known to inhibit the crystallization of supersaturated drugs. Moreover, assistance in the initial dissolution process is also an important role for the polymeric excipients. The other purpose is to physically stabilize the amorphous state of the drugs. Possible stabilization mechanisms include steric hindrance for crystallization of drugs, increased viscosity to retard crystal growth, direct interactions with drug molecules, adsorption onto the growth surface of crystals, and influence on the nucleation process. The amount of polymer included in marketed ASDs is generally more than double the amount of the drug, resulting in a significant increase in formulation volume. A more detailed understanding of the crystallization inhibition mechanism is required to reduce the amount of excipients, which is one of the focus of this chapter.

## 2.2 Nucleation and Crystal Growth

Crystallization is generally considered to comprise two steps: nucleation and crystal growth. However, nucleation step remains poorly understood. Classical nucleation theory explains nucleation by considering the competition between the Gibbs energy change of transformation, $\Delta G_t$, and the interfacial energy to suppress the growth of nuclei, $\Delta G_i$. Equations for each contribution are expressed as follows:

$$\Delta G_t = \frac{4}{3}\pi r^3 \Delta G_v \tag{2.1}$$

$$\Delta G_i = 4\pi r^2 \sigma \tag{2.2}$$

Here $r$ and $s$ represent the radius of the nuclei and interfacial energy, respectively. $\Delta G_v$ is the difference in Gibbs energy between the crystal and liquid/glass states. The sum of these contributions reaches a maximum for nuclear formation at the critical size. If thermal fluctuations allow the formation of nuclei larger than a critical size, they will exhibit further growth. Although classical nucleation theory provides a phenomenological understanding of the nucleation behavior in amorphous drugs, its practical application remains challenging. One of the main challenges in advancing crystallization theory is the difficulty in the experimental detection of nuclei. Moreover, their structure is not well understood. Therefore, the evaluation of the crystallization process generally relies on the observation of crystal growth process, from which the nucleation process is inferred. Various solid-state reaction equations have been applied to approximate crystal growth. Among them, the Avrami–Erofeev equation, as shown below, generally offers a good approximation for the time-dependent crystallinity *(X)* evolution of low-molecular-weight organic compounds [10, 11].

$$\mathrm{X}(\%) = 100\left[1 - \exp\left\{-k_c (t-d)^n\right\}\right] \tag{2.3}$$

where $k_c$ and $d$ are the crystallization rate and induction time, respectively, and $n$ is the Avrami exponent, which is generally considered to be dependent on the dimensions of crystal growth and nucleation mechanisms (Table 2.1). Thus, this equation provides insights into the nucleation process. However, the best fit may not provide

**Table 2.1** Interpretation of Avrami constant under constant rate of crystal growth

| Dimension of crystal growth | Homogeneous nucleation (Constant nucleation rate) | Heterogeneous nucleation (Constant nuclei number) |
|---|---|---|
| 1D | 2 | 1 |
| 2D | 3 | 2 |
| 3D | 4 | 3 |

appropriate results because $n$ and $k$ play compensatory roles in the fitting procedure. Thus, the validity of the results must be examined. Some material properties, such as porosity and density, are likely to influence the dimensions of crystal growth [12]. Moreover, changes in crystallization mechanisms and kinetics during the crystallization process are not considered in this simple equation.

Because the nucleation process cannot be observed directly, several indirect methods have been employed to determine the nucleation rate and/or temperature. Microscopic observation, where the number of growing crystals is counted under a microscope after annealing at the nucleation temperature, is the most conventional method for this purpose. A systematic study of indomethacin glass was reported by Andronis et al. [13], where the temperatures corresponding to the maximum nucleation rates for forms $\gamma$ and $\alpha$ were identified to be 40 and 60 °C, respectively. Although this method remains the gold standard for evaluating nucleation, it has several limitations and problems. Observation by the human eye, or even the use of image analysis, has limitations in counting all crystals, differentiation of similar crystal habits, identification of minor crystals, and evaluation of crystal form transformation. In fact, indomethacin glass is likely to favor nucleation to form δ at 60 °C [14], which has a morphology similar to that of form α. As form δ melts at 125 °C and immediately recrystallizes to form α, the observation using a microscope above this temperature results in misassignment to form α, although the nuclei formed during the annealing were of form δ. Also, the same crystal form may look differently [15]. Other issues for microscopic observation include difficulty in controlling atmospheric conditions such as moisture level and temperature. In addition, the samples frequently experience a confined environment during observation, which can affect the crystallization behavior.

Problems associated with the microscopic method may be overcome using differential scanning calorimetry (DSC) analysis, where cold crystallization is investigated after annealing at nucleation temperatures [16–18]. Figure 2.2a shows typical DSC heating curves of ibuprofen glass after annealing at various temperatures for 1 h. Ibuprofen forms relatively stable glass, which is difficult to crystallize without annealing to create nuclei before heating. The optimum nucleation temperatures for forms I (stable form) and II (metastable form) are ca. −15 °C [18] and below −105 °C [19], respectively. During DSC heating, annealing in advance at 0, −10, and −30 °C provides cold crystallization peak, of which the temperature depends on the annealing condition, followed by its melting at 76 °C. The lower onset crystallization temperature and larger crystallization enthalpy suggest the formation of a larger number of nuclei during annealing. Figure 2.2b presents the DSC curve of the IBP glass that experienced equilibration at −80 °C, followed by annealing at 0 °C for 1 h. The DSC curve exhibits an endothermic peak characteristic to form II at 10 °C, indicating nucleation to form II occurred at −80 °C.

If the probability of cold crystallization is high, the optimum nucleation temperature can be determined by identifying the annealing temperature which gives the lowest onset crystallization temperature and the largest crystallization enthalpy [17]. However, if the probability of crystallization is low, the probability of crystallization, rather than the crystallization temperature/enthalpy, provides a clearer

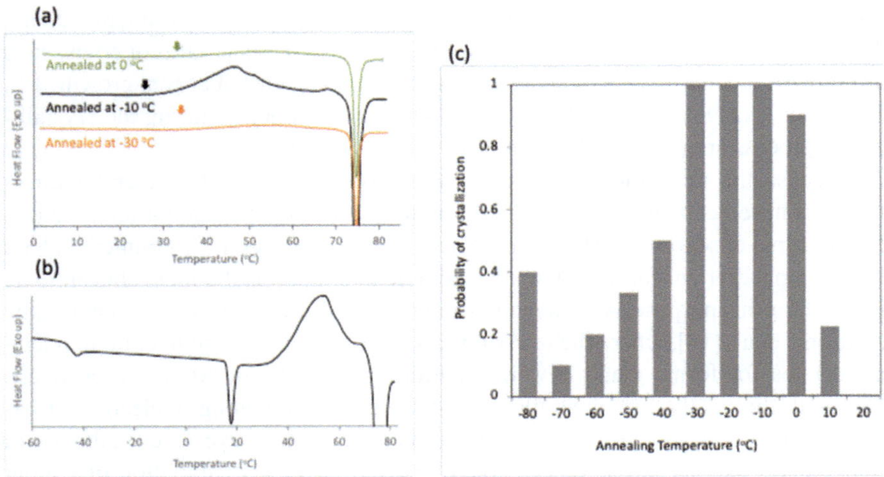


**Fig. 2.2** (**a**) Representative DSC heating curves of ibuprofen glass annealed at various temperatures for 1 h. The annealing temperatures are indicated in the figure. The heating rate was 10 °C / min. (**b**) A representative DSC heating curve of ibuprofen glass equilibrated at −80 °C, followed by annealing at 0 h for 1 h. Then, the sample was cooled to −80 °C again to acquire the heating curve at 10 °C/min. (**c**) Probability of cold crystallization of ibuprofen glass after the annealing at various temperatures for 1 h. (Figures (a) and (c) are adapted with modifications under CC-BY 4.0 license [18])

indication of the nucleation temperature. Figure 2.2c shows the probability of cold crystallization of ibuprofen glass after annealing at various temperatures for 1 h. Crystallization consistently occurred after annealing at −10, −20, and −30 °C because of nucleation for form I. By reducing the annealing period, the optimum nucleation temperature was identified as −15 °C [18]. Figure 2.2c also shows higher probability of crystallization after annealing at −80 °C relative to annealing at higher temperature such as −60 °C and −70 °C, due to nucleation for form II. DSC offers precise temperature control and atmosphere with a low-moisture level compared to microscopic observations. Moreover, the crystallization process is free from the confinement effect.

Nucleation at low temperatures is sometimes debated because the formation of cracks is suspected to accelerate crystallization [20]. Discriminating between the effects of nucleation and crack formation is difficult, as both is supposed to occur in the same temperature range. However, cracks are healed when the glass is heated above the glass transition temperature ($T_g$), where cold crystallization is observed in DSC. Moreover, in the case of celecoxib glass, the optimum nucleation temperature was observed at −50 °C, and annealing at lower temperature was less effective [17], although cracks were formed in the same manner. There observation suggest nucleation at −50 °C.

Table 2.2 summarizes examples of reported nucleation temperatures, which are typically slightly above $T_g$ [21] or in the freezing temperature range [22]. The reason to find nucleation temperatures in these temperature ranges requires further

**Table 2.2** Examples of nucleation temperature

| Compounds | $T_g$ (°C) | $T_m$ (°C) | $T_g/T_m$ | Nucleation temperature (°C) |
|---|---|---|---|---|
| Acetaminophen | 23 | 169 | 0.67 | 50~60[a] |
| D-Arabitol | −12 | 102 | 0.70 | 5[b], 15[c] |
| Celecoxib | 58 | 162 (III) | 0.76 | −50[d] |
| Fenofibrate | −20 | 80 | 0.72 | −20~10[e] |
| Fluconazole | 32 | 140 | 0.74 | 50[f] |
| Ibuprofen | −47 | 76 (I) | 0.65 | −30[b], −15[g] |
| | | 17 (II) | 0.94 | <−105[h], <−80[i] |
| Indomethacin | 45 | 161 (γ) | 0.73 | 40[j] |
| | | 155 (α) | 0.74 | 50[j] |
| Loratadine | 34 | 135 | 0.75 | −40[k] |
| Nifedipine | 46 | 137 (γ′) | 0.78 | 100[l] |
| | | 164 (β′) | 0.73 | 80[l] |
| D-Sorbitol | -4 | 80 | 0.76 | 15[c] |

[a]Trasi et al. [28]
[b]Descamps et al. [16]
[c]Huang et al. [21]
[d]Kawakami [17]
[e]Unpublished result
[f]Zhang et al. [29]
[g]Kawakami et al. [18]
[h]Dudognon et al. [19]
[i]See text
[j]Kawakami [14]
[k]Chang et al. [30]
[l]Gui et al. [31]

investigation; however, one assumption for nucleation above $T_g$ is related to molecular cooperativity. For ibuprofen glass, the characteristic length of the cooperatively rearranged region [23, 24] agreed with the critical size of the nuclei determined from Eqs. (2.1) and (2.2) [18]. A similar idea was provided by Fokin et al., who assumed the formation of "structural units" to explain the slower diffusion rate than expected, which is required to justify the observation of nucleation temperatures slightly above $T_g$. [25]. Presumably the nucleation in this temperature range is general for any compounds; however, it is difficult to detect if the temperature range overlaps with that for crystal growth. Nucleation at low temperatures is likely associated with local molecular mobility, observed as a secondary relaxation. Crystallization of indomethacin glass was enhanced after annealing at −10 °C, where the contribution of $\beta$ relaxation was suggested [26]. Suppression of $\gamma$ relaxation in celecoxib glass was found to significantly retard its crystallization [27], as discussed later, further indicating a relationship between nucleation and local molecular mobility detected as secondary relaxation.

Nuclei are expected to grow into large crystals at temperatures higher than those for nucleation. As molecular diffusion plays an important role in crystal growth, it occurs basically above $T_g$. The growth process can be captured visually using optical microscopy, where a quantitative evaluation of the growth rate is possible, although the use of a cover glass may influence the growth process. In addition, microscopic observation enables discrimination between bulk and surface growth, where the latter is generally faster [32–34].

Kinetic information obtained under a microscope cannot be directly correlated with the crystallization of bulk materials. The behavior of bulk materials is better captured by other techniques, including X-ray powder diffraction (XRPD) and DSC. Although XRPD is generally not recommended for quantitative analysis owing to issues related to the preferred orientation of crystals, the results provide an intuitive comprehension of crystallinity. DSC provides quantitative information on crystallinity, and the procedure used to calculate it depends on the properties of the compound. Figure 2.3 depicts various options for calculating crystallinity. If crystallization occurs during DSC heating, crystallization enthalpy may be used for the calculation. Crystallization enthalpy is expected to be proportional to the amount of amorphous part. In this procedure, attention is required to the difference in crystallization temperature. As can be found in Fig. 2.1, the crystallization enthalpy is the distance between the supercooled liquid and crystal lines. Because the supercooled liquid line is steeper than that of the crystal, crystallization enthalpy increases with increasing temperature. Thus, crystallinity must be determined using the following equation:

$$\mathrm{X}(\%) = 100\left\{\Delta H / \left(\Delta H_q - \int_T^{Tq} \Delta Cp dT\right)\right\} \tag{2.4}$$

where $\Delta H$ and $\Delta H_q$ are the crystallization enthalpies of the sample of interest and control glass, respectively. $T$ and $T_q$ are their crystallization temperatures, and $\Delta C_p$ is the difference in heat capacity between the amorphous and crystalline states.

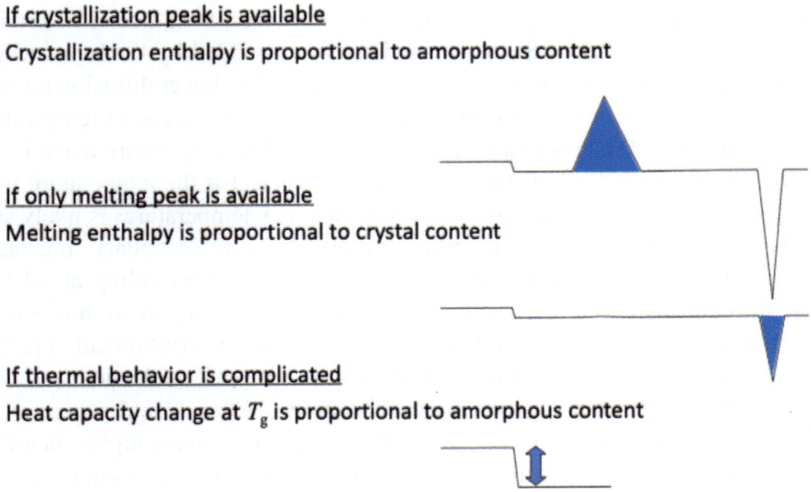


**Fig. 2.3** Various procedures to calculate crystallinity from DSC curves

Unless crystallization occurs during DSC heating, the melting enthalpy is proportional to the amount of crystal part. However, crystallization and melting enthalpies cannot be used for determining crystallinity if the DSC curve exhibits complicated behavior, including the overlapping of the peak of interest with other peaks. $\Delta C_p$ is always available for estimating the amount of the amorphous part. However, the drawback of this procedure is that the determined crystallinity value is sensitive to the analysis procedure. If the preparation of the glass sample and the annealing study to observe the crystallization rate are performed in an identical DSC program, a shift in the baseline after annealing enables the determination of crystallinity. Crystallinity must be calculated by considering the thermal behavior of the compound.

The crystal growth rate varies significantly depending on the compound, as presented in Tables 2.3, 2.4 and 2.5. The classification into three groups is explained in the next section. Growth is generally active between $T_m$ and $T_g$, and the growth rate is expected to exhibit a bell-shaped dependence on temperature, where the low- and high-temperature sides are controlled by kinetics and thermodynamics, respectively. However, unusually fast growth is sometimes observed just below $T_g$, which is explained by diffusionless crystal growth [35–39]. Figure 2.4 shows the growth rate of celecoxib [40], which exhibited an unusually fast growth rate below $T_g$. The growth rate at 45 °C is faster than that at the higher temperature near $T_g$. Because this is a one-dimensional comparison, the crystallization rate of the bulk material has much larger difference [41]. Possible mechanisms for this observation include crystal growth by coalescence of embryos, reduced interfacial tension at the liquid/crystal surface, and the incorporation of nearby molecules using local molecular motion [39].

In crystallization experiments, data are sometimes poorly reproducible. This happens especially for compounds, of which the nucleation is heterogeneous, and the crystal growth rate is fast, as the fast crystal growth stresses the presence or absence of nuclei that appear stochastically in individual samples. Figure 2.5 shows the evolution of crystallinity of acetaminophen glass, where the large variation in data is evident. This observation was made using samples in DSC pans that usually provide highly reproducible results. Thus, data variation is anticipated to be much larger for samples prepared by practical methods. Acetaminophen possesses a fast crystal growth rate as shown later. The Avrami parameter for this fitting was ca. 2, indicating heterogeneous nucleation, which can also be supported by observation using optical microscopy. The reproducibility in the crystallization study requires more attention and should be analyzed from a viewpoint of statistical thermodynamics, as it is closely related to the crystallization mechanism.

**Table 2.3** Examples of Class 1 compounds

| Compounds | $M_w$ (Da) | $T_m$ (°C) | $T_g$ (°C) | $T_g/T_m$ | $\Delta C_p$ (J/mol/K) | $\Delta H_m$ (kJ/mol) | $\Delta S_m$ (J/mol/K) | Growth rate (μm/s) |
|---|---|---|---|---|---|---|---|---|
| Antipyrin | 188 | 111 | −25 | 0.65 | 90.8 | 25.2 | 65.6 | – |
| Anthranilic acid | 137 | 147 | 5 | 0.66 | – | 22.8 | 54.3 | – |
| Atenolol | 266 | 153 | 22 | 0.69 | – | 37.5 | 88.0 | – |
| Atovaquone | 367 | 219 | – | – | – | 33.5 | 68.1 | – |
| Benzamide | 121 | 127 | −10 | 0.66 | – | 21.7 | 54.2 | – |
| Benzocaine | 165 | 89 | −31 | 0.67 | – | 22.6 | 62.4 | – |
| Caffeine | 194 | 237 | 72 | 0.68 | – | 20.8 | 40.8 | – |
| Carbamazepine | 236 | 192 | 61 | 0.72 | – | 25.5 | 55.9 | 45.9 |
| Chlorpropamide | 277 | 118 | 17 | 0.74 | 133 | 27.4 | 70.0 | – |
| Chlorzoxazone | 170 | 191 | 38 | 0.67 | – | 25.6 | 55.2 | – |
| Clofibric acid | 215 | 121 | – | – | – | 29.0 | 73.6 | – |
| Diflunisal | 250 | 213 | – | – | – | 35.6 | 73.2 | – |
| Felbinac | 212 | 164 | 24 | 0.68 | – | 29.8 | 68.2 | – |
| Flufenamic acid | 281 | 135 | 17 | 0.71 | – | 27.1 | 66.4 | – |
| Griseofulvin | 353 | 218 | 89 | 0.74 | 154 | 39.1 | 79.6 | 66.6 |
| Haloperidol | 376 | 152 | 33 | 0.72 | – | 54.3 | 128 | – |
| Indoprofen | 281 | 212 | 50 | 0.67 | – | 36.0 | 74.2 | – |
| Lidocaine | 234 | 68 | −39 | 0.69 | – | 16.7 | 49.0 | – |
| Mefenamic acid | 241 | 231 | – | – | – | 39.4 | 78.2 | – |
| Naproxen | 230 | 157 | 56 | 0.77 | – | 32.4 | 75.3 | – |
| Nepafenac | 254 | 183 | – | – | – | 42.8 | 93.8 | – |
| Phenacetin | 179 | 136 | 2 | 0.67 | – | 31.5 | 77.0 | – |
| Piroxicam | 331 | 201 | – | – | – | 35.6 | 75.1 | 51.3 |
| Probenecid | 285 | 199 | – | – | – | 40.4 | 85.6 | – |
| Saccharin | 183 | 228 | – | – | – | 29.5 | 58.9 | – |
| Salicylic acid | 138 | 159 | – | – | – | 24.9 | 57.6 | – |
| Theophylline | 180 | 272 | 94 | 0.67 | – | 29.6 | 54.3 | – |
| Tolbutamide | 270 | 128 | 5 | 0.69 | 117 | 26.2 | 64.6 | 20.6 |
| Tolfenamic acid | 262 | 213 | 63 | 0.69 | – | 38.8 | 79.8 | – |
| Average | 237 | 172 | 27 | 0.69 | 124 | 31.1 | 69.9 | 46.1 |

$M_w$ molecular weight, $\Delta H_m$ melting enthalpy, $\Delta S_m$ melting entropy, $\Delta C_p$ heat capacity change at $T_g$. Tables 2.2, 2.3, and 2.4 were created by updating previously reported table [46] using our data. Crystal growth rate was adapted from literature [47] except for our data*

## 2.3 Crystallization Tendency

The ease of crystallization depends on the compound and is known as the crystallization tendency. Alternatively, the term “glass-forming ability” describes the opposite meaning, the resistance to crystallization. Compounds with low crystallization tendencies are good candidates for ASD technology. Therefore, investigating crystallization tendency is the first step in the development of ASDs.

**Table 2.4** Examples of Class 2 compounds

| Compounds | $M_w$ (Da) | $T_m$ (°C) | $T_g$ (°C) | $T_g/T_m$ | $\Delta C_p$ (J/mol/K) | $\Delta H_m$ (kJ/mol) | $\Delta S_m$ (J/mol/K) | Growth rate (μm/s) |
|---|---|---|---|---|---|---|---|---|
| Acetaminophen | 151 | 169 | 23 | 0.67 | 103 | 27.2 | 61.5 | 100 |
| Aripiprazole | 448 | 141 | 33 | 0.74 | 233 | 40.9 | 98.9 | – |
| Bifonazole | 310 | 149 | 16 | 0.68 | 133 | 39.2 | 92.9 | 4.0 |
| Cinnarizine | 369 | 120 | 7 | 0.71 | 143 | 40.9 | 104 | 1.8 |
| Clofoctol | 365 | 88 | −4 | 0.75 | 256 | 35.2 | 97.5 | – |
| Dibucaine | 343 | 65 | −35 | 0.70 | 134 | 29.2 | 86.4 | 2.7 |
| Droperidol | 379 | 143 | 29 | 0.73 | 197 | 40.0 | 96.1 | 8.2 |
| Fluconazole | 306 | 140 | 31 | 0.74 | 150 | 32.4 | 78.4 | – |
| Flurbiprofen | 244 | 115 | −5 | 0.69 | 100 | 27.4 | 70.6 | 7.9 |
| Nifedipine | 346 | 172 | 46 | 0.72 | 144 | 38.2 | 85.8 | – |
| Phenobarbital | 233 | 174 | 42 | 0.70 | 126 | 28.7 | 64.2 | – |
| Phenylbutazone | 308 | 106 | −6 | 0.70 | 132 | 27.6 | 72.8 | – |
| Progesterone | 314 | 130 | 10 | 0.70 | 126 | 27.3 | 67.6 | – |
| Tamoxifen | 372 | 101 | −13 | 0.69 | 141 | 34.5 | 92.2 | – |
| Tolazamide | 311 | 172 | 18 | 0.65 | 171 | 43.4 | 97.5 | 5.9 |
| Average | 320 | 132 | 13 | 0.70 | 153 | 34.1 | 84.4 | 18.6 |

In the field of physics, the most widely used method for determining crystallization tendency has been to identify the critical cooling rate required to achieve vitrification. This procedure was adapted by Blaabjerg et al. for evaluating pharmaceutical glasses: compounds that crystallize even at 750 °C/min were noted as Class 1, those that can be vitrified at ca. 10–20 °C/min were classified as Class 2, and compounds requiring very slow cooling rates below 2 °C/min for crystallization were categorized as Class 3 [43, 44].

An alternative method was provided by Baird et al. [45], which enabled classification using a cooling/heating cycle in DSC. In this procedure, compounds that crystallize during cooling from the melt at 20 °C /min are classified as Class 1. Those that do not crystallize during cooling from the melt but crystallize during subsequent reheating at 10 °C/min are classified as Class 2 compounds. Class 3 compounds do not crystallize during the cooling/reheating cycle mentioned above. As this procedure requires only one measurement per compound and can be performed using conventional DSC, it is frequently employed in the field of pharmaceutical science. However, one drawback of this method is the lack of interest in the lowest temperature achieved at the end of the cooling process. In the case of celecoxib glass, cold crystallization was not observed during DSC heating after cooling to room temperature, whereas crystallization was observed during heating after deeper cooling [17]. The availability of cold crystallization was influenced by the temperature achieved during cooling, as the nucleation temperature is in the freezing temperature range. Therefore, it is necessary to define the lowest temperature achieved after cooling. Finding of the nucleation temperatures slightly above $T_g$ is the general observation. Thus, it should be sufficient for Class 2 compounds to

**Table 2.5** Examples of Class 3 compounds

| Compounds | $M_w$ (Da) | $T_m$ (°C) | $T_g$ (°C) | $T_g/T_m$ | $\Delta C_p$ (J/mol/K) | $\Delta H_m$ (kJ/mol) | $\Delta S_m$ (J/mol/K) | Growth rate (μm/s) |
|---|---|---|---|---|---|---|---|---|
| Aceclofenac | 354 | 153 | 10 | 0.66 | 170 | 42.3 | 99.3 | 7.4 |
| Carvediol | 406 | 114 | 36 | 0.80 | 300 | 48.0 | 124 | 0.07 |
| Celecoxib | 381 | 163 | 58 | 0.76 | 171 | 37.4 | 85.8 | 55.3 |
| Clotrimazole | 345 | 141 | 28 | 0.73 | 114 | 33.3 | 80.4 | 0.02 |
| Curcumin | 368 | 182 | 62 | 0.74 | 190 | 50.1 | 110 | – |
| Emtricitabine | 247 | 154 | 66 | 0.78 | 148 | 29.4 | 68.8 | – |
| Efavirenz | 316 | 138 | 33 | 0.74 | 90.4 | 15.3 | 37.1 | 60.7 |
| Felodipine | 384 | 147 | 45 | 0.76 | 142 | 31.0 | 73.8 | 0.86 |
| Fenofibrate | 361 | 80 | −19 | 0.72 | 157 | 33.0 | 93.4 | – |
| Ibuprofen | 206 | 76 | −44 | 0.66 | 74.6 | 26.5 | 75.9 | 1.50* |
| Indomethacin | 358 | 161 | 45 | 0.73 | 149 | 37.6 | 86.6 | 0.47 |
| Itraconazole | 706 | 168 | 58 | 0.75 | 302 | 57.6 | 131 | – |
| Ketoconazole | 531 | 147 | 44 | 0.75 | 215 | 52.9 | 126 | – |
| Ketoprofen | 254 | 95 | −3 | 0.73 | 120 | 28.3 | 76.9 | 0.11 |
| Loratadine | 383 | 134 | 35 | 0.76 | 113 | 27.3 | 67.1 | 0.84 |
| Miconazole | 417 | 86 | 1 | 0.76 | – | 32.8 | 91.3 | 0.03 |
| Naftopidil | 393 | 126 | 22 | 0.74 | 259 | 53.8 | 135 | – |
| Nilutamide | 317 | 155 | 33 | 0.72 | – | 31.0 | 72.4 | 50.0 |
| Nimesulide | 308 | 150 | 21 | 0.70 | 139 | 33.4 | 78.9 | 216 |
| Pimozide | 462 | 219 | 54 | 0.66 | 171 | 42.7 | 86.8 | 92.4 |
| Probucol | 517 | 126 | 27 | 0.75 | 154 | 39.3 | 98.5 | 2.0 |
| Procaine | 236 | 61 | −39 | 0.70 | 155 | 26.2 | 78.4 | – |
| Ribavirin | 244 | 168 | 56 | 0.75 | 173 | 45.7 | 104 | – |
| Ritonavir | 721 | 122 | 47 | 0.81 | 339 | 65.3 | 165 | – |
| Simvastatin | 419 | 139 | 29 | 0.73 | 172 | 34.4 | 83.4 | – |
| Stiripentol | 234 | 76 | −25 | 0.71 | 128 | 30.5 | 87.4 | – |
| Sulfamerazine | 264 | 237 | 60 | 0.65 | 145 | 44.4 | 87.0 | – |
| Terfenadine | 472 | 152 | 55 | 0.77 | 245 | 52.4 | 123 | 1.02* |
| Average | 379 | 138 | 28 | 0.73 | 174 | 38.6 | 93.8 | 30.5 |

crystallize if they are cooled to 10 °C below $T_g$. Based on this definition, celecoxib is categorized as a Class 3 compound. Examples of compounds in each class, based on this revised definition, are presented in Tables 2.3, 2.4, and 2.5.

Several factors affect the crystallization tendencies of certain compounds. As molecular complexity is an important factor, larger values were found for molecular weight and melting entropy of compounds in the higher classes. Differences in nucleation and crystal growth temperatures are generally considered to influence crystallization tendencies. As discussed in the previous section, the nucleation temperature for the stable form is usually slightly above $T_g$. It easily overlaps with the temperature region of crystal growth, especially when $T_g/T_m$ is small; thus, compounds in the lower class tend to exhibit a small $T_g/T_m$. The crystal growth rate is another important factor in determining crystallization tendencies. Class 1

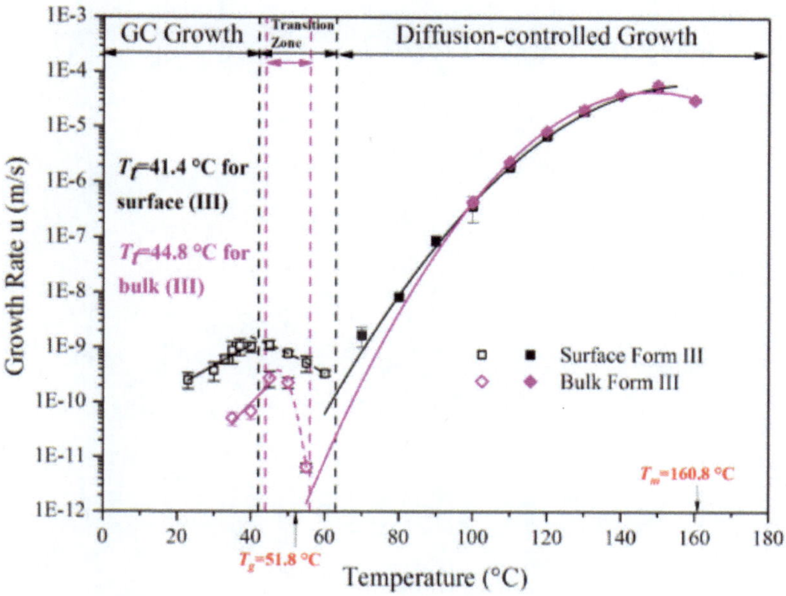


**Fig. 2.4** Surface and bulk crystal growth rates of celecoxib in supercooled liquid and glass states. Crystal forms are presented in the figure [40]. (The figure was adapted with permission from American Chemical Society)

compounds tend to have relatively fast growth rates, whereas those in higher classes exhibit low growth rates, with some exceptions, including celecoxib, efavirenz, nimesulide, and pimozide. Therefore, as shown in Fig. 2.6, a rough correlation exists between the crystal growth rate and melting entropy [48]. The exceptionally rapid growth in some Class 3 compounds can be attributed to their well-separated nucleation and crystal growth temperatures. Most compounds marketed as ASDs belong to Class 3 [8].

## 2.4 Possible Correlation Between Relaxation Time and Physicochemical Properties of Glasses

The enthalpy of the amorphous material generally lies on the supercooled liquid line above $T_g$ (Fig. 2.1) because the equilibrium state can be easily achieved owing to its sufficiently high molecular mobility. Below $T_g$, the enthalpy of the amorphous material is assumed to lie on the fresh glass line indicated in the figure, if it is prepared by cooling the melt. It consumes excess enthalpy relative to the metastable glass line to reduce its Gibbs energy, a process called as structural relaxation ($\alpha$-relaxation). The Kohlrausch–Williams–Watts (KWW) equation shown below is frequently used to describe this relaxation process [49, 50].

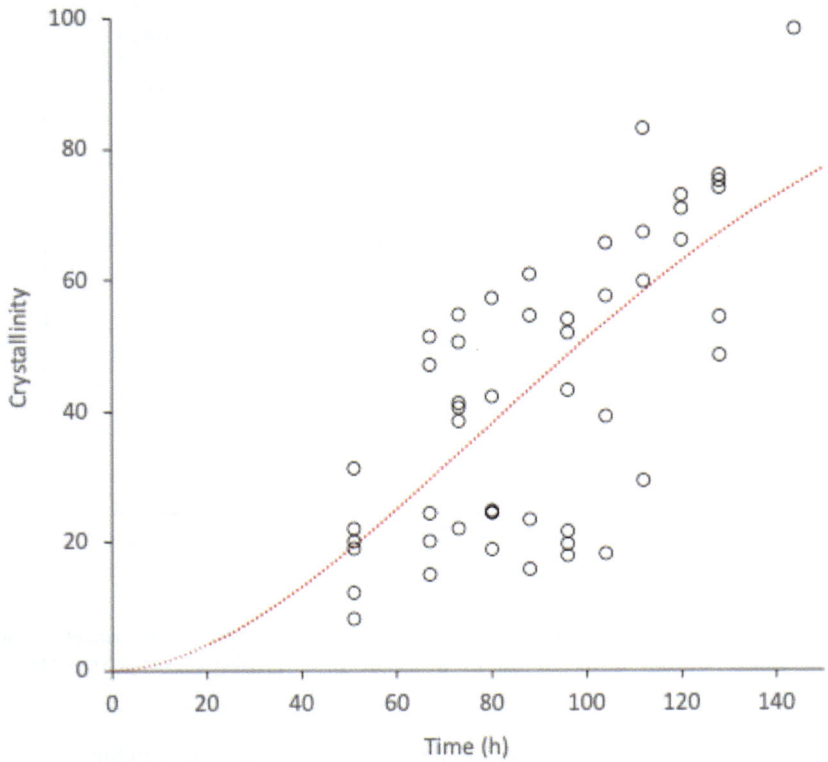


**Fig. 2.5** Evolution of crystallinity of amorphous acetaminophen as a function of time at 30 °C. The experimental procedure can be found in literature [42]. The red line shows the best-fitting by Avrami–Erofeev equation. Unpublished result

$$\phi(t,T) = \exp\left[-\left(\frac{1}{\tau(T)}\right)\right]^{\beta}, 0 < \beta \le 1 \quad (2.5)$$

where $\phi$ is the fraction of the excess enthalpy remaining with respect to the metastable glass state. The relaxation time $\tau_\alpha$ represents the molecular mobility of the glass and is regarded as a function of temperature $T$. The fitting parameter $\beta$, which takes values between 0 and 1, practically works as a fitting parameter and is believed to reflect the distribution of relaxation times.

During storage or annealing experiments, fresh glass (point *a* in Fig. 2.1) consumes excess enthalpy to reach state *b*. If this glass is subjected to DSC heating, it changes its state to *c*, where the slope of the *b-c* line may be approximated as parallel to the fresh glass line. Subsequently, the state jumps to *d* to return to equilibrium because of the acquisition of sufficient molecular mobility at $T_g$, followed by a state change along the supercooled liquid line with increasing temperature. Figure 2.7a shows the DSC curves of the ritonavir glass after annealing at 40 °C for various time

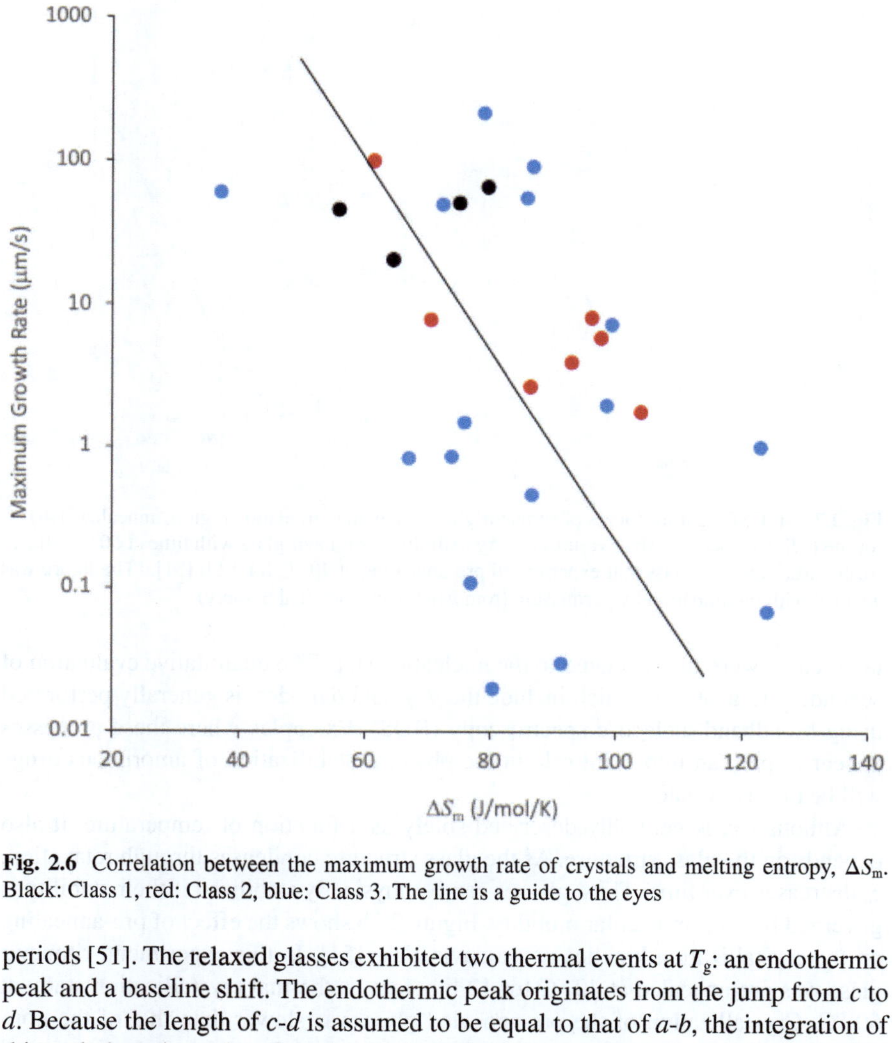


**Fig. 2.6** Correlation between the maximum growth rate of crystals and melting entropy, $\Delta S_m$. Black: Class 1, red: Class 2, blue: Class 3. The line is a guide to the eyes

periods [51]. The relaxed glasses exhibited two thermal events at $T_g$: an endothermic peak and a baseline shift. The endothermic peak originates from the jump from *c* to *d*. Because the length of *c-d* is assumed to be equal to that of *a-b*, the integration of this endothermic peak, referred to as the enthalpy recovery peak, provides quantitative information on the degree of relaxation occurring during storage/annealing. The baseline shift results from a change in heat capacity. Because heat capacity is the derivative of enthalpy with respect to temperature, the change in heat capacity at $T_g$ can be found as a change in slope, as illustrated in Fig. 2.1. The growth of the recovery peak describes proceeding of relaxation during storage/annealing at 40 °C. Application of the KWW equation provides relaxation parameters to acquire information on molecular mobility and its distribution from τ and β, respectively.

As $\tau_\alpha$ reflects global molecular mobility, it is expected to correlate with both physical and chemical stability. Some observations revealed a correlation between $\tau_{a\alpha}$ and crystallization onset time [52]; however, the contributions of other factors, especially local molecular mobility, also appear to play a role [26, 27, 37, 53]. The

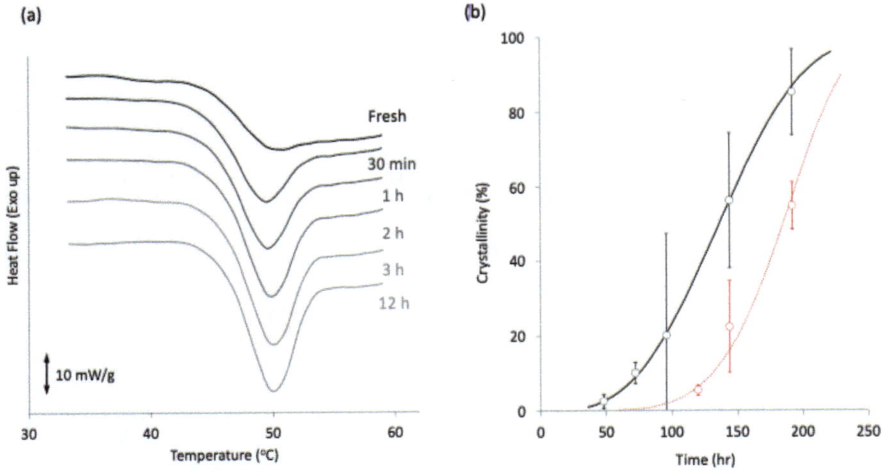


**Fig. 2.7** (**a**) DSC heating curves of ritonavir glass in the glass transition region, annealed at 40 °C for the indicated periods. (**b**) Evolution of crystallinity of ritonavir glass with time at 60 °C. Black: fresh glass. Red: the glass that experienced pre-annealing at 40 °C for 12 h [51]. (The figure was adapted with modification by permission from American Chemical Society)

latter cases were likely related to the nucleation step. The quantitative evaluation of secondary relaxations, which include the $\beta$, $\gamma$, and $\delta$ modes, is generally performed using broadband dielectric spectroscopy (BDS). Examples where these processes appear to play an important role in the physical stabilization of amorphous drugs will be discussed later.

Although $\tau_\alpha$ is generally described solely as a function of temperature, it also depends on the glass structure. As the glass structure condenses through relaxation, $\tau_\alpha$ decreases over time. Thus, glasses can be physically stabilized if their stability is governed by their molecular mobility. Figure 2.7b shows the effect of pre-annealing on the crystallization kinetics of ritonavir glass [51]. In this experiment, ritonavir glass was pre-annealed at 40 °C for 12 h before isothermal crystallization study at 60 °C. Crystallization of the pre-annealed glass was slower than that of the non-annealed glass. Pair-distribution function analysis and Fourier transform infrared spectroscopy measurements indicated a rearrangement of hydrogen bonding during pre-annealing at 40 °C, which was also considered to be the origin of stabilization.

The correlation between $\tau_\alpha$ and chemical stability should also be observed if the decomposition of interest is governed by molecular mobility. Both good and poor correlations were observed, depending on the decomposition mechanism [54]. Chemical stability can also be improved by annealing because of a decrease in molecular mobility [55].

## 2.5 Temperature-Dependence of Amorphous Properties

The temperature dependence of the physicochemical properties of general materials can be described using the Arrhenius equation; however, this model does not apply to amorphous materials over a wide range of temperatures. Their non-Arrhenius behavior can be quantified using the concept of fragility. Because fragile glasses frequently exhibit large changes in specific heat capacity at $T_g$, the degree of the change in the heat capacity can be used as a measure of fragility, which is known as "thermodynamic fragility" [56]. However, "dynamic fragility" is more widely used to describe the non-Arrhenius behavior of liquids and glasses.

Various methods have been proposed to calculate dynamic fragility [57]. One method focuses on configurational entropy $S_c$, of which temperature-dependence can be described as

$$S_c = \int_{T_K}^{T} \left( \frac{C_p^{conf}(T)}{T} \right) dT \tag{2.6}$$

where $C_p^{conf}$ denotes the configurational heat capacity. Given that $C_p^{conf}(T)$ is constant [57], the following equation can be obtained by integrating Eq. (2.6) up to $T_m$ using the configurational heat capacity at $T_g$.

$$\frac{1}{T_K} = \frac{1}{T_m} \left( 1 + \frac{\Delta H_m}{C_p^{conf}(T_g)T_g} \right) \tag{2.7}$$

where $\Delta H_m$ is the melting enthalpy. The strength parameter, $D$, can be calculated using the Vogel–Tammann–Fulcher (VTF) equation:

$$\tau(T) = \tau_0 \exp\left( \frac{DT_0}{T - T_0} \right) \tag{2.8}$$

where $\tau_0$ is the timescale of the vibrational motion (ca. $10^{-14}$ s), $T_0$ is a constant analogous to $T_k$, and fragility $m$ can be determined using the following equation.

$$m = \frac{DT_0 / T_g}{2.303\left(1 - T_0 / T_g\right)^2} \tag{2.9}$$

Dependence of $T_g$ on the ramp rate, $q$, in DSC measurements also enables calculation of fragility. $D$ and $m$ are given by the following equations.

$$m = \frac{d\log\tau}{d\left(T_g / T\right)}\bigg|_{T=Tg} = \frac{1}{2.303T_g}\frac{\Delta E_{Tg}}{R}, \tag{2.10}$$

$$\text{where } -\frac{\Delta E_{Tg}}{R} = \frac{d(\ln q)}{d(1/T_g)} \tag{2.11}$$

The heating rate to observe $T_g$ must be equal to the cooling rate of the melt used during glass formation [58]. However, this is possible only for a part of Class 3 compounds in the classification system of crystallization tendency, as hot crystallization occurs under slow cooling rates from the melt for other compounds. From a practical standpoint, the use of a constant cooling rate does not significantly affect the calculated fragility values [59].

Another method for calculating fragility involves analyzing the glass transition width, for which the following equation is applied.

$$\frac{1}{T_g^{\text{onset}} - T_g^{\text{end}}} = \frac{C\Delta E_{Tg}}{R} \tag{2.12}$$

where $T_g^{\text{onset}}$ and $T_g^{\text{end}}$ denote the onset and endpoint temperatures of $T_g$, respectively. Calculations using this equation require a material-dependent constant $C$. When this value is not available, a universal value of 5 is sometimes applied. Fragility values obtained from different calculation methods may not be consistent because each method is based on different assumptions [60].

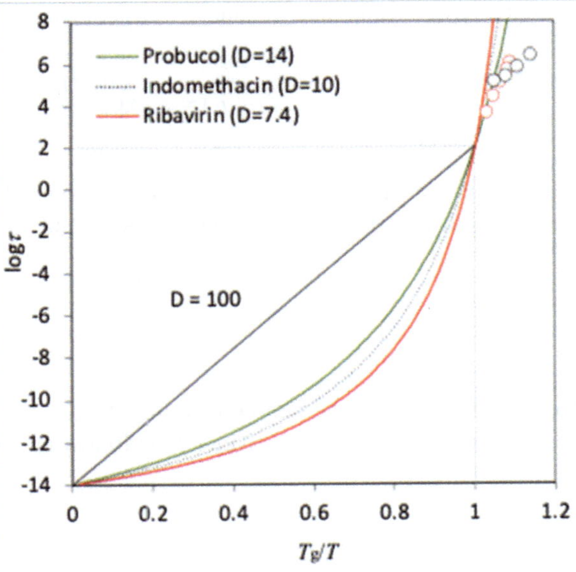


**Fig. 2.8** Relaxation times of liquid/glass with various fragility values. The relaxation time data below $T_g$ were taken from the literature. Red and black circles represent indomethacin [63] and ribavirin [64], respectively

Figure 2.8 demonstrates molecular mobility (relaxation time, $\tau$) of liquid/glass with various fragility values. A liquid/glass with small fragility ($m < 40$, $D > 30$) is regarded as strong, whereas one with large fragility ($m > 75$, $D < 10$) is called fragile [57, 61]. As illustrated in Fig. 2.8, fragile systems exhibit significant nonlinearity in molecular mobility above $T_g$.

However, the above discussion is only valid under equilibrium conditions. Thus, below $T_g$, the VTF equation is applicable only when the state of the glass is on the extrapolation of the supercooled liquid line. If not, as is typically true for real glasses, molecular mobility approaches Arrhenius behavior [62]. Also plotted in Fig. 2.8 is the relaxation times of indomethacin and ribavirin glasses below $T_g$, which do not follow the VTF curves.

The overall crystallization behavior is frequently reported to correlate with molecular mobility. Figure 2.9 shows the initiation time for the crystallization of Class 1 and 2 compounds as a function of $T_g/T$ [42]. Regardless of the diversity of their chemical structures, the initiation times for crystallization occurred on a straight line. The initiation times for class 3 compounds were longer than prediction from this line, because nucleation serves as the rate-limiting step in their crystallization process.

The chemical stability of a glass is occasionally governed by its molecular mobility [54, 65]. Thus, the nonlinearity of the decomposition rate may be observed below and above $T_g$ [65], raising concerns about the applicability of accelerated chemical stability studies for pharmaceutical glasses. However, from a practical perspective, the temperature range of interest for pharmaceutical development is relatively narrow, and behavior below $T_g$ can likely be approximated by the Arrhenius rule. Figure 2.10 shows the chemical stability of albendazole ASDs prepared by spray drying or hot-melt extrusion as a function of storage temperature, where the

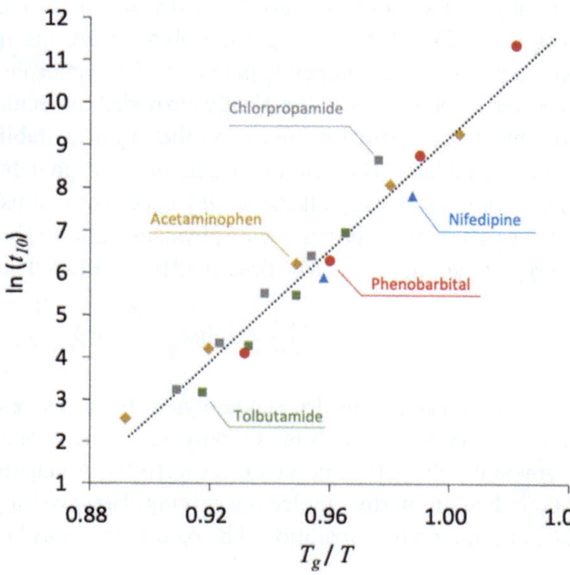


**Fig. 2.9** (a) Initiation time for crystallization of class I/II compounds as a function of $T_g/T$. $t_{10}$ represents the time to reach 10% crystallinity in minutes. (The figure was adapted with modification by permission from American Chemical Society [42])

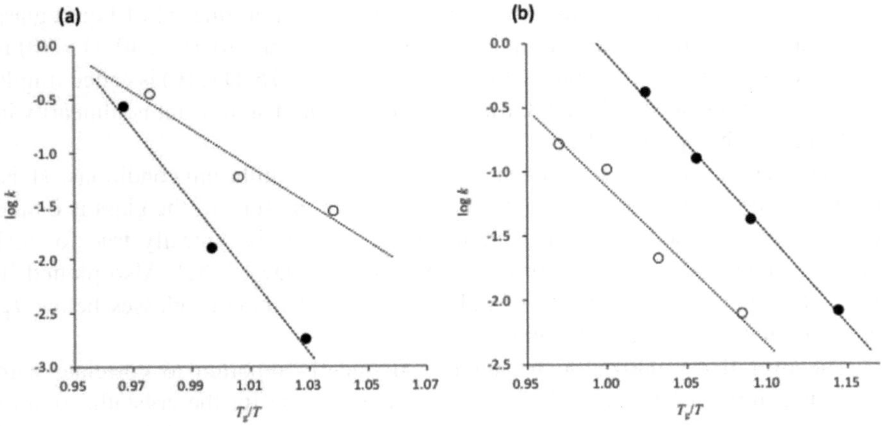


**Fig. 2.10** Decomposition rates (per month) of albendazole (ALZ) ASDs as a function of temperature. (**a**) Spray-dried, (**b**) Hot-melt extruded ASDs. Compositions for open and closed symbols are ALZ/PVPVA = 2/8 and ALZ/Soluplus = 2/8, respectively. Unpublished results

decomposition rate is linearly dependent on temperature. Therefore, accelerated stability tests can provide practical information on chemical stability unless crystallization occurs.

## 2.6 Physical Stabilization of Amorphous Drug by Mixing with Polymers

The physical stability of amorphous drugs can be improved by mixing them with excipients, for which hydrophilic polymers are the most commonly used. If the polymer has a much higher $T_g$ than that of the amorphous drug, physical stability is expected to be improved drastically, provided molecular mobility is coupled. Other possible mechanisms for improving the physical stability of amorphous drugs will be discussed later. The amount of drug must be smaller than its solid solubility limit in the polymer matrix; otherwise, any excess drug must be phase-separated to form a drug-rich phase that has lower physical stability than expected [60]. The Gibbs energy of their mixing can be described by the Flory-Huggins equation shown below:

$$\frac{\Delta G}{kT} = \phi_d \ln \phi_d + \frac{\phi_p}{N} \ln \phi_p + \chi \phi_d \phi_p, \tag{2.13}$$

where $\phi_d$ and $\phi_p$ are the drug and polymer fractions, respectively. $\chi$ is the interaction parameter between the drug and polymer. $k$ and $N$ are the Boltzmann constant and segment number of the polymer, respectively. A negative $\chi$ always yields a negative $\Delta G$, indicating perfect molecular mixing. However, a positive $\chi$ may offer positive $\Delta G$, causing phase separation. The parameter $\chi$ has been determined using various methodologies, including systematic investigations of the $T_g$ in binary mixtures [60] and observations of melting temperature depression [66, 67].

However, the mixing state of ASDs depends on the preparation method because of the slow dynamics of amorphous solids, and the resulting ASD may not be in the equilibrium state. If an ASD is prepared under mechanical stress at elevated temperatures, as in hot-melt extrusion, higher thermodynamic miscibility is expected. This is because the mixing of the drug and the excipient polymer is generally an endothermic process. In such cases, phase separation may proceed slowly at storage temperatures over time [60]. Spray drying is a rapid process to form ASD particles, making the achievement of equilibrium unlikely during preparation. The diffusion rates of the polymer and drug during drying of sprayed droplets are different because of a large difference in their molecular weights [68], which may allow spontaneous formation of a core-shell structure in the dried particles. Figure 2.11 shows a comparison of the DSC curves of spray- and freeze-dried ASDs composed of psicose and PVP [66]. When the proportion of psicose was below 40%, only one $T_g$ was observed for the freeze-dried ASDs, suggesting homogeneous structure. However, multiple $T_g$ values were observed in the spray-dried ASDs regardless of mixing ratio, indicating structural heterogeneity. A similar phase separation was reported for a miconazole-PVPVA ASD, where the mixing state varied depending on the collection site (cyclone, collector, etc.) in the spray dryer [69]. Solvent selection also influences the mixing state in spray drying, as the miscibility between two solutes, a polymer and a drug, depends on the solvent type used. Interactions between the two solutes can be more attractive if they exhibit low miscibility with the solvent [70]. Therefore, the addition of an antisolvent to the spray-dried solvent may improve the mixing of the components [71]. The features of each preparation method must be well understood to predict the performance of ASDs.

The possible stabilization mechanisms of amorphous drugs using polymeric excipients include a decrease in molecular mobility of the drug, direct interaction between drug and molecules to inhibit crystallization, steric hindrance and increased viscosity that inhibit crystal growth, adsorption of polymer molecules onto nuclei

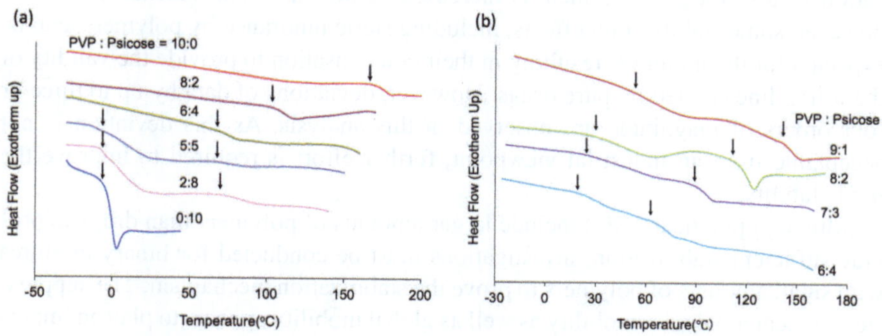


**Fig. 2.11** DSC curves of ASDs prepared by (**a**) freeze-drying and (**b**) spray-drying [66]. ASDs are composed of PVP and psicose. Their mixing ratios are indicated in the figure. $T_g$ values are indicated by arrows. (The figure was adapted with permission from Elsevier)

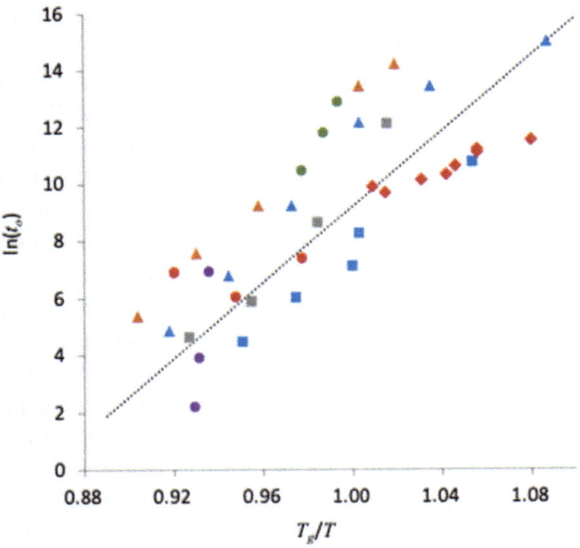


**Fig. 2.12** Initiation time, $t_0$ (in minutes), for crystallization of binary ASDs as a function of $T_g/T$. (▲, ■,●) Nifedipine/PVP ASDs prepared by melt-quench [72, 73] followed by milling [74]. (▲,●) Phenobarbital/PVP ASDs prepared by melt-quench [72, 73]. (■,◆) Sanofi-Aventis compound/hydroxypropyl methylcellulose phthalate ASDs prepared by spray-drying stored under dried and humid conditions, respectively [75]. (●) Fenofibrate ASDs prepared by spray-drying or hot-melt extrusion using PVPVA or HPMCAS as excipients [18]. Definition of initiation time for crystallization is slightly different depending on literature, but it does not impact this analysis. (The figure was reproduced with modification under CC-BY 4.0 license [46])

and/or crystal surfaces to suppress their growth, and an impact on the nucleation process. If molecular mobility is the dominant factor for crystallization, as same as for pure drugs, *the* $T_g/T$ of ASDs is expected to correlate with the initiation time of crystallization, as presented in Fig. 2.9. Figure 2.12 shows their relationship for drug/polymer binary mixtures. Manufacturing processes include several destabilization factors for glasses, such as increased surface area and mechanical stress; however, some stabilization effects, including steric hindrance by polymers, can be expected for the mixtures, resulting in their compensation to provide the validity of the fitting line created by pure drugs. However, deviations of data by up to three or four orders of magnitude are observed in this analysis. As this deviation is not acceptable from an industrial viewpoint, further effort is required to improve the predictability.

Although practical ASDs include larger amounts of polymers than drugs to provide sufficient stabilization, investigations must be conducted for binary mixtures with small amounts of polymers to prove the stabilization mechanism. The suppression of local molecular mobility as well as global mobility appears to play an important role in inhibiting crystallization. For instance, the stability of celecoxib glass was significantly improved in the presence of 10% octaacetylmaltose, where $\gamma$-relaxation was found to be suppressed using BDS [27]. The $\gamma$-process of celecoxib glass is related to the mobility of the sulfonamide group, which can form hydrogen bonds with the nitrogen atom of the pyrazole ring of another celecoxib molecule during crystallization [40]. Thus, it is reasonable to assume that suppression of $\gamma$-relaxation effectively inhibits crystallization of celecoxib glass.

The addition of polymers is known to alter the nucleation temperature. The inclusion of 10% poly(acrylic acid) and HPMCAS was reported to decrease the nucleation temperature of acetaminophen glass by ca. 50 and 30 °C, respectively [28]. Nucleation of fluconazole glass was accelerated, and the temperature for the fastest nucleation rate was decreased by the addition of 10% poly(ethylene oxide) [29]. The optimum temperature for nucleation of ibuprofen glass is −15 °C; however, it decreased to −50 °C by adding 2% Eudragit L100, whereas the addition of 2% HPMCAS appeared to widen nucleation temperature range [18]. The temperature range for nucleation of acetaminophen glass is also reported to be widened in the presence of 10% polymer [28]. Although further examination is required to explain these observations, they may be attributed to changes in molecular cooperativity and glass/nucleus interfacial tension [18].

## 2.7 Physical Stabilization of Amorphous Drug by Mixing with Mesoporous Materials

Another type of excipient that has attracted interest for stabilizing amorphous drugs is mesoporous materials. When amorphous drugs are filled in mesopores, their crystallization can be suppressed by two mechanisms: direct interaction with the pore surface and size-confinement effects in the pores. The effect of confinement in nanopores is still under debate, with both accelerated and suppressed mobilities being reported. The diffusion of water molecules in a graphene slab with a pore size of 3.1 nm has been reported to be accelerated [76]. This is because the molecular motion parallel to the pore wall, which was the major mode of movement, was likely to be faster than that perpendicular to the surface [77]. However, suppression of the diffusion of water molecules in 0.7 nm pores was also observed, most likely due to restricted translational motion. Similar observations have been made for pharmaceutical compounds. BDS and NMR studies of ibuprofen glass in MCM-41 with 3.6 nm pore indicated its higher molecular mobility near and below $T_g$ than that of bulk ibuprofen [78, 79]. The same has been reported for flufenamic acid glass [80]. Studies on prednisolone glass in 1.60 and 2.16 nm pores revealed that larger pores provided greater stabilization [81]. In contrast, when flufenamic acid glass was entrapped in 3.2, 7.1, and 29 nm pores, crystallization was found only in the largest pores [80]. Nifedipine glass in pores smaller than 12 nm was reported to offer better stability than that in pores of 50–198 nm range [82]. These investigations suggest the presence of a specific pore size range for stabilization, which should depend on the interaction between the porous materials and the drug, as well as the physicochemical properties of the drug.

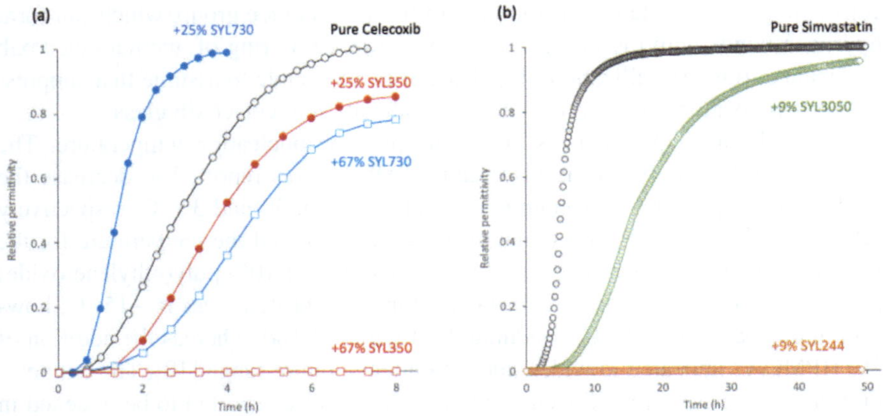


**Fig. 2.13** Changes in relative permittivity of (**a**) celecoxib glass at 95 °C [83] and (**b**) simvastatin glass at 90 °C [84] as a function of time in the presence of mesoporous silica, as observed by BDS. Relative permittivity is understood as crystallinity, if dependence of permittivity on molecular state (free or in the pore) is ignored. Pore and particle sizes of mesoporous silica are as follows. SYL 350: pore 21 nm, particle 4 μm; SYL730: pore 2.5 nm, particle 4 μm; SYL3050: pore 23 nm, particle 59 μm; SYL244: pore 23 nm, particle 2.5–3.7 μm

From a practical standpoint, the addition of large amounts of mesoporous material to capture all amorphous drugs in the pores is not ideal owing to concerns about pill burden. Thus, stabilization of entrapped molecules alone is insufficient. However, the stabilization effect appears to be effective even for overloaded drugs. For instance, crystallization of celecoxib glass was retarded in the presence of 25% mesoporous silica with 21 nm pores (Fig. 2.13a), although more than 80% silica is required for complete loading of the drug [83]. This stabilization effect was not observed for mesoporous silica with 2.5 nm pores, suggesting that a certain pore size is required to exert a stabilization effect on both internal and external drug molecules. The addition of 67% silica with larger pores completely inhibited crystallization. The entrapment study of simvastatin glass, using mesoporous silica with an identical pore size of 23 nm but different particle sizes of 3 and 59 μm, revealed that mesoporous particles with smaller particle sizes were more effective for stabilization (Fig. 2.13b) [84]. This observation was explained by the inhibition of crystal growth because of the larger number for the smaller particles, and larger surface area enabling easier exchange of molecules inside and outside the particles. The latter idea is consistent with the assumption made for explaining the observed effect of pore size. The addition of mesoporous materials appears to offer a double-sided stabilization effect, potentially accelerating or decelerating crystallization depending on the properties of the mesoporous materials [83, 84] and mixing ratios [85]. The stabilization mechanisms of mesoporous materials are completely different from those of polymeric excipients. Further investigations in this field may broaden the design strategies available for ASDs.

## 2.8 Summary and Outlook

This chapter discusses the physical stabilities of amorphous drugs and their mixtures with polymeric and mesoporous materials. Glass theory still involves many unsolved problems, such as the nucleation theory, origin of the glass transition, and the Kauzmann paradox. Among these, the nucleation theory is closely related to the quality and performance of ASDs; thus, a deeper understanding is highly demanded. Physical stabilization can be achieved by adding large amounts of excipients; however, this is expected to be reduced, for which the current interpretation of the stabilization mechanism using excipients is not sufficient. It is meaningful to elucidate these stabilization mechanisms at the molecular level by observing the relationship between local molecular mobility and crystallization behavior. Although ASD technology is applicable to any type of poorly soluble drug, challenges such as pill burden remain. Further improvements in ASD technology should be pursued by focusing on the fundamental aspects of glass theory.

## References


1. Brown D. Unfinished business: target-based drug discovery. Drug Discov Today. 2007;12(23–24):1007–12.
2. Kawakami K. Modification of physicochemical characteristics of active pharmaceutical ingredients and application of supersaturatable dosage forms for improving bioavailability of poorly absorbed drugs. Adv Drug Deliv Rev. 2012;64(6):480–95.
3. Brouwers J, Brewster ME, Augustijns P. Supersaturating drug delivery systems: the answer to solubility-limited oral bioavailability? J Pharm Sci. 2009;98(8):2549–72.
4. Newman A, Knipp G, Zografi G. Assessing the performance of amorphous solid dispersions. J Pharm Sci. 2012;101(4):1355–77.
5. Paudel A, Worku ZA, Meeus J, Guns S, Van den Mooter G. Manufacturing of solid dispersions of poorly water soluble drugs by spray drying: formulation and process considerations. Int J Pharm. 2013;453(1):253–84.
6. Porat D, Dahan A. Active intestinal drug absorption and the solubility-permeability interplay. Int J Pharm. 2018;537(1–2):84–93.
7. Kawakami K. Current status of amorphous formulation and other special dosage forms as formulations for early clinical phases. J Pharm Sci. 2009;98(9):2875–85.
8. Wyttenbach N, Kuentz M. Glass-forming ability of compounds in marketed amorphous drug products. Eur J Pharm Biopharm. 2017;112:204–8.
9. Newman A, Zografi G. What we need to know about solid-state isothermal crystallization of organic molecules from the amorphous state below the glass transition temperature. Mol Pharm. 2020;17(6):1761–77.
10. Kawakami K, Miyoshi K, Tamura N, Yamaguchi T, Ida Y. Crystallization of sucrose glass under ambient conditions: evaluation of crystallization rate and unusual melting behavior of resultant crystals. J Pharm Sci. 2006;95(6):1354–63.
11. Zhou D, Schmitt EA, Zhang GG, Law D, Vyazovkin S, Wight CA, et al. Crystallization kinetics of amorphous nifedipine studied by model-fitting and model-free approaches. J Pharm Sci. 2003;92(9):1779–92.
12. Kawakami K. Surface effects on the crystallization of ritonavir glass. J Pharm Sci. 2015;104(1):276–9.

13. Andronis V, Zografi G. Crystal nucleation and growth of indomethacin polymorphs from the amorphous state. J Non-Cryst Solids. 2000;271(3):236–48.
14. Kawakami K. Rigid nuclei and flexible nuclei: appearance and disappearance of nuclei in indomethacin glass revealed by a long-term annealing study. J Phys Chem B. 2023;127(26):5967–77.
15. Wu T, Yu L. Origin of enhanced crystal growth kinetics near Tg probed with indomethacin polymorphs. J Phys Chem B. 2006;110(32):15694–9.
16. Descamps M, Dudognon E. Crystallization from the amorphous state: nucleation–growth decoupling, polymorphism interplay, and the role of interfaces. J Pharm Sci. 2014;103(9):2615–28.
17. Kawakami K. Nucleation and crystallization of celecoxib glass: impact of experience of low temperature on physical stability. Thermochim Acta. 2019;671:43–7.
18. Kawakami K, Ohyama K. Crystal nucleation in ibuprofen glass: possible relevance between the characteristic length of the cooperatively rearranging region and the size of crystal nuclei. J Phys Chem B. 2025;129(7):2096–104.
19. Dudognon E, Danède F, Descamps M, Correia NT. Evidence for a new crystalline phase of racemic Ibuprofen. Pharm Res. 2008;25(12):2853–8.
20. Su Y, Yu L, Cai T. Enhanced crystal nucleation in glass-forming liquids by tensile fracture in the glassy state. Cryst Growth Des. 2019;19(1):40–4.
21. Huang C, Chen Z, Gui Y, Shi C, Zhang GGZ, Yu L. Crystal nucleation rates in glass-forming molecular liquids: D-sorbitol, D-arabitol, D-xylitol, and glycerol. J Chem Phys. 2018;149(5):054503.
22. Abyzov AS, Fokin VM, Yuritsyn NS, Nascimento MLF, Schmelzer JWP, Zanotto ED. Crystal nucleation in a glass during relaxation well below T(g). J Chem Phys. 2023;158(6):064501.
23. Donth E. The size of cooperatively rearranging regions at the glass transition. J Non-Cryst Solids. 1982;53(3):325–30.
24. Donth E-J. The glass transition: relaxation dynamics in liquids and disordered materials. Berlin/Heidelberg: Springer; 2001.
25. Fokin VM, Abyzov AS, Zanotto ED, Cassar DR, Rodrigues AM, Schmelzer JWP. Crystal nucleation in glass-forming liquids: variation of the size of the "structural units" with temperature. J Non-Cryst Solids. 2016;447:35–44.
26. Vyazovkin S, Dranca I. Effect of physical aging on nucleation of amorphous indomethacin. J Phys Chem B. 2007;111(25):7283–7.
27. Grzybowska K, Chmiel K, Knapik-Kowalczuk J, Grzybowski A, Jurkiewicz K, Paluch M. Molecular factors governing the liquid and glassy states recrystallization of celecoxib in binary mixtures with excipients of different molecular weights. Mol Pharm. 2017;14(4):1154–68.
28. Trasi NS, Taylor LS. Effect of polymers on nucleation and crystal growth of amorphous acetaminophen. CrystEngComm. 2012;14(16):5188–97.
29. Zhang J, Liu Z, Wu H, Cai T. Effect of polymeric excipients on nucleation and crystal growth kinetics of amorphous fluconazole. Biomater Sci. 2021;9(12):4308–16.
30. Chang R, Fu Q, Li Y, Wang M, Du W, Chang C, et al. Crystallization and relaxation dynamics of amorphous loratadine under different quench-cooling temperatures. CrystEngComm. 2017;19(2):335–45.
31. Gui Y, Huang C, Shi C, Stelzer T, Zhang GGZ, Yu L. Polymorphic selectivity in crystal nucleation. J Chem Phys. 2022;156(14):144504.
32. Cai T, Zhu L, Yu L. Crystallization of organic glasses: effects of polymer additives on bulk and surface crystal growth in amorphous nifedipine. Pharm Res. 2011;28(10):2458–66.
33. Wang K, Sun CC. Correction to crystal growth of celecoxib from amorphous state: polymorphism, growth mechanism, and kinetics. Cryst Growth Des. 2019;19(8):4894.
34. Wu H, Yao X, Gui Y, Hao H, Yu L. Surface enhancement of crystal nucleation in amorphous acetaminophen. Cryst Growth Des. 2022;22(9):5598–606.
35. Hatase M, Hanaya M, Hikima T, Oguni M. Discovery of homogeneous-nucleation-based crystallization in simple glass-forming liquid of toluene below its glass-transition temperature. J Non-Cryst Solids. 2002;307–310:257–63.

36. Hatase M, Hanaya M, Oguni M. Studies of homogeneous-nucleation-based crystal growth: significant role of phenyl ring in the structure formation. J Non-Cryst Solids. 2004;333(2):129–36.
37. Hikima T, Hanaya M, Oguni M. Microscopic observation of a peculiar crystallization in the glass transition region and β-process as potentially controlling the growth rate in triphenylethylene. J Mol Struct. 1999;479(2):245–50.
38. Konishi T, Tanaka H. Possible origin of enhanced crystal growth in a glass. Phys Rev B. 2007;76(22):220201.
39. Sun Y, Xi H, Chen S, Ediger MD, Yu L. Crystallization near glass transition: transition from diffusion-controlled to diffusionless crystal growth studied with seven polymorphs. J Phys Chem B. 2008;112(18):5594–601.
40. Wang K, Sun CC. Crystal growth of celecoxib from amorphous state: polymorphism, growth mechanism, and kinetics. Cryst Growth Des. 2019;19(6):3592–600.
41. Han X, Dai K, Kawakami K. Influence of nucleation on relaxation, molecular cooperativity, and physical stability of celecoxib glass. Mol Pharm. 2024;21(4):1794–803.
42. Kawakami K, Harada T, Miura K, Yoshihashi Y, Yonemochi E, Terada K, et al. Relationship between crystallization tendencies during cooling from melt and isothermal storage: toward a general understanding of physical stability of pharmaceutical glasses. Mol Pharm. 2014;11(6):1835–43.
43. Blaabjerg LI, Lindenberg E, Löbmann K, Grohganz H, Rades T. Glass forming ability of amorphous drugs investigated by continuous cooling and isothermal transformation. Mol Pharm. 2016;13(9):3318–25.
44. Blaabjerg LI, Lindenberg E, Rades T, Grohganz H, Löbmann K. Influence of preparation pathway on the glass forming ability. Int J Pharm. 2017;521(1–2):232–8.
45. Baird JA, Van Eerdenbrugh B, Taylor LS. A classification system to assess the crystallization tendency of organic molecules from undercooled melts. J Pharm Sci. 2010;99(9):3787–806.
46. Kawakami K. Crystallization tendency of pharmaceutical glasses: relevance to compound properties, impact of formulation process, and implications for design of amorphous solid dispersions. Pharmaceutics. 2019;11(5):202.
47. Trasi NS, Baird JA, Kestur US, Taylor LS. Factors influencing crystal growth rates from undercooled liquids of pharmaceutical compounds. J Phys Chem B. 2014;118(33):9974–82.
48. Ediger MD, Harrowell P, Yu L. Crystal growth kinetics exhibit a fragility-dependent decoupling from viscosity. J Chem Phys. 2008;128(3):034709.
49. Angell CA, Ngai KL, McKenna GB, McMillan PF, Martin SW. Relaxation in glassforming liquids and amorphous solids. J Appl Phys. 2000;88(6):3113–57.
50. Kawakami K, Pikal MJ. Calorimetric investigation of the structural relaxation of amorphous materials: evaluating validity of the methodologies. J Pharm Sci. 2005;94(5):948–65.
51. Tominaka S, Kawakami K, Fukushima M, Miyazaki A. Physical stabilization of pharmaceutical glasses based on hydrogen bond reorganization under sub-Tg temperature. Mol Pharm. 2017;14(1):264–73.
52. Bhardwaj SP, Arora KK, Kwong E, Templeton A, Clas SD, Suryanarayanan R. Correlation between molecular mobility and physical stability of amorphous itraconazole. Mol Pharm. 2013;10(2):694–700.
53. Mehta M, Ragoonanan V, McKenna GB, Suryanarayanan R. Correlation between molecular mobility and physical stability in pharmaceutical glasses. Mol Pharm. 2016;13(4):1267–77.
54. Shamblin SL, Hancock BC, Pikal MJ. Coupling between chemical reactivity and structural relaxation in pharmaceutical glasses. Pharm Res. 2006;23(10):2254–68.
55. Luthra SA, Hodge IM, Utz M, Pikal MJ. Correlation of annealing with chemical stability in lyophilized pharmaceutical glasses. J Pharm Sci. 2008;97(12):5240–51.
56. Huang D, McKenna GB. New insights into the fragility dilemma in liquids. J Chem Phys. 2001;114(13):5621–30.
57. Crowley KJ, Zografi G. The use of thermal methods for predicting glass-former fragility. Thermochim Acta. 2001;380(2):79–93.

58. Moynihan CT, Easteal AJ, Wilder J, Tucker J. Dependence of the glass transition temperature on heating and cooling rate. J Phys Chem. 1974;78(26):2673–7.
59. Kawakami K, Harada T, Yoshihashi Y, Yonemochi E, Terada K, Moriyama H. Correlation between glass-forming ability and fragility of pharmaceutical compounds. J Phys Chem B. 2015;119(14):4873–80.
60. Kawakami K, Bi Y, Yoshihashi Y, Sugano K, Terada K. Time-dependent phase separation of amorphous solid dispersions: implications for accelerated stability studies. J Drug Deliv Sci Technol. 2018;46:197–206.
61. Yu L. Amorphous pharmaceutical solids: preparation, characterization and stabilization. Adv Drug Deliv Rev. 2001;48(1):27–42.
62. Shamblin SL, Tang X, Chang L, Hancock BC, Pikal MJ. Characterization of the time scales of molecular motion in pharmaceutically important glasses. J Phys Chem B. 1999;103(20):4113–21.
63. Hancock BC, Shamblin SL, Zografi G. Molecular mobility of amorphous pharmaceutical solids below their glass transition temperatures. Pharm Res. 1995;12(6):799–806.
64. Kawakami K. Dynamics of ribavirin glass in the sub-Tg temperature region. J Phys Chem B. 2011;115(39):11375–81.
65. Yoshioka S, Aso Y. Correlations between molecular mobility and chemical stability during storage of amorphous pharmaceuticals. J Pharm Sci. 2007;96(5):960–81.
66. Kawakami K, Hasegawa Y, Deguchi K, Ohki S, Shimizu T, Yoshihashi Y, et al. Competition of thermodynamic and dynamic factors during formation of multicomponent particles via spray drying. J Pharm Sci. 2013;102(2):518–29.
67. Marsac PJ, Li T, Taylor LS. Estimation of drug-polymer miscibility and solubility in amorphous solid dispersions using experimentally determined interaction parameters. Pharm Res. 2009;26(1):139–51.
68. Vehring R. Pharmaceutical particle engineering via spray drying. Pharm Res. 2008;25(5):999–1022.
69. Worku ZA, Aarts J, Singh A, Van den Mooter G. Drug-polymer miscibility across a spray dryer: a case study of naproxen and miconazole solid dispersions. Mol Pharm. 2014;11(4):1094–101.
70. Bergfeldt K, Piculell L, Linse P. Segregation and association in mixed polymer solutions from Flory–Huggins model calculations. J Phys Chem. 1996;100(9):3680–7.
71. Paudel A, Van den Mooter G. Influence of solvent composition on the miscibility and physical stability of naproxen/PVP K 25 solid dispersions prepared by cosolvent spray-drying. Pharm Res. 2012;29(1):251–70.
72. Aso Y, Yoshioka S, Kojima S. Molecular mobility-based estimation of the crystallization rates of amorphous nifedipine and phenobarbital in poly(vinylpyrrolidone) solid dispersions. J Pharm Sci. 2004;93(2):384–91.
73. Caron V, Bhugra C, Pikal MJ. Prediction of onset of crystallization in amorphous pharmaceutical systems: phenobarbital, nifedipine/PVP, and phenobarbital/PVP. J Pharm Sci. 2010;99(9):3887–900.
74. Kothari K, Ragoonanan V, Suryanarayanan R. The role of polymer concentration on the molecular mobility and physical stability of nifedipine solid dispersions. Mol Pharm. 2015;12(5):1477–84.
75. Greco S, Authelin JR, Leveder C, Segalini A. A practical method to predict physical stability of amorphous solid dispersions. Pharm Res. 2012;29(10):2792–805.
76. Martí J, Sala J, Guàrdia E. Molecular dynamics simulations of water confined in graphene nanochannels: from ambient to supercritical environments. J Mol Liq. 2010;153(1):72–8.
77. Martí J, Nagy G, Guàrdia E, Gordillo MC. Molecular dynamics simulation of liquid water confined inside graphite channels: dielectric and dynamical properties. J Phys Chem B. 2006;110(47):23987–94.
78. Brás A, Fonseca I, Dionísio M, Schönhals A, Affouard F, Correia N. Influence of nanoscale confinement on the molecular mobility of ibuprofen. J Phys Chem C. 2014;118:13857–68.

79. Guenneau F, Panesar K, Nossov A, Springuel-Huet M-A, Azaïs T, Babonneau F, et al. Probing the mobility of ibuprofen confined in MCM-41 materials using MAS-PFG NMR and hyperpolarised-129Xe NMR spectroscopy. Phys Chem Chem Phys. 2013;15(43):18805–8.
80. Nartowski KP, Malhotra D, Hawarden LE, Sibik J, Iuga D, Zeitler JA, et al. (19) F NMR spectroscopy as a highly sensitive method for the direct monitoring of confined crystallization within nanoporous materials. Angew Chem Int Ed Engl. 2016;55(31):8904–8.
81. Nishiwaki A, Watanabe A, Higashi K, Tozuka Y, Moribe K, Yamamoto K. Molecular states of prednisolone dispersed in folded sheet mesoporous silica (FSM-16). Int J Pharm. 2009;378(1):17–22.
82. Cheng S, McKenna GB. Nanoconfinement effects on the glass transition and crystallization behaviors of nifedipine. Mol Pharm. 2019;16(2):856–66.
83. Han X, Kawakami K. Influence of pore size of mesoporous silica on physical stability of over-loaded celecoxib glass. Mol Pharm. 2025;22(5):2556–67.
84. Knapik-Kowalczuk J, Kramarczyk D, Chmiel K, Romanova J, Kawakami K, Paluch M. Importance of mesoporous silica particle size in the stabilization of amorphous pharmaceuticals-the case of simvastatin. Pharmaceutics. 2020;12(4):384.
85. Kramarczyk D, Knapik-Kowalczuk J, Klimontko J, Kurek M, Jachowicz R, Paluch M. Tuning the physical state of aripiprazole by mesoporous silica. Mol Pharm. 2024;21(5):2315–26.

# Chapter 3
# Polymer Selection for the Successful Preparation of Amorphous Solid Dispersions

**Jelena Djuris, Djordje Medarevic, Aleksandar Pavic, and Svetlana Ibric**

**Abstract** The increasing prevalence of poorly water-soluble drug candidates has positioned amorphous solid dispersions (ASDs) as a key enabling technology in modern pharmaceutical development. While the amorphization of active pharmaceutical ingredients is central to this approach, the long-term stability, manufacturability, and biopharmaceutical performance of ASDs are critically governed by the properties of the polymeric carrier. This chapter provides a structured and mechanistic perspective on rational polymer selection for the successful development of ASDs, with a particular focus on antifungal drugs as model compounds.

Key physicochemical principles underlying polymer function: miscibility, intermolecular interactions, glass transition behavior, and steric effects are discussed in relation to their impact on supersaturation generation, crystallization inhibition, and physical stability. Practical decision criteria for polymer selection are outlined, integrating compatibility screening, predictive and data-driven tools, manufacturability constraints, and regulatory considerations. Established, next-generation, and naturally derived polymer classes are reviewed in the context of their advantages and limitations.

Case studies involving azole and non-azole antifungal agents illustrate how polymer chemistry, grade selection, and processing route jointly determine formulation performance and clinical relevance. Finally, emerging early de-risking strategies, including advanced in vitro methodologies and new approach methodologies (NAMs), are highlighted as essential components of future ASD development work-

J. Djuris (✉) · D. Medarevic · S. Ibric
Department of Pharmaceutical Technology, Faculty of Pharmacy, University of Belgrade, Belgrade, Serbia
e-mail: jelena.djuris@pharmacy.bg.ac.rs; djordje.medarevic@pharmacy.bg.ac.rs; svetlana.ibric@pharmacy.bg.ac.rs

A. Pavic
Institute of Molecular Genetics and Genetic Engineering, University of Belgrade, Belgrade, Serbia
e-mail: aleksandar.pavic@imgge.bg.ac.rs



I. Nikolakakis, I. Partheniadis (eds.), *Amorphous Solid Dispersions*, AAPS Introductions in the Pharmaceutical Sciences 5,
https://doi.org/10.1007/978-3-032-24883-1_3

flows. Collectively, this chapter emphasizes that polymers are not inert excipients but active design elements whose informed selection is pivotal for translating ASD technologies into robust and clinically reliable drug delivery systems.

**Keywords** Antifungal drugs · Supersaturation · Crystallization inhibition · Biopharmaceutical performance · Hot-melt extrusion · Spray drying · Early de-risking · New approach methodologies

## 3.1 Introduction

The increasing prevalence of poorly water-soluble drug candidates continues to pose a major challenge in pharmaceutical development. It is estimated that a substantial proportion of marketed drugs, as well as the vast majority of new chemical entities entering development pipelines, fall into Biopharmaceutics Classification System (BCS) classes II or IV, where solubility and dissolution rate represent critical barriers to oral bioavailability. In this context, amorphous solid dispersions (ASDs) have emerged as one of the most successful formulation strategies for enabling oral delivery of such compounds, and their relevance continues to grow alongside the expanding chemical complexity of modern drug candidates.

ASDs improve drug performance by dispersing an active pharmaceutical ingredient (API) within a polymeric matrix in an amorphous, high-energy state, thereby increasing apparent solubility and dissolution rate [1–3]. However, despite their long-standing use and numerous commercial successes, the development of robust ASD formulations remains far from trivial [4]. The stability, manufacturability, and in vivo performance of an ASD are not dictated by the API alone, but are critically dependent on the properties of the polymeric carrier and its interactions with the drug.

Historically, polymer selection in ASD development has often been guided by empirical considerations, regulatory familiarity, or prior formulation experience. While this approach has yielded effective products, it has also contributed to an underestimation of the active role that polymers play in determining both therapeutic performance and biological behavior. Polymers are frequently treated as functionally inert excipients, yet they directly influence molecular mobility, supersaturation maintenance, precipitation kinetics, and, in some cases, biological exposure and tolerability [5]. These effects become particularly relevant as formulation strategies evolve toward higher drug loadings, complex polymer architectures, and multifunctional carrier systems [6, 7].

Importantly, the development of advanced drug delivery systems introduces challenges that are not necessarily apparent during the development of the drug substance itself. While the safety and efficacy of an API are typically well characterized early in the pipeline, formulation-driven changes in drug presentation, release kinetics, or tissue exposure may give rise to new, formulation-specific risks. This discrepancy underscores the need for a more integrated development paradigm, in which physicochemical optimization is complemented by early consideration of biological performance and safety at the level of the complete delivery system.

In recent years, this need has been reflected in the growing adoption of new approach methodologies (NAMs), which aim to support earlier, more informative decision-making while reducing reliance on late-stage, resource-intensive testing. When applied thoughtfully, such tools can help identify formulation-dependent effects on drug behavior, reveal unanticipated safety concerns related to carrier systems, and ultimately de-risk development before costly scale-up and clinical evaluation [8]. Within this framework, antifungal drugs represent particularly instructive model compounds, as they combine pronounced solubility limitations with narrow therapeutic windows and clinically relevant safety concerns [9–13].

Based on this rationale, the present chapter focuses on polymer selection as a central design decision in the development of ASDs. Using antifungal drugs as representative case studies, we integrate physicochemical principles, processing considerations, regulatory constraints, and emerging early-stage evaluation strategies to provide a structured and application-oriented perspective on rational polymer selection for successful ASD development.

## 3.2 Amorphous Solid Dispersions: Concept and Biopharmaceutical Rationale

ASDs represent a formulation strategy in which an API is molecularly dispersed within a polymeric carrier in a non-crystalline, high-energy state. In contrast to crystalline solids, amorphous materials lack long-range molecular order, eliminating the energetic barrier associated with crystal lattice disruption during dissolution. As a result, ASDs are capable of generating substantially higher apparent solubility and faster dissolution rates than their crystalline counterparts.

From a biopharmaceutical perspective, the primary advantage of ASDs lies in their ability to induce and maintain a supersaturated drug concentration upon contact with gastrointestinal fluids. This behavior is often described using the "spring–parachute" concept, where rapid drug release from the amorphous phase generates a supersaturation "spring," followed by a "parachute" effect mediated by the polymer, which delays precipitation and prolongs the availability of dissolved drug for absorption [1, 14], as represented in Fig. 3.1. The effectiveness of this mechanism is governed not only by the intrinsic properties of the API but also by the physicochemical characteristics of the polymeric matrix.

Despite these advantages, the amorphous state is inherently metastable. Without adequate stabilization, amorphous drugs are prone to recrystallization during storage or upon exposure to aqueous media, leading to loss of dissolution advantage and, ultimately, therapeutic failure. Polymers used in ASD systems therefore serve multiple, interrelated functions: they reduce molecular mobility through elevation of the system glass transition temperature, establish specific intermolecular

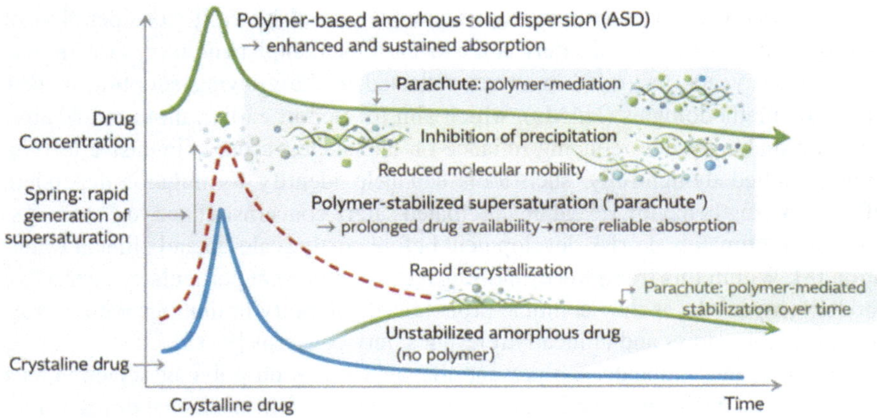


**Fig. 3.1** Schematic illustration of the *spring–parachute* concept highlighting the biopharmaceutical rationale of polymer-based amorphous solid dispersions (ASDs). Rapid dissolution of amorphous drug forms generates a transient supersaturated state (*spring*), which rapidly collapses in the absence of a polymer due to recrystallization. In contrast, polymer-based ASDs stabilize the supersaturated state (*parachute*) by inhibiting precipitation and reducing molecular mobility, thereby prolonging drug availability and enabling improved and more reliable absorption compared with crystalline and unstabilized amorphous drug forms

interactions with the drug, and introduce steric barriers that hinder nucleation and crystal growth. Consequently, polymer selection is not merely a matter of enabling amorphization, but a central determinant of long-term stability and in vivo performance.

### *3.2.1 Antifungal Drugs as Model Compounds for ASD Formulation Design*

Antifungal agents used to treat human fungal infections constitute a therapeutically important yet formulation-challenging class of drugs, making them particularly suitable model compounds for evaluating the capabilities and limitations of ASD systems [15]. Many clinically used antifungals exhibit pronounced hydrophobicity, high melting points, and limited aqueous solubility, placing them predominantly in BCS class II or IV. As a consequence, oral bioavailability is frequently dissolution-limited and highly sensitive to formulation variables.

Azole antifungals, such as itraconazole and posaconazole, are among the most extensively studied examples. These compounds display extremely low intrinsic solubility and strong dependence on gastrointestinal conditions, including pH and food intake, resulting in notable inter- and intra-patient variability. For such drugs, ASD-based formulations have demonstrated the ability to mitigate food effects and increase exposure; however, achieving consistent performance requires careful selection of polymeric carriers capable of sustaining supersaturation under physiologically relevant conditions [8, 16].

In contrast, polyene antifungals, most notably amphotericin B, exhibit a distinct set of challenges. Beyond poor solubility and permeability, their tendency to form large, toxic aggregates in aqueous environments introduces an additional safety dimension to formulation design. While amorphization and polymer-based encapsulation can improve apparent solubility, changes in aggregation behavior and biological exposure may profoundly influence toxicity and therapeutic profiles, underscoring the need for formulation strategies that balance efficacy enhancement with safety considerations [17, 18].

Echinocandins and newer antifungal agents further illustrate the diversity of formulation barriers encountered within this therapeutic class. High molecular weight, limited permeability, and structural complexity often restrict the applicability of conventional ASD approaches, necessitating hybrid or multifunctional delivery systems [19]. Collectively, these characteristics position antifungal drugs as a stringent test case for polymer performance, revealing not only the solubilization potential of ASD systems but also their limitations and formulation-driven risks.

Polymers prevent recrystallization by raising the glass transition temperature ($T_g$) of the system, reducing molecular mobility, and forming specific intermolecular interactions (e.g., hydrogen bonding, ionic, or van der Waals forces) with the drug. These interactions hinder nucleation and crystal growth, thereby extending the supersaturation plateau and improving oral absorption [20].

## 3.3 Mechanistic Basis of Polymer Function in Amorphous Solid Dispersions

### *3.3.1 Molecular Miscibility and Phase Behavior*

A prerequisite for the successful formation and long-term stability of ASDs is molecular-level miscibility between the drug and the polymeric carrier. When miscibility is insufficient, phase separation may occur either during storage or upon exposure to moisture, leading to localized drug enrichment, accelerated molecular mobility, and ultimately recrystallization [21].

Thermodynamic approaches, such as the use of Hansen solubility parameters or Flory–Huggins interaction parameters, are commonly employed as initial screening tools to estimate drug–polymer compatibility [22]. A small difference in solubility parameters is generally associated with favorable mixing behavior; however, these models provide only an approximation and may fail to capture specific intermolecular interactions or kinetic effects relevant to ASD performance [23].

Experimental techniques remain essential for confirming miscibility. Differential scanning calorimetry is frequently used to detect the presence of a single, composition-dependent glass transition temperature, indicative of a homogeneous amorphous phase. Complementary methods, including film casting and powder X-ray diffraction, provide practical insight into phase stability and crystallization tendencies under controlled conditions. Together, these approaches highlight that miscibility is not a binary attribute, but a dynamic property influenced by composition, processing history, and environmental exposure [24, 25].

### 3.3.2 Intermolecular Interactions and Inhibition of Crystallization

Beyond thermodynamic miscibility, the ability of a polymer to inhibit drug crystallization is governed by the nature and strength of intermolecular interactions within the ASD system. Hydrogen bonding, ionic interactions, and dipole–dipole forces between functional groups of the drug and polymer can significantly reduce the tendency of drug molecules to self-associate and form a crystalline lattice [26].

These interactions may act at different stages of the crystallization process. Some polymers primarily suppress nucleation by stabilizing individual drug molecules in a dispersed state, whereas others slow crystal growth by adsorbing onto emerging crystal surfaces and disrupting lattice propagation. The effectiveness of crystallization inhibition, therefore, depends not only on interaction strength but also on the spatial and temporal accessibility of interacting functional groups [27].

Importantly, the presence of specific interactions does not guarantee physical stability under all conditions. Environmental factors such as humidity or temperature fluctuations may weaken drug–polymer interactions, increase molecular mobility, and promote phase separation or recrystallization. Consequently, a mechanistic understanding of intermolecular interactions must be considered in conjunction with polymer mobility and environmental sensitivity [28].

The key mechanisms by which polymers stabilize amorphous solid dispersions are schematically summarized in Fig. 3.2.

Long-term physical stability of amorphous solid dispersions is a prerequisite for ensuring both adequate shelf-life and consistent therapeutic performance. The selected polymer plays a central role in suppressing drug crystallization during storage and handling by reducing molecular mobility, stabilizing supersaturated states, and mitigating the impact of environmental stressors such as temperature and humidity [28, 29].

### 3.3.3 Glass Transition Temperature, Molecular Mobility, and Physical Stability

The glass transition temperature ($T_g$) of an amorphous solid dispersion is a critical parameter governing molecular mobility and, by extension, physical stability. In general, an increase in $T_g$ is associated with reduced molecular motion and a lower probability of nucleation and crystal growth. Polymers with high intrinsic $T_g$ values

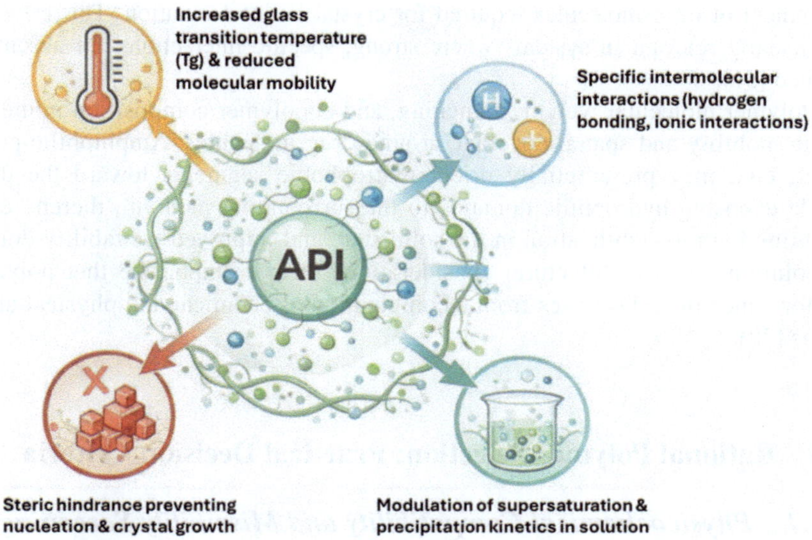


**Fig. 3.2** Schematic overview of the key mechanisms by which polymers stabilize amorphous solid dispersions (ASDs). Polymer incorporation increases the glass transition temperature and reduces molecular mobility, promotes specific intermolecular interactions with the drug (e.g., hydrogen bonding or ionic interactions), provides steric hindrance that suppresses nucleation and crystal growth, and modulates supersaturation and precipitation kinetics in solution. Acting in concert, these mechanisms prevent recrystallization and enable the maintenance of a physically and biopharmaceutically relevant amorphous state

are therefore often favored as ASD carriers, particularly for drugs with strong recrystallization tendencies [29].

However, the $T_g$ of an ASD system is not solely determined by the polymer itself, but reflects the combined contributions of the drug, polymer, and any absorbed moisture. Many commonly used ASD polymers are hygroscopic, and water acts as a potent plasticizer, lowering the $T_g$ and increasing molecular mobility even at relatively low humidity levels. This effect can negate the stabilizing influence of a high-$T_g$ polymer and accelerate physical degradation during storage [30].

As a result, $T_g$ should be viewed as a contextual, rather than absolute, indicator of stability. Reliable ASD design requires consideration of polymer hygroscopicity, environmental exposure, and formulation composition, as well as the margin between storage temperature and the effective $T_g$ of the system. These factors collectively determine whether a polymer can provide durable stabilization of the amorphous drug under realistic conditions [7].

### 3.3.4 *Steric Effects and Polymer Architecture*

In addition to specific interactions and thermal effects, the physical architecture of polymer chains plays an important role in stabilizing amorphous drugs. Long, flexible polymer chains introduce steric hindrance that can physically impede the alignment of drug molecules required for crystal lattice formation. This effect is particularly relevant in systems where strong, specific interactions are absent or limited [31].

Polymer molecular weight, branching, and copolymer composition influence chain mobility and spatial coverage around drug molecules. Amphiphilic polymers, e.g., may preferentially orient hydrophobic segments toward the drug while exposing hydrophilic domains to the surrounding medium, thereby contributing both to stabilization in the solid state and improved wettability during dissolution. Such architectural considerations further emphasize that polymer performance in ASDs arises from a combination of chemical and physical attributes [32].

## 3.4 Rational Polymer Selection: Practical Decision Criteria

### *3.4.1 Physicochemical Compatibility and Miscibility Screening*

The first step in rational polymer selection for amorphous solid dispersions is the assessment of physicochemical compatibility between the drug and candidate polymers. Early identification of incompatible systems is critical, as inadequate miscibility represents a primary cause of phase separation and long-term instability in ASDs [33].

Theoretical approaches, including Hansen solubility parameters and related thermodynamic models, are widely used as initial screening tools [34]. Small differences in solubility parameters are generally indicative of favorable mixing behavior and can help narrow the pool of potential polymers. However, such calculations provide only a coarse approximation and do not account for specific drug–polymer interactions or kinetic effects introduced during processing [35].

Experimental verification is therefore indispensable. Differential scanning calorimetry is commonly employed to detect a single, composition-dependent glass transition temperature, which supports the presence of a homogeneous amorphous phase. Simple yet informative techniques, such as film casting from a common solvent, allow rapid visual assessment of phase separation tendencies, while powder X-ray diffraction confirms the absence of crystalline domains. Used in combination, these methods enable efficient prioritization of polymer candidates before resource-intensive formulation development is undertaken [36–38].

### *3.4.2 Structure- and Property-Based Prediction of Drug–Polymer Compatibility*

In addition to experimental screening, contemporary approaches to polymer selection increasingly rely on structure- and property-based prediction of drug–polymer compatibility. Advances in molecular descriptor calculation and data-driven modeling have enabled the use of drug and polymer structural features to anticipate miscibility, interaction potential, and stabilization capacity prior to experimental evaluation [39].

Such approaches typically integrate information on functional group complementarity, hydrogen bond donor and acceptor capacity, polarity, molecular flexibility, and hydrophobic–hydrophilic balance. For polymers, repeat unit chemistry, degree of substitution, and copolymer composition are particularly relevant, while for APIs, aromaticity, ionizable groups, and conformational rigidity often play a decisive role. These features can be translated into quantitative descriptors that support the estimation of interaction strength and mixing behavior [40].

Thermodynamic models and mixing rules, including extensions of Flory–Huggins theory and glass transition-based models, further contribute to the prediction of phase behavior in ASD systems. More recently, machine learning techniques have been applied to relate molecular descriptors of drugs and polymers to experimental outcomes such as single glass transition temperature formation, physical stability, and dissolution performance. By learning from existing datasets, these models offer a powerful means of narrowing the design space and prioritizing promising drug–polymer combinations [41–43].

It is important to emphasize that predictive approaches are not intended to replace experimental validation. Rather, they serve as decision-support tools that reduce empirical trial-and-error, accelerate early development, and guide targeted experimental screening. When used in conjunction with classical physicochemical and stability assessments, structure- and property-based predictions represent an increasingly valuable component of rational polymer selection strategies.

### 3.4.3 Physical Stability and Shelf-Life Considerations

Beyond initial miscibility, the long-term physical stability of an ASD is a decisive criterion in polymer selection [44]. A formulation that remains amorphous immediately after preparation may nevertheless undergo gradual phase separation or recrystallization during storage, particularly under conditions of elevated temperature or humidity [35].

Polymer hygroscopicity plays a central role in this context. Many polymers commonly used in ASD systems readily absorb moisture, which acts as a plasticizer and lowers the effective glass transition temperature of the dispersion. This increase in molecular mobility can dramatically accelerate crystallization processes, even when high-$T_g$ polymers are employed. Consequently, stability assessments must consider not only the intrinsic properties of the polymer, but also its interaction with environmental moisture and the intended storage conditions [38].

Drug loading further modulates stability behavior. Higher drug loadings reduce the relative proportion of polymer available to immobilize drug molecules and inhibit crystallization, narrowing the stability margin of the system [45]. Rational polymer selection, therefore, requires balancing desired drug loading with acceptable stability risk, supported by accelerated and, where possible, real-time stability studies.

### 3.4.4 Process Compatibility and Manufacturability

The suitability of a polymer for ASD development is also determined by its compatibility with the intended manufacturing process [46]. Polymers that perform well from a physicochemical standpoint may prove unsuitable for large-scale processing due to unfavorable thermal, rheological, or solubility characteristics [3].

For solvent-free techniques such as hot-melt extrusion, polymers must exhibit adequate thermal stability and melt viscosity within a processing window that does not compromise the integrity of the drug. In contrast, spray drying requires sufficient solubility of the polymer in volatile organic solvents and the ability to form stable droplets during atomization. These process-dependent requirements can significantly constrain polymer choice and must be evaluated early in development to avoid late-stage reformulation [47].

Importantly, manufacturability considerations extend beyond the primary ASD production step. The downstream behavior of ASD intermediates, including powder flow, compressibility, and suitability for tableting or encapsulation, is strongly influenced by polymer properties and should be integrated into polymer selection decisions from the outset [48].

### 3.4.5 Regulatory Acceptance and Safety Considerations

Regulatory status represents an essential, and often decisive, constraint in polymer selection. Polymers with established use in oral pharmaceutical products, particularly those listed in regulatory excipient databases or compendial monographs, are generally favored due to their well-characterized safety profiles and reduced regulatory burden [49]. However, reliance on regulatory familiarity alone may limit formulation innovation. Emerging polymers and modified polymer architectures have demonstrated superior performance in terms of drug loading, stability, or dissolution behavior, yet their adoption introduces additional safety and regulatory challenges. While such materials may offer clear physicochemical advantages, their biological behavior—either alone or in combination with a drug—may not be fully predictable based on existing excipient data [50]. This tension highlights the importance of integrating safety considerations earlier in the formulation design process. Rather than deferring safety evaluation to late development stages, rational polymer selection increasingly benefits from early-stage strategies aimed at identifying formulation-specific risks associated with novel or multifunctional carrier systems. Such an approach supports informed decision-making and reduces the likelihood of costly redevelopment driven by unforeseen safety concerns [51].

## 3.5 Polymer Classes Used in Amorphous Solid Dispersion Systems

### *3.5.1 Established Synthetic and Semi-Synthetic Polymers*

Synthetic and semi-synthetic polymers remain the backbone of most commercially successful ASD formulations, primarily due to their well-characterized physicochemical properties, regulatory acceptance, and reproducible performance. These materials have been extensively studied across a wide range of drugs and manufacturing platforms, making them the first-line choice in many formulation development programs.

Vinyl polymers, particularly polyvinylpyrrolidone (PVP) and its copolymer with vinyl acetate (PVPVA), are among the most widely used ASD carriers [52]. Their high glass transition temperatures and strong hydrogen bonding capacity enable effective stabilization of amorphous drugs. However, their pronounced hygroscopicity represents a significant limitation, as moisture uptake can lead to plasticization, reduced physical stability, and accelerated recrystallization under humid storage conditions [53].

Cellulose derivatives, including hydroxypropyl methylcellulose (HPMC), hydroxypropyl methylcellulose phthalate (HPMCP), and hydroxypropyl methylcellulose acetate succinate (HPMCAS), occupy a central position in ASD technology. Among these, HPMCAS has emerged as a particularly versatile polymer due to its amphiphilic character and pH-dependent solubility. By remaining insoluble under gastric conditions and dissolving at higher intestinal pH values, HPMCAS-based ASDs can delay drug release and promote sustained supersaturation at sites favorable for absorption [54–57].

Methacrylate-based polymers, such as those within the Eudragit® family, further expand formulation flexibility by offering tailored solubility profiles and release characteristics. Their diverse functionalization allows adaptation to immediate-release, delayed-release, or enteric delivery strategies, although careful consideration of processing conditions and polymer–drug compatibility is required [58, 59].

### *3.5.2 Functional and Next-Generation Synthetic Polymers*

To address the limitations of traditional ASD carriers, particularly with respect to drug loading, processability, and dissolution performance, a number of functional and next-generation synthetic polymers have been developed, among which Soluplus® and Apinovex™ have received the majority of attention. These materials are often designed with specific formulation challenges in mind, rather than broad applicability [60].

Soluplus® represents a prominent example of a purpose-designed ASD polymer. As an amphiphilic graft copolymer, it combines hydrophilic and lipophilic domains within a single macromolecular structure, facilitating both drug stabilization in the solid state and enhanced wettability upon dissolution. Its favorable processing characteristics have made it particularly suitable for hot-melt extrusion, contributing to its widespread adoption in ASD formulations [46].

More recently, modified polymer systems have been introduced to support higher drug loadings while maintaining physical stability. Polymers such as Apinovex™ have been engineered to provide optimized rheological properties and reduced crystallization propensity, enabling drug loadings that significantly exceed those achievable with conventional carriers. While these materials offer clear formulation advantages, their use necessitates careful evaluation of safety and regulatory considerations due to their more limited history of pharmaceutical application [61, 62].

### 3.5.3 Natural and Naturally Derived Polymers

Naturally derived polymers have attracted increasing interest as ASD carriers, driven in part by their perceived biocompatibility and alignment with sustainability considerations. Polysaccharides such as chitosan, alginate, and pullulan, as well as cyclodextrin-based polymers, have been investigated for their potential to enhance solubility and modulate drug release [63, 64].

Chitosan, a cationic polysaccharide, offers additional functional attributes such as mucoadhesion, which may prolong gastrointestinal residence time and enhance local drug exposure [65]. Alginate-based systems have demonstrated rapid drug release and favorable dissolution profiles for certain poorly soluble compounds [66, 67]. Cyclodextrin polymers, particularly crosslinked or polymerized forms, are of interest for drugs prone to aggregation, as they can encapsulate hydrophobic moieties and modulate drug presentation in aqueous environments [68, 69].

Despite these advantages, natural polymers present specific challenges in ASD applications. Batch-to-batch variability, limited thermal stability, and less predictable processing behavior can complicate scale-up and reproducibility. Consequently, their use in ASD systems often requires more extensive characterization and tighter control strategies than established synthetic alternatives [70].

### 3.5.4 Polymer Blends, Ternary, and Multifunctional Systems

In many cases, a single polymer is insufficient to simultaneously meet the requirements of miscibility, physical stability, dissolution performance, and manufacturability. This has led to the development of polymer blends and ternary ASD systems, in which two polymers or a polymer–surfactant combination are employed to achieve complementary functionality [18, 71].

Dual polymer systems can combine the stabilizing capacity of one polymer with the dissolution-enhancing properties of another, resulting in improved performance relative to binary dispersions. Similarly, the incorporation of surfactants into ternary systems may enhance wettability and delay precipitation, although excessive surfactant content can introduce new stability or safety concerns [72, 73].

As system complexity increases, so does the potential for unintended interactions and formulation-specific risks. Rational design of such multifunctional systems, therefore, requires a clear understanding of the role of each component and careful justification of added complexity, supported by both physicochemical and early biological evaluation.

## 3.6 Manufacturing Technologies and Their Impact on Polymer Choice

### *3.6.1 Hot-Melt Extrusion*

Hot-melt extrusion (HME) is one of the most widely applied manufacturing techniques for amorphous solid dispersions, owing to its solvent-free nature, continuous processing capability, and scalability. In this process, the drug and polymer are mixed under controlled temperature and shear conditions to form a homogeneous molten mass, which solidifies upon cooling into an amorphous dispersion [74, 75].

The applicability of HME imposes specific constraints on polymer selection. Suitable polymers must exhibit sufficient thermal stability within the processing temperature window and possess melt viscosities that allow effective mixing without excessive shear stress. Polymers such as PVPVA and Soluplus® are frequently employed due to their favorable softening behavior and compatibility with moderate processing temperatures [76–78].

However, high-viscosity polymers, including certain cellulose derivatives, may present challenges during extrusion, particularly when high drug loadings are targeted. In such cases, the use of plasticizers or polymer blends may be required to lower processing temperatures and improve melt flow, introducing additional formulation complexity. These considerations highlight that polymer performance in HME-based ASDs reflects not only solid-state stabilization capacity, but also rheological behavior under dynamic processing conditions [79].

### *3.6.2 Spray Drying*

Spray drying represents an alternative, widely used approach for the production of amorphous solid dispersions, particularly for thermolabile drugs that cannot tolerate the elevated temperatures associated with melt-based processing. In spray drying, the drug and polymer are co-dissolved in a suitable solvent system and rapidly dried, resulting in kinetic trapping of the drug in an amorphous state [3, 80, 81].

Polymer selection for spray-dried ASDs is strongly influenced by solvent compatibility and solution properties. Polymers must exhibit adequate solubility in volatile organic solvents and support stable atomization during the drying process. These requirements often favor polymers such as HPMC, HPMCAS, and PVP, which dissolve readily in common spray-drying solvents [82–84].

The rapid solvent evaporation characteristic of spray drying typically produces porous, low-density particles with high specific surface area. While this morphology can enhance dissolution performance, it may also introduce challenges related to powder flow, compressibility, and downstream processing. Consequently, polymer choice must account not only for solid-state stabilization and dissolution behavior, but also for the physical attributes of the resulting ASD intermediate [85].

### *3.6.3 Emerging and Alternative Manufacturing Technologies*

Beyond conventional hot-melt extrusion and spray drying, several emerging manufacturing technologies have been explored to expand the applicability of ASD systems. Techniques such as KinetiSol® processing, electrospinning, and supercritical fluid-based methods aim to overcome limitations associated with thermal stress, solvent use, or drug loading capacity [52, 86, 87].

These approaches introduce distinct constraints on polymer properties, including ultra-fast melting behavior, rapid solidification kinetics, or compatibility with supercritical media. While such technologies offer promising solutions for specific formulation challenges, their adoption remains limited by equipment availability, scale-up considerations, and regulatory familiarity. As a result, polymer selection for these processes requires careful alignment between material properties, processing feasibility, and development timelines.

### *3.6.4 Integrating Polymer and Process Selection*

Taken together, manufacturing considerations underscore the need to integrate polymer selection with process selection from the earliest stages of ASD development. A polymer that offers excellent stabilization potential may be impractical for a given manufacturing platform, while a process-efficient polymer may compromise long-term stability or dissolution performance [47].

Rational ASD design, therefore, requires iterative evaluation of material properties and processing behavior, rather than sequential decision-making. By aligning polymer characteristics with the intended manufacturing strategy, formulation scientists can reduce development risk, streamline scale-up, and improve the likelihood of successful translation from laboratory-scale screening to commercial production, as represented in Fig. 3.3.

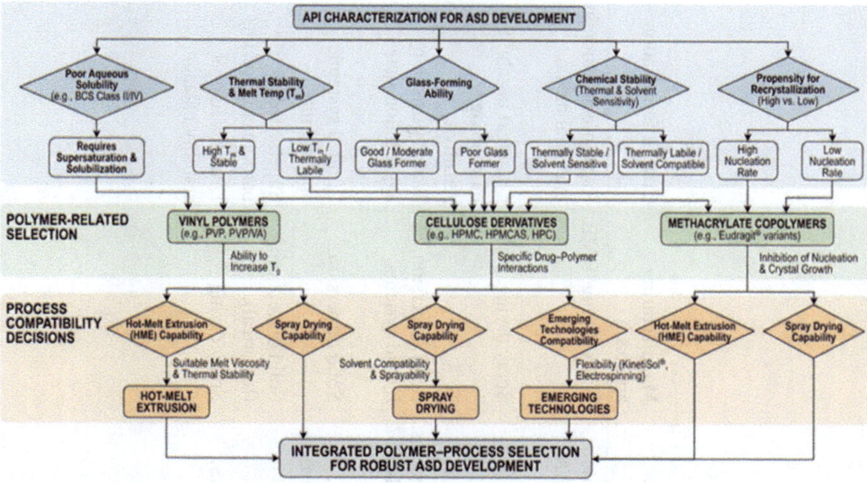


**Fig. 3.3** Decision tree illustrating the integrated selection of polymers and manufacturing technologies for amorphous solid dispersions (ASDs) based on key active pharmaceutical ingredient (API) properties. The scheme highlights how poor aqueous solubility, thermal and chemical stability, glass-forming ability, and recrystallization propensity guide polymer class selection, which in turn constrains compatible processing routes such as hot-melt extrusion, spray drying, and emerging technologies. The figure emphasizes that robust ASD development requires the simultaneous consideration of API characteristics, polymer functionality, and process-specific requirements

## 3.7 Case Studies: Polymer Selection in Antifungal Amorphous Solid Dispersions

Antifungal drugs represent biopharmaceutically diverse model compounds for rational polymer selection in amorphous solid dispersions. Key formulation challenges and corresponding polymer strategies for selected antifungal agents are summarized in Table 3.1.

### *3.7.1 Azole Antifungals: Managing Supersaturation and Food Effects*

Azole antifungals represent the most extensively studied class of drugs formulated as amorphous solid dispersions and provide instructive examples of rational polymer selection. Compounds such as itraconazole and posaconazole are characterized by extremely low aqueous solubility and strong sensitivity to gastrointestinal conditions, resulting in variable oral bioavailability.

**Table 3.1** Antifungal drug examples illustrating polymer selection strategies in ASD systems

| API | Antifungal class | Key biopharmaceutical challenge | Dominant formulation objective | Polymer classes typically used | Manufacturing approach | Critical formulation risk |
|---|---|---|---|---|---|---|
| Itraconazole | Triazole | Extremely low solubility; pH-dependent precipitation | Maintenance of supersaturation after gastric emptying | HPMCAS (various grades), PVP/VA | Spray drying, HME | Rapid precipitation in intestinal media |
| Posaconazole | Triazole | Low solubility; exposure variability | Stabilization of supersaturation and controlled release | HPMCAS | Spray drying, HME | Chemical instability; morphology sensitivity |
| Amphotericin B | Polyene | Aggregation; formulation-related toxicity | Modulation of aggregation; localized exposure | PEG-based systems, polymer blends | Solid dispersion, hybrid systems | Toxicity driven by aggregation state |
| Nystatin | Polyene | Poor solubility; topical relevance | Rapid dissolution and local availability | Maltodextrins, hydrophilic carriers | Solid dispersion | Limited systemic applicability |
| Griseofulvin | Benzofuran | Dissolution-limited absorption | Reduction of crystallinity and improved wettability | PEG, PVP | Solid dispersion (historical) | Superseded by particle engineering |
| Terbinafine HCl | Allylamine | Low solubility; pediatric formulations | Stable supersaturation at low dose | PVP, HPMCAS, Soluplus® | Spray drying | Drug loading vs. stability balance |

#### 3.7.1.1 Itraconazole as an Exemplar of Rational Polymer Selection in Amorphous Solid Dispersions

Itraconazole is widely considered a representative compound for the development and evaluation of ASD strategies due to its extremely low aqueous solubility, pronounced lipophilicity, and strong dependence of absorption on gastrointestinal conditions. Classified as a BCS class II drug, itraconazole exhibits dissolution-limited oral bioavailability and a pronounced pH-dependent solubility, whereby the drug dissolved under gastric conditions readily precipitates upon transition to the intestinal environment. These properties make itraconazole a test case for assessing whether ASD technology can not only enhance apparent solubility but also effectively control supersaturation and precipitation kinetics under physiologically relevant conditions [88, 89].

Mechanistic studies have demonstrated that polymer selection for itraconazole ASDs cannot be guided solely by glass transition temperature or empirical dissolution enhancement. Investigations into molecular mobility have revealed that different polymers stabilize itraconazole through distinct mechanisms. Bhardwaj et al. (2014) showed that hydroxypropyl methylcellulose acetate succinate (HPMCAS) significantly retards global molecular mobility, as reflected by prolonged α-relaxation times, thereby delaying the onset of crystallization [90]. In contrast, polyvinylpyrrolidone (PVP), despite its high glass transition temperature, did not measurably affect α-relaxation or crystallization onset, but instead influenced crystal growth kinetics, suggesting a fundamentally different stabilization pathway. These findings underscore that suppression of nucleation and inhibition of crystal growth represent separable targets in ASD design, and that polymer choice dictates which of these mechanisms predominates [90].

Beyond polymer identity, substitution patterns within a given polymer class further modulate itraconazole performance. Comparative studies of different HPMCAS grades (L, M, and H) have demonstrated that variations in acetyl and succinoyl content translate into distinct balances between hydrophobic drug–polymer interactions and polymer–water affinity. Lower substitution grades tend to promote higher levels of supersaturation and drug flux across biomimetic membranes, whereas more hydrophobic grades provide stronger precipitation inhibition but may compromise overall release and permeation. These results illustrate that stronger drug–polymer interactions do not necessarily yield superior biopharmaceutical performance, and that grade selection represents a critical, yet often underappreciated, design variable [91].

The complexity of itraconazole behavior has also motivated the exploration of ternary ASD systems incorporating surfactants alongside polymers. Such approaches aim to enhance wetting, modulate supersaturation kinetics, or mitigate precipitation following pH shift. However, systematic studies have revealed that polymer–surfactant interactions can be highly drug- and medium-specific. In some cases, surfactants compete with the drug for polymer binding, diminishing the stabilizing effect of the polymer and accelerating precipitation. Consequently, conventional screening metrics such as critical micelle concentration or equilibrium solubility have proven insufficient predictors of in situ performance. Instead, maintenance of high supersaturation over time emerges as the most relevant criterion for polymer–surfactant selection in itraconazole ASDs [88]. Nevertheless, well-designed ternary systems can offer tangible benefits when polymer choice and surfactant functionality are aligned. Formulations combining enteric polymers such as HPMCAS with surfactants like D-α-tocopheryl polyethylene glycol succinate (TPGS) have demonstrated improved dissolution profiles under biorelevant conditions and enhanced systemic exposure in vivo. These outcomes highlight that ternary strategies can be effective for itraconazole, provided that polymer selection remains the primary determinant of supersaturation control and that surfactants are introduced to support polymer function [89].

Itraconazole further exemplifies the extent to which manufacturing technology modulates polymer performance in ASD systems. Comparative studies employing identical drug–polymer compositions but different processing routes have shown pronounced differences in amorphicity, particle morphology, and dissolution behavior. High-speed electrospinning of itraconazole–PVPVA dispersions produced highly amorphous fibrous matrices with rapid and near-complete drug release, whereas spray drying yielded slower dissolution and solvent-cast films retained residual crystallinity due to prolonged solvent evaporation and phase separation. These findings demonstrate that polymer suitability cannot be decoupled from processing kinetics, as the same polymer may succeed or fail depending on the manufacturing route [92]. More recent work integrating continuous manufacturing approaches has reinforced this conclusion. Comparative evaluation of hot-melt extrusion and spray drying, supported by compatibility screening based on solubility parameters, film casting, and molecular modeling, has shown that process selection directly constrains viable polymer candidates and optimal drug loading ranges. Such integrated workflows exemplify how rational polymer selection for itraconazole increasingly relies on the simultaneous consideration of material properties and process feasibility rather than sequential decision-making [93].

The relevance of these formulation principles is underscored by their successful translation into clinically approved products. A prominent example is SUBA™ itraconazole, marketed as TOLSURA® (and LOZANOC® in some regions), which employs a spray-dried ASD of itraconazole in an HPMCAS-based polymer matrix. The use of an enteric, amphiphilic polymer enables pH-dependent drug release in the intestine, improved maintenance of supersaturation, and reduced sensitivity to gastric pH and food intake. Clinically, this formulation achieves higher and more consistent bioavailability at lower doses compared with conventional itraconazole capsules, and is therefore not interchangeable with earlier products such as Sporanox®.

#### 3.7.1.2 Posaconazole as an Exemplar of Mechanism-Informed Polymer Selection and Processing Effects in Amorphous Solid Dispersions

Posaconazole is a second-generation triazole antifungal and, like itraconazole, serves as a highly instructive model compound for ASD development due to its poor aqueous solubility and formulation-sensitive bioperformance. In addition to classic BCS class II limitations, posaconazole exhibits pronounced sensitivity to gastrointestinal conditions (including pH and fed/fasted state), and therefore strongly benefits from ASD strategies designed to generate and sustain supersaturation at physiologically relevant sites [13].

A defining feature of posaconazole ASD development is the repeated emergence of HPMCAS as a high-performing carrier polymer, yet its effectiveness cannot be reduced to a single property (e.g., $T_g$ or pH-dependent solubility). Mechanistic work demonstrates that polymer solution behavior under dissolving pH, including self-association/aggregation and drug–polymer affinity, can be central to maintaining supersaturation and inhibiting crystallization. In a detailed study using posaconazole as a model drug, stronger supersaturation maintenance and crystallization inhibition correlated with higher HPMCAS aggregation numbers (affected by grade and pH) and with the extent of HPMCAS co-precipitation with posaconazole, interpreted as a proxy for drug–polymer affinity [94]. This provides a mechanistic explanation for why HPMCAS-based ASDs frequently deliver robust performance: HPMCAS can act not only through solid-state stabilization, but also through controlling the solution-phase events that govern precipitation kinetics.

Posaconazole also highlights that polymer selection and processing route are interdependent decisions. A recent comparative study prepared posaconazole-HPMCAS ASDs by HME and two spray drying configurations, showing that processing affects particle morphology and chemical stability, even when drug loading and polymer identity are held constant. Specifically, spray-dried samples exhibited reduced chemical stability under stress conditions relative to HME, attributed to morphological features such as smaller particle size and hollow structures facilitating oxygen penetration; meanwhile, the different ASDs showed broadly similar supersaturation induction and de-supersaturation kinetics under the tested conditions. These findings reinforce a key decision principle: for posaconazole-HPMCAS systems, process selection can become a primary determinant of developability (stability, robustness, downstream handling), not merely a manufacturing preference [11].

The clinical relevance of these polymer selection principles is underscored by the availability of marketed ASD-based posaconazole products. Noxafil® delayed-release tablets represent a prominent example and incorporate HPMCAS as a key excipient in the formulation. The use of an enteric/amphiphilic polymer aligns with the broader mechanistic rationale for posaconazole: enabling controlled release beyond the stomach and supporting sustained supersaturation in intestinal conditions where absorption occurs [95]. Taken together, posaconazole demonstrates that successful ASD development depends on (i) selecting polymers that control both solid-state stability and solution-phase precipitation events, (ii) recognizing that polymer grade and pH-dependent polymer behavior can influence supersaturation outcomes, and (iii) integrating process choice early, as processing can materially alter stability and manufacturability even within the same drug–polymer system.

#### 3.7.1.3 Amphotericin B, Nystatin, Griseofulvin, and Terbinafine: Expanding the Scope of ASD Strategies in Antifungal Therapy

While triazole antifungals such as itraconazole and posaconazole have become archetypal examples for mechanistic polymer selection in ASD, other antifungal agents present distinct formulation challenges that have also motivated the application of ASD-based approaches. Amphotericin B, nystatin, griseofulvin, and terbinafine span a wide physicochemical spectrum, from highly lipophilic small molecules to large, amphoteric polyenes, and collectively demonstrate how ASD strategies can be adapted to different therapeutic and biopharmaceutical objectives.

Amphotericin B (AmB) and nystatin are polyene antifungals characterized by large molecular size, amphiphilicity, and extremely low aqueous solubility. Their clinical use has historically relied on formulations designed to mitigate toxicity or enable local rather than systemic exposure. For AmB, marketed products such as Fungizone® (amphotericin B deoxycholate) and lipid-based formulations including AmBisome®, Abelcet®, and Amphotec® are not ASDs but rather micellar or lipid-associated systems developed to reduce nephrotoxicity and improve tolerability. Nevertheless, solid dispersion strategies have been explored to improve amphotericin B solubility, dissolution behavior, and antifungal efficacy, particularly in localized applications. Studies employing hydrophilic carriers such as maltodextrins, polyethylene glycols, or polymeric matrices have demonstrated partial amorphization of amphotericin B and enhanced interaction with fungal biofilms, supporting the concept that even highly complex molecules can benefit from dispersion-based formulation approaches [17, 18].

Nystatin, which is primarily used for topical or oral mucosal infections and is not systemically absorbed, provides a particularly illustrative example of how ASD concepts can be repurposed toward local efficacy rather than bioavailability enhancement. Solid dispersions of nystatin with hydrophilic carriers such as maltodextrin have been shown to substantially reduce crystallinity, accelerate dissolution, and markedly enhance antifungal and antibiofilm activity against *Candida* species. Importantly, in these systems, the benefit of the carrier extends beyond solubility enhancement to include hyperosmotic effects and improved penetration into biofilm matrices, underscoring that ASD performance metrics may differ depending on the therapeutic target [96].

Griseofulvin represents one of the earliest antifungal drugs for which solid dispersion strategies were investigated, long before the widespread adoption of the ASD concept in modern formulation science. As a poorly water-soluble, BCS class II compound, griseofulvin exhibits dissolution-limited absorption and substantial variability in oral bioavailability. Early studies demonstrated that dispersing griseofulvin in hydrophilic carriers such as polyethylene glycols, polyvinylpyrrolidone, or sugars could reduce crystallinity, decrease particle size, and significantly enhance dissolution rate. Although many of these formulations predate contemporary mechanistic frameworks, they embody core ASD principles: disruption of crystal lattice energy, improved wettability, and increased effective surface area. Notably, marketed griseofulvin products historically relied on particle size reduction (micronized or ultramicronized griseofulvin) rather than true ASDs; nevertheless, the griseofulvin case highlights how solid dispersion concepts laid foundational groundwork for later ASD technologies applied to more complex antifungal agents [9, 97].

Terbinafine hydrochloride is a lipophilic allylamine antifungal with high permeability but low aqueous solubility, classified as a BCS class II drug. Unlike amphotericin B and nystatin, terbinafine is intended for systemic oral therapy, making it a more direct candidate for ASD-based bioavailability enhancement. Recent studies employing a rational design framework have demonstrated that careful screening of polymers using solubility parameters, equilibrium solubility measurements, and dose number calculations can guide carrier selection for terbinafine ASDs. In particular, polymers such as Soluplus®, HPMCAS, and high-molecular-weight polyvinylpyrrolidones have shown favorable interactions with terbinafine, enabling the formation of stable spray-dried ASDs with prolonged supersaturation and enhanced apparent solubility. These findings are especially relevant in the context of pediatric formulation development, where dose flexibility, stability, and excipient safety are critical considerations. While marketed terbinafine products (e.g., Lamisil® tablets) are not based on ASD technology, the growing body of preclinical evidence positions terbinafine as a strong candidate for future ASD-enabled oral formulations [98].

## 3.8 Early De-Risking and Advanced Evaluation Tools in ASD Development

### *3.8.1 Rationale for Early De-Risking in ASD Design*

Despite significant progress in amorphous solid dispersion technologies, late-stage formulation failure remains common. In many cases, these failures are not caused by insufficient solubility enhancement, but by formulation-driven risks that emerge only after scale-up or biological evaluation. These include rapid precipitation following supersaturation, high variability in systemic exposure, and formulation-induced safety concerns.

Antifungal drugs are particularly sensitive to such risks due to narrow therapeutic windows, aggregation-prone structures, and concentration-dependent toxicity. As demonstrated by the case studies in this chapter, improvements in apparent solubility do not necessarily translate into improved therapeutic performance. Early de-risking strategies are therefore essential to identify formulation liabilities at stages where design modifications are still feasible [95]. A structured overview of early de-risking questions, tools, and decision outputs across ASD development stages is provided in Table 3.2.

### 3.8.2 Predictive and Data-Driven Tools for Early Formulation Screening

In silico and data-driven approaches are increasingly used to support early decision-making in ASD development. Molecular descriptor analysis, thermodynamic modeling, and machine learning—guided frameworks enable prioritization of polymer candidates based on predicted performance, interaction profiles, and biological relevance.

Notably, recent studies applying machine learning to antifungal polymer design demonstrate how polymer structure can be linked to antifungal efficacy and safety outcomes. Although such approaches do not replace experimental validation, they effectively reduce formulation uncertainty by narrowing the design space and excluding high-risk candidates before extensive experimental work [99].

### 3.8.3 Advanced In Vitro Tools Beyond Conventional Dissolution Testing

Conventional compendial dissolution tests provide limited insight into the dynamic behavior of ASD systems under physiological conditions. In particular, they fail to capture supersaturation generation, precipitation kinetics, and polymer-mediated stabilization following gastrointestinal pH shifts.

Advanced in vitro methodologies, including pH-shift dissolution experiments, biphasic dissolution systems, and dissolution–permeation setups, have therefore gained prominence as early de-risking tools. These approaches allow simultaneous assessment of drug release, precipitation tendencies, and absorption potential, which is particularly relevant for antifungal ASDs where sustained supersaturation is often the primary determinant of in vivo exposure.

### 3.8.4 Integrated Biological Models and New Approach Methodologies

As ASD formulations increasingly modulate drug exposure profiles, early evaluation of biological effects becomes essential—not only to assess efficacy but also to identify formulation-specific safety risks. New approach methodologies (NAMs),

**Table 3.2** Early de-risking framework and advanced evaluation tools for polymer-based amorphous solid dispersions (ASDs)

| Development question | Typical risk addressed | Representative tools (examples) | Key information obtained | Implications for polymer selection | Implications for process choice |
|---|---|---|---|---|---|
| Is the API suitable for ASD formulation? | Poor glass-forming ability; rapid recrystallization | DSC, hot-stage microscopy, stress testing | Glass-forming tendency; intrinsic physical stability | Need for polymers with strong stabilizing capacity (high $T_g$, strong interactions) | May exclude thermally demanding processes |
| Is the API miscible with the polymer? | Phase separation; physical instability | Film casting, DSC (single $T_g$), XRPD | Drug–polymer miscibility and homogeneity | Selection or exclusion of polymer classes or grades | Guides feasible drug loading and processing window |
| Can supersaturation be generated and maintained in solution? | Rapid precipitation after dissolution | Supersaturation and pH-shift dissolution tests | Duration and robustness of supersaturation | Preference for precipitation-inhibiting polymers (e.g. HPMCAS, amphiphilic systems) | Favors processes yielding optimal polymer distribution |
| Is the system physically stable under relevant storage conditions? | Moisture-induced plasticization and crystallization | DVS, accelerated stability studies | Sensitivity to humidity and temperature | Need for less hygroscopic polymers or polymer blends | Influences choice of process and packaging strategy |
| Is the formulation compatible with the intended manufacturing route? | Thermal or solvent-induced degradation; poor processability | Thermal analysis, solvent compatibility screening | Processability limits and degradation risk | Polymer selection constrained by thermal or solvent stability | Selection between HME, spray drying, or emerging technologies |

including alternative biological models, offer valuable opportunities for early-stage assessment while reducing reliance on traditional animal testing.

Among these, the zebrafish (*Danio rerio*) embryo model has emerged as a versatile platform for early de-risking in drug and formulation development. Owing to its optical transparency, rapid development, and conserved molecular pathways, the zebrafish model enables simultaneous observation of pharmacodynamic effects, developmental toxicity, and morphological changes following exposure to drug formulations.

In the context of ASD development, zebrafish embryos can serve as an intermediate evaluation platform to compare different formulation variants, assess dose–response relationships, and identify formulation-driven toxicity that may not be evident from physicochemical characterization alone. Importantly, this model is particularly well suited for testing antifungal agents, where alterations in solubility, aggregation state, or release kinetics may significantly influence both efficacy and safety [100, 101]. By positioning such models as screening and prioritization tools, rather than definitive predictors of clinical outcome, NAM approaches complement advanced in vitro and in silico methods and contribute to a more integrated early de-risking strategy.

### *3.8.5 Translational Value of Early De-Risking Strategies*

The integration of predictive modeling, advanced in vitro testing, and biologically relevant screening models represents a shift toward iterative, knowledge-driven ASD development. Rather than uncovering formulation shortcomings at late preclinical stages, early de-risking enables informed decision-making while formulation flexibility remains high.

For antifungal ASDs, where polymer selection and processing choices strongly influence both exposure and safety, such integrated strategies improve robustness and translational success. Early de-risking therefore represents a critical component in the evolution of ASD technology from a solubility-enhancement tool into a comprehensive drug delivery platform.

## 3.9 Conclusions and Future Perspectives

Amorphous solid dispersions remain one of the most powerful formulation strategies for addressing the solubility- and dissolution-related limitations of antifungal drugs. However, as illustrated throughout this chapter, the success of ASD-based systems depends far more on rational polymer selection and integrated design strategies than on amorphization alone. Polymers are not inert excipients but active design elements that govern miscibility, molecular mobility, supersaturation maintenance, and, ultimately, therapeutic performance. Through mechanistic analysis and case studies, this chapter highlights that polymer selection must be approached as a multidimensional decision-making process. Factors such as polymer chemistry, grade, molecular architecture, and hygroscopicity interact with drug properties and manufacturing conditions to define both opportunities and risks. For antifungal agents, where narrow therapeutic windows, precipitation-prone supersaturation, and formulation-sensitive toxicity are common, these considerations become particularly critical.

The case studies of itraconazole and posaconazole demonstrate how mechanistic understanding of polymer function translates into clinically successful formulations when aligned with appropriate manufacturing technologies. In contrast, examples involving polyene antifungals and other non-azole agents underscore the limitations of applying ASD strategies without a clear appreciation of biological context and therapeutic objectives. Together, these examples reinforce that there is no universally optimal polymer or formulation architecture; rather, effective ASD design emerges from aligning polymer properties with the dominant biopharmaceutical challenges of each drug.

Looking forward, the development of ASD systems is expected to become increasingly data-informed and integrative. Advances in predictive modelling, molecular descriptor analysis, and machine learning offer opportunities to rationalize early-stage polymer selection and reduce empirical trial-and-error. At the same time, the growing adoption of advanced in vitro tools and NAMs enables earlier identification of formulation-specific risks related to precipitation behavior, exposure, and safety.

Future progress in ASD technology will therefore depend on the convergence of mechanistic insight, smart material selection, process-aware design, and early de-risking strategies. By embedding these principles into formulation development workflows, ASD-based delivery systems can continue to evolve from solubility-enhancement tools into robust, clinically reliable platforms for antifungal therapy and beyond.

## References


1. Jermain SV, Brough C, Williams RO 3rd. Amorphous solid dispersions and nanocrystal technologies for poorly water-soluble drug delivery – an update. Int J Pharm. 2018;535(1–2):379–92.
2. Qian F, Huang J, Hussain MA. Drug–polymer solubility and miscibility: stability consideration and practical challenges in amorphous solid dispersion development. J Pharm Sci. 2010;99(7):2941–7.
3. Vasconcelos T, Marques S, das Neves J, Sarmento B. Amorphous solid dispersions: rational selection of a manufacturing process. Adv Drug Deliv Rev. 2016;100:85–101.
4. Oliveira VDS, de Almeida AS, Albuquerque IDS, Duarte FÍC, Queiroz B, Converti A, et al. Therapeutic applications of solid dispersions for drugs and new molecules: in vitro and in vivo activities. Pharmaceutics. 2020;12(10):933.

5. Lehmkemper K, Kyeremateng SO, Heinzerling O, Degenhardt M, Sadowski G. Long-term physical stability of PVP- and PVPVA-amorphous solid dispersions. Mol Pharm. 2017;14(1):157–71.
6. Meng F, Trivino A, Prasad D, Chauhan H. Investigation and correlation of drug polymer miscibility and molecular interactions by various approaches for the preparation of amorphous solid dispersions. Eur J Pharm Sci. 2015;71:12–24.
7. Shi Q, Chen H, Wang Y, Wang R, Xu J, Zhang C. Amorphous solid dispersions: role of the polymer and its importance in physical stability and in vitro performance. Pharmaceutics. 2022;14(8):1747.
8. Tambe S, Jain D, Meruva SK, Rongala G, Juluri A, Nihalani G, et al. recent advances in amorphous solid dispersions: preformulation, formulation strategies, technological advancements and characterization. Pharmaceutics. 2022;14(10):2203.
9. Al-Obaidi H, Kowalczyk RM, Kalgudi R, Zariwala MG. Griseofulvin solvate solid dispersions with synergistic effect against fungal biofilms. Colloids Surf B Biointerfaces. 2019;184:110540.
10. Anklam E, Bahl MI, Ball R, Beger RD, Cohen J, Fitzpatrick S, et al. Emerging technologies and their impact on regulatory science. Exp Biol Med (Maywood). 2022;247(1):1–75.
11. He R, Lamm MS, Brunskill A, Axnanda S, Li Y. Impact of processing methods on the physico-chemical properties of posaconazole amorphous solid dispersions. Pharm Res. 2024;41(1):141–51.
12. Madabushi R, Seo P, Zhao L, Tegenge M, Zhu H. Review: role of model-informed drug development approaches in the lifecycle of drug development and regulatory decision-making. Pharm Res. 2022;39(8):1669–80.
13. Patel K, Kevlani V, Shah S. A novel posaconazole oral formulation using spray dried solid dispersion technology: in-vitro and in-vivo study. Drug Deliv Transl Res. 2024;14(5):1253–76.
14. Li Y, Mann AKP, Zhang D, Yang Z. Processing impact on in vitro and in vivo performance of solid dispersions-a comparison between hot-melt extrusion and spray drying. Pharmaceutics. 2021;13(8):1307.
15. Wu S, Song R, Liu T, Li C. Antifungal therapy: novel drug delivery strategies driven by new targets. Adv Drug Deliv Rev. 2023;199:114967.
16. O'Shea JP, Holm R, O'Driscoll CM, Griffin BT. Food for thought: formulating away the food effect – a PEARRL review. J Pharm Pharmacol. 2019;71(4):510–35.
17. Mahfud MAS, Syahirah NA, Akram M, Mahfufah U, Saputra MD, Elim D, et al. Solid dispersion incorporated into dissolving microneedles for improved antifungal activity of amphotericin B: in vivo study in a Fungal Keratitis model. Mol Pharm. 2023;20(12):6246–61.
18. Smith-Craven MM, Dening TJ, Basra AK, Hageman MJ. Enhanced dissolution of amphotericin B through development of amorphous solid dispersions containing polymer and surfactants. J Pharm Sci. 2024;113(8):2454–63.
19. Hüttel W. Echinocandins: structural diversity, biosynthesis, and development of antimycotics. Appl Microbiol Biotechnol. 2021;105(1):55–66.
20. Pendam D, Tomake P, Debaje S, Guleria K, Saha A, Thakran P, et al. Advances in formulation strategies and stability considerations of amorphous solid dispersions. J Drug Deliv Sci Technol. 2025;108:106922.
21. Baghel S, Cathcart H, O'Reilly NJ. Polymeric amorphous solid dispersions: a review of amorphization, crystallization, stabilization, solid-state characterization, and aqueous solubilization of biopharmaceutical classification system class II drugs. J Pharm Sci. 2016;105(9):2527–44.
22. Marsac PJ, Li T, Taylor LS. Estimation of drug-polymer miscibility and solubility in amorphous solid dispersions using experimentally determined interaction parameters. Pharm Res. 2009;26(1):139–51.
23. Tian Y, Qian K, Jacobs E, Amstad E, Jones DS, Stella L, et al. The investigation of Flory-Huggins interaction parameters for amorphous solid dispersion across the entire temperature and composition range. Pharmaceutics. 2019;11(8):420.

24. Medarević D, Djuriš J, Barmpalexis P, Kachrimanis K, Ibrić S. Analytical and computational methods for the estimation of drug-polymer solubility and miscibility in solid dispersions development. Pharmaceutics. 2019;11(8):372.
25. Qu B, Wu H, Ding Z, Bu R, Zhang H, Han J, et al. Amorphous solid dispersion formulation of pimobendan for bioavailability enhancement: a comprehensive study on miscibility, interactions, and in vitro dissolution behavior. J Drug Deliv Sci Technol. 2025;108:106884.
26. Bookwala M, Wildfong PLD. The implications of drug-polymer interactions on the physical stability of amorphous solid dispersions. Pharm Res. 2023;40(12):2963–81.
27. Joshi P, Sangamwar AT. Stabilizing supersaturated drug-delivery system through mechanism of nucleation and crystal growth inhibition of drugs. Ther Deliv. 2018;9(12):873–85.
28. Lin X, Hu Y, Liu L, Su L, Li N, Yu J, et al. Physical stability of amorphous solid dispersions: a physicochemical perspective with thermodynamic, kinetic and environmental aspects. Pharm Res. 2018;35(6):125.
29. Wu J, Mooter GV. The influence of hydrogen bonding between different crystallization tendency drugs and PVPVA on the stability of amorphous solid dispersions. Int J Pharm. 2023;646:123440.
30. Xiang T-X, Anderson BD. Effects of molecular interactions on miscibility and mobility of ibuprofen in amorphous solid dispersions with various polymers. J Pharm Sci. 2019;108(1):178–86.
31. Mohapatra S, Samanta S, Kothari K, Mistry P, Suryanarayanan R. Effect of polymer molecular weight on the crystallization behavior of indomethacin amorphous solid dispersions. Cryst Growth Des. 2017;17(6):3142–50.
32. Nair AR, Lakshman YD, Anand VSK, Sree KSN, Bhat K, Dengale SJ. Overview of extensively employed polymeric carriers in solid dispersion technology. AAPS PharmSciTech. 2020;21(8):309.
33. Zhang Z, Dong L, Guo J, Li L, Tian B, Zhao Q, et al. Prediction of the physical stability of amorphous solid dispersions: relationship of aging and phase separation with the thermodynamic and kinetic models along with characterization techniques. Expert Opin Drug Deliv. 2021;18(2):249–64.
34. Louwerse MJ, Maldonado A, Rousseau S, Moreau-Masselon C, Roux B, Rothenberg G. Revisiting Hansen solubility parameters by including thermodynamics. ChemPhysChem. 2017;18(21):2999–3006.
35. Rumondor ACF, Ivanisevic I, Bates S, Alonzo DE, Taylor LS. Evaluation of drug-polymer miscibility in amorphous solid dispersion systems. Pharm Res. 2009;26(11):2523–34.
36. Dedroog S, Pas T, Vergauwen B, Huygens C, Van den Mooter G. Solid-state analysis of amorphous solid dispersions: why DSC and XRPD may not be regarded as stand-alone techniques. J Pharm Biomed Anal. 2020;178:112937.
37. Liu X, Feng X, Williams RO, Zhang F. Characterization of amorphous solid dispersions. J Pharm Investig. 2018;48(1):19–41.
38. Ma X, Williams RO. Characterization of amorphous solid dispersions: an update. J Drug Deliv Sci Technol. 2019;50:113–24.
39. Thakore SD, Akhtar J, Jain R, Paudel A, Bansal AK. Analytical and computational methods for the determination of drug-polymer solubility and miscibility. Mol Pharm. 2021;18(8):2835–66.
40. Sharma A, Shah S, Wagh S, Pandey G, Pradhan AK, Shukla S, et al. Ever-increasing role of computational tools in solid-state pharmaceutics: advancing drug development with enhanced molecular understanding and risk assessment. Mol Pharm. 2025;22(9):5165–92.
41. Bhatta HP, Han HK, Maharjan R, Jeong SH. Recent techniques to improve amorphous dispersion performance with quality design, physicochemical monitoring, molecular simulation, and machine learning. Pharmaceutics. 2025;17(10):1249.
42. Liao H-C, Lin Y-H, Peng C-H, Li Y-P. Directed message passing neural networks for accurate prediction of polymer–solvent interaction parameters. ACS Eng Au. 2025;5(5):530–9.

43. Walden DM, Bundey Y, Jagarapu A, Antontsev V, Chakravarty K, Varshney J. Molecular simulation and statistical learning methods toward predicting drug-polymer amorphous solid dispersion miscibility, stability, and formulation design. Molecules. 2021;26(1):182.
44. Wegiel LA, Mauer LJ, Edgar KJ, Taylor LS. Crystallization of amorphous solid dispersions of resveratrol during preparation and storage-impact of different polymers. J Pharm Sci. 2013;102(1):171–84.
45. Bhugra C, Telang C, Schwabe R, Zhong L. Reduced crystallization temperature methodology for polymer selection in amorphous solid dispersions: stability perspective. Mol Pharm. 2016;13(9):3326–33.
46. Saha SK, Joshi A, Singh R, Dubey K. Review of industrially recognized polymers and manufacturing processes for amorphous solid dispersion based formulations. Pharm Dev Technol. 2023;28(7):678–96.
47. Iyer R, Petrovska Jovanovska V, Berginc K, Jaklič M, Fabiani F, Harlacher C, et al. Amorphous solid dispersions (ASDs): the influence of material properties, manufacturing processes and analytical technologies in drug product development. Pharmaceutics. 2021;13(10):1682.
48. Szabó E, Démuth B, Galata DL, Vass P, Hirsch E, Csontos I, et al. Continuous formulation approaches of amorphous solid dispersions: significance of powder flow properties and feeding performance. Pharmaceutics. 2019;11(12):654.
49. Abrantes CG, Duarte D, Reis CP. An overview of pharmaceutical excipients: safe or not safe? J Pharm Sci. 2016;105(7):2019–26.
50. Kim S, Kim JH, Jeon O, Kwon IC, Park K. Engineered polymers for advanced drug delivery. Eur J Pharm Biopharm. 2009;71(3):420–30.
51. Umekar M, Raut N, Mahore J, Malode D, Koche S, Manghani M, et al. Comparative insights into polymers for solid dispersions: mechanisms, sustainability, regulatory and clinical perspectives. Polym Plast Technol Eng. 2025;64(16):2429–51.
52. Rusdin A, Mohd Gazzali A, Ain Thomas N, Megantara S, Aulifa DL, Budiman A, et al. Advancing drug delivery paradigms: polyvinyl pyrolidone (PVP)-based amorphous solid dispersion for enhanced physicochemical properties and therapeutic efficacy. Polymers (Basel). 2024;16(2):286.
53. Van Duong T, Van den Mooter G. The role of the carrier in the formulation of pharmaceutical solid dispersions. Part II: amorphous carriers. Expert Opin Drug Deliv. 2016;13(12):1681–94.
54. Butreddy A. Hydroxypropyl methylcellulose acetate succinate as an exceptional polymer for amorphous solid dispersion formulations: a review from bench to clinic. Eur J Pharm Biopharm. 2022;177:289–307.
55. Papageorgiou GZ, Bikiaris D, Kanaze FI, Karavas E, Stergiou A, Georgarakis E. Tailoring the release rates of fluconazole using solid dispersions in polymer blends. Drug Dev Ind Pharm. 2008;34(3):336–46.
56. Purohit HS, Taylor LS. Miscibility of itraconazole-hydroxypropyl methylcellulose blends: insights with high resolution analytical methodologies. Mol Pharm. 2015;12(12):4542–53.
57. Sarode AL, Obara S, Tanno FK, Sandhu H, Iyer R, Shah N. Stability assessment of hypromellose acetate succinate (HPMCAS) NF for application in hot melt extrusion (HME). Carbohydr Polym. 2014;101:146–53.
58. Dos Santos J, da Silva GS, Velho MC, Beck RCR. Eudragit(®): a versatile family of polymers for hot melt extrusion and 3D printing processes in pharmaceutics. Pharmaceutics. 2021;13(9):1424.
59. Schmied FP, Bernhardt A, Moers C, Meier C, Endres T, Klein S. A novel aminomethacrylate-based copolymer for solubility enhancement-from radical polymer synthesis to manufacture and characterization of amorphous solid dispersions. Polymers (Basel). 2022;14(7):1281.
60. Schittny A, Huwyler J, Puchkov M. Mechanisms of increased bioavailability through amorphous solid dispersions: a review. Drug Deliv. 2020;27(1):110–27.
61. Kommarajula P, Prithipaul V, Kunda NK. Enhanced solubility and dissolution of Alectinib via amorphous solid dispersion with a novel polymer, Apinovex™. Int J Pharm. 2026;690:126551.

62. Zeneli E, Bohets H, Mebenga FN, Tistaert C, Holm R, Kuentz M. Guiding excipient selection for amorphous solid dispersions by combining an in vitro-in-silico approach – II: supersaturation and drug release. Int J Pharm. 2025;682:125976.
63. Qin ZY, Jia XW, Liu Q, Kong BH, Wang H. Fast dissolving oral films for drug delivery prepared from chitosan/pullulan electrospinning nanofibers. Int J Biol Macromol. 2019;137:224–31.
64. Roy S, Min S-J, Biswas D, Rhim J-W. Pullulan/chitosan-based functional film incorporated with curcumin-integrated chitosan nanoparticles. Colloids Surf A Physicochem Eng Asp. 2023;660:130898.
65. Madera-Santana TJ, Herrera-Méndez CH, Rodríguez-Núñez JR. An overview of the chemical modifications of chitosan and their advantages. Green Mater. 2018;6(4):131–42.
66. Guan J, Liu Q, Zhang X, Zhang Y, Chokshi R, Wu H, et al. Alginate as a potential diphase solid dispersion carrier with enhanced drug dissolution and improved storage stability. Eur J Pharm Sci. 2018;114:346–55.
67. Luo C, Wu W, Lou S, Zhao S, Yang K. Improving the in vivo bioavailability and in vitro anti-inflammatory activity of tanshinone IIA by alginate solid dispersion. J Drug Deliv Sci Technol. 2020;60:101966.
68. Mai NNS, Nakai R, Kawano Y, Hanawa T. Enhancing the solubility of curcumin using a solid dispersion system with hydroxypropyl-β-cyclodextrin prepared by grinding, freeze-drying, and common solvent evaporation methods. Pharmacy (Basel). 2020;8(4):203.
69. Schoeman C, van Niekerk S, Liebenberg W, Hamman J. Cyclodextrin inclusion complex and amorphous solid dispersions as formulation approaches for enhancement of curcumin's solubility and nasal epithelial membrane permeation. Futur J Pharm Sci. 2024;10(1):85.
70. Budiman A, Ivana H, Huang KA, Huang SA, Nadhira MS, Rusdin A, et al. Biocompatible natural polymer-based amorphous solid dispersion system improving drug physicochemical properties, stability, and efficacy. Polymers. 2025; 17(15):2059.
71. Solanki NG, Gumaste SG, Shah AV, Serajuddin ATM. Effects of surfactants on itraconazole-hydroxypropyl methylcellulose acetate succinate solid dispersion prepared by hot melt extrusion. II: rheological analysis and extrudability testing. J Pharm Sci. 2019;108(9):3063–73.
72. Baghel S, Cathcart H, O'Reilly NJ. Investigation into the solid-state properties and dissolution profile of spray-dried ternary amorphous solid dispersions: a rational step toward the design and development of a multicomponent amorphous system. Mol Pharm. 2018;15(9):3796–812.
73. Borchardt-Setter KA, Yu J, Stelzer T, Yu L. Surface enrichment and depletion of components in a ternary drug-surfactant-polymer amorphous solid dispersion. Mol Pharm. 2024;21(8):4074–81.
74. Kyeremateng SO, Voges K, Dohrn S, Sobich E, Lander U, Weber S, et al. A hot-melt extrusion risk assessment classification system for amorphous solid dispersion formulation development. Pharmaceutics. 2022;14(5):1044.
75. Tran PHL, Lee BJ, Tran TTD. Recent studies on the processes and formulation impacts in the development of solid dispersions by hot-melt extrusion. Eur J Pharm Biopharm. 2021;164:13–9.
76. Darwich M, Mohylyuk V, Kolter K, Bodmeier R, Dashevskiy A. Enhancement of itraconazole solubility and release by hot-melt extrusion with Soluplus®. J Drug Deliv Sci Technol. 2023;81:104280.
77. Li S, Zhang Z, Gu W, Gallas M, Jones D, Boulet P, et al. Hot melt extruded high-dose amorphous solid dispersions containing lumefantrine and Soluplus. Int J Pharm. 2024;665:124676.
78. Zhao Y, Xie X, Zhao Y, Gao Y, Cai C, Zhang Q, et al. Effect of plasticizers on manufacturing ritonavir/copovidone solid dispersions via hot-melt extrusion: preformulation, physicochemical characterization, and pharmacokinetics in rats. Eur J Pharm Sci. 2019;127:60–70.
79. Alshahrani SM, Lu W, Park JB, Morott JT, Alsulays BB, Majumdar S, et al. Stability-enhanced hot-melt extruded amorphous solid dispersions via combinations of Soluplus® and HPMCAS-HF. AAPS PharmSciTech. 2015;16(4):824–34.

80. Duarte Í, Santos JL, Pinto JF, Temtem M. Screening methodologies for the development of spray-dried amorphous solid dispersions. Pharm Res. 2015;32(1):222–37.
81. Ziaee A, Albadarin AB, Padrela L, Faucher A, O'Reilly E, Walker G. Spray drying ternary amorphous solid dispersions of ibuprofen – an investigation into critical formulation and processing parameters. Eur J Pharm Biopharm. 2017;120:43–51.
82. Huang S, Williams RO 3rd. Effects of the preparation process on the properties of amorphous solid dispersions. AAPS PharmSciTech. 2018;19(5):1971–84.
83. Li J, Patel D, Wang G. Use of spray-dried dispersions in early pharmaceutical development: theoretical and practical challenges. AAPS J. 2017;19(2):321–33.
84. Yu D, Nie H, Hoag SW. Comprehensive evaluation of polymer types and ratios in spray-dried dispersions: compaction, dissolution, and physical stability. Int J Pharm. 2024;650:123674.
85. Vehring R. Pharmaceutical particle engineering via spray drying. Pharm Res. 2008;25(5):999–1022.
86. Ellenberger DJ, Miller DA, Williams RO. Expanding the application and formulation space of amorphous solid dispersions with KinetiSol®: a review. AAPS PharmSciTech. 2018;19(5):1933–56.
87. Tran P, Park JS. Application of supercritical fluid technology for solid dispersion to enhance solubility and bioavailability of poorly water-soluble drugs. Int J Pharm. 2021;610:121247.
88. Deshpande TM, Shi H, Pietryka J, Hoag SW, Medek A. Investigation of polymer/surfactant interactions and their impact on itraconazole solubility and precipitation kinetics for developing spray-dried amorphous solid dispersions. Mol Pharm. 2018;15(3):962–74.
89. Feng D, Peng T, Huang Z, Singh V, Shi Y, Wen T, et al. Polymer⁻Surfactant system based amorphous solid dispersion: precipitation inhibition and bioavailability enhancement of itraconazole. Pharmaceutics. 2018;10(2):53.
90. Bhardwaj SP, Arora KK, Kwong E, Templeton A, Clas SD, Suryanarayanan R. Mechanism of amorphous itraconazole stabilization in polymer solid dispersions: role of molecular mobility. Mol Pharm. 2014;11(11):4228–37.
91. Adhikari A, Polli JE. Characterization of grades of HPMCAS spray dried dispersions of itraconazole based on supersaturation kinetics and molecular interactions impacting formulation performance. Pharm Res. 2020;37(10):192.
92. Nagy ZK, Balogh A, Démuth B, Pataki H, Vigh T, Szabó B, et al. High speed electrospinning for scaled-up production of amorphous solid dispersion of itraconazole. Int J Pharm. 2015;480(1–2):137–42.
93. Kshirsagar MM, Chatale BC, Dyawanapelly S, Vora LK, Amin PD. Continuous processing strategies for amorphous solid dispersions of itraconazole: impact of polymer selection and manufacturing techniques. Pharmaceutics. 2025;17(9):1090.
94. Wang S, Liu C, Chen Y, Zhu AD, Qian F. Aggregation of hydroxypropyl methylcellulose acetate succinate under its dissolving pH and the impact on drug supersaturation. Mol Pharm. 2018;15(10):4643–53.
95. Mugheirbi NA, O'Connell P, Serrano DR, Healy AM, Taylor LS, Tajber L. A comparative study on the performance of inert and functionalized spheres coated with solid dispersions made of two structurally related antifungal drugs. Mol Pharm. 2017;14(11):3718–28.
96. Benavent C, Torrado-Salmerón C, Torrado-Santiago S. Development of a solid dispersion of nystatin with maltodextrin as a carrier agent: improvements in antifungal efficacy against Candida spp. biofilm infections. Pharmaceuticals (Basel). 2021;14(5):397.
97. Rahman M, Ahmad S, Tarabokija J, Parker N, Bilgili E. Spray-dried amorphous solid dispersions of griseofulvin in HPC/Soluplus/SDS: elucidating the multifaceted impact of SDS as a minor component. Pharmaceutics. 2020;12(3):197.
98. Boza IAF, da Silva SL, Guedes NB, Bazzo GC, Stulzer HK. Pediatric formulation optimization using a rational design: exploring amorphous solid dispersion technology with terbinafine hydrochloride as a case study. AAPS PharmSciTech. 2025;26(1):40.

99. Roh J, Huang KT, Choudhury A, Quan T, Huang CJ, Kumar P, et al. Machine learning-guided design of high-performance antifungal polymers against Candida albicans. ACS Polym Au. 2025;5(5):557–69.
100. Maciel-Magalhães M, Medeiros RJ, Guedes N, Brito TM, Souza GF, Canabarro BR, et al. Amphotericin B encapsulation in polymeric nanoparticles: toxicity insights via cells and zebrafish embryo testing. Pharmaceutics. 2025;17(1):116.
101. Pavic A, Stojanovic Z, Pekmezovic M, Veljović Đ, O'Connor K, Malagurski I, et al. Polyenes in medium chain length polyhydroxyalkanoate (mcl-PHA) biopolymer microspheres with reduced toxicity and improved therapeutic effect against Candida infection in zebrafish model. Pharmaceutics. 2022;14(4):696.

# Part II
# Manufacturing and Engineering Technologies

# Chapter 4
# Hot Melt Extrusion (HME) as an Innovative Technology for the Design and Development of Amorphous Solid Dispersions

**Siva S. Kolipaka and Dennis Douroumis**

**Abstract** Hot melt extrusion (HME) is increasingly employed as a continuous manufacturing technology for producing amorphous solid dispersions (ASDs) capable of overcoming solubility and bioavailability limitations associated with modern drug candidates. By enabling intimate molecular mixing of drug and polymer under controlled thermal and mechanical conditions, HME transforms crystalline active pharmaceutical ingredients into kinetically stable amorphous systems with enhanced dissolution and supersaturation behaviour. The quality and performance of these extrudates are driven by defined critical quality attributes (CQAs), including glass transition temperature, degree of crystallinity, dissolution kinetics and impurity profiles. These are influenced by critical material attributes (CMAs) such as polymer characteristics, drug loading, particle size, and hygroscopicity, as well as critical process parameters (CPPs), including barrel temperature, screw speed, residence time, and specific mechanical energy. A quality by design (QbD) framework supports rational formulation and process development by linking CQAs, CMAs and CPPs to a justified design space. Contemporary approaches integrating Process Analytical Technology (PAT) and real-time release testing (RTRT) enable continuous monitoring of mixing, amorphization and chemical stability, aligning with the expectations of FDA, EMA and ICH guidelines. Recent case studies demonstrate the versatility of HME beyond solubility enhancement, including taste masked paediatric dosage forms, abuse deterrent designs, fixed dose combinations and gastroretentive systems. Altogether, these advances highlight HME as a scalable, solvent-free and regulatory aligned technology for developing high-performance oral drug products with improved patient outcomes.

S. S. Kolipaka
Centre for Research Innovation (CRI), University of Greenwich, Chatham, UK

D. Douroumis (✉)
School of Life and Health Sciences, Department of Health Sciences,
University of Nicosia -Athens, Athens, Greece
e-mail: douroumis.d@unic.ac.cy

 
I. Nikolakakis, I. Partheniadis (eds.), *Amorphous Solid Dispersions*, AAPS Introductions in the Pharmaceutical Sciences 5,
https://doi.org/10.1007/978-3-032-24883-1_4

**Keywords** Thermomechanical processing · Twin-screw extrusion · Continuous manufacturing · Process analytical technology (PAT) · Residence time distribution (RTD)

## 4.1 Introduction

Hot melt extrusion (HME) is a versatile and efficient technology that has been widely used in pharmaceutical manufacturing to enhance the solubility and bioavailability of poorly water-soluble drugs. During the extrusion process, the melting/mixing of active pharmaceutical ingredients and polymeric matrix leads to the production of solid dispersions that improve drug solubility. Since the 1930s, when HME was first developed for commercial use in the plastics and food sectors, it has been utilized in pharmaceutical manufacturing [1]. HME has been largely acknowledged as a suitable technique for processing pharmaceuticals, aiming to enhance the physicochemical properties of challenging drug candidates, such as biopharmaceutics classification system II (BCS class-II) molecules via the formation of solid dispersions (SDs). In recent decades, HME technologies have evolved into a sophisticated platform for the production of different drug delivery systems. Since its beginnings in the early 1970s, HME has been used to develop several drug delivery methods such as enhancing solubility, solid dispersions, implants, oral disintegration mechanisms, ophthalmic delivery, controlled, sustained release drug delivery, granules, cocrystals, taste masking, concealing flavour, and formulating films and strips using environmentally friendly technology that avoids organic solvents, minimizes manufacturing steps, and is easily scalable [2].

The HME process entails the feeding of a physical blend into the barrel, where it is conveyed, mixed, melted, dispersed, and subjected to torque and pressure to produce SDs [3]. Feeder, barrel, screw, and a die are the main components of the extruder (Fig. 4.1). HME is a continuous pharmaceutical process that entails propelling polymeric materials using a rotating screw at temperatures exceeding their glass transition temperature ($T_g$) and occasionally surpassing the melting temperature ($T_m$) to facilitate molecular-level mixing of active compounds and thermoplastic binders (materials that soften when heated and solidify upon cooling), without experiencing any chemical changes such as Hypromellose, hydroxypropyl cellulose, polyethylene oxide. Hot melt extruders are also feasible for wet granulation using solvents, liquified APIs, with a shear force similar to that of a standard high shear granulator [4]. In (HME), twin-screw wet granulation offers an advantage over traditional granulators, such as high-shear mixers, due to its unique capability for continuous manufacturing [5].

This cutting-edge technology also promotes continuous manufacturing [6] and process monitoring with integration of downstream auxiliary equipment or instruments such as process analytical tool (PAT), such as Raman spectroscopy and near-infrared spectroscopy (NIR) for evaluating the manufacturing process and product quality [7, 8].

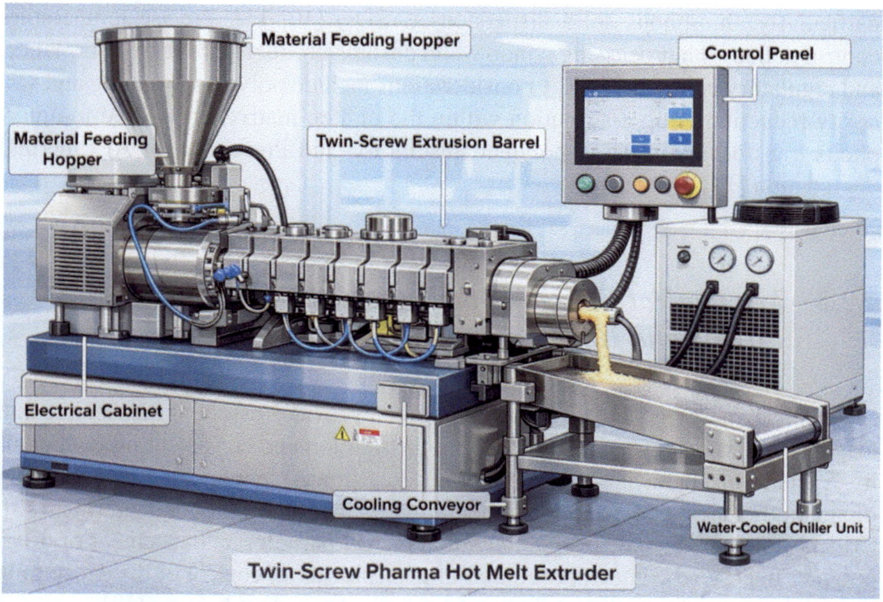


**Fig. 4.1** Hot melt extrusion process

This chapter provides a comprehensive assessment of HME technology, encompassing theoretical principles, excipient and equipment considerations, thermomechanical processing, characterization, stability evaluation, dosage form development, regulatory implications, and novel applications.

## 4.2 Overview of Hot Melt Extrusion

Extrusion is the process of changing the physicochemical properties of substances by passing them through a heated barrel under pressure with rotating screws. During this process, the material is softened or melted, mixed, conveyed, and forced through a die (feed rate, screw speed, and temperature) to form strands, pellets, granules, films, or filaments according to the materials' melting and miscibility characteristics, followed by extrusion of the processed material [1, 9].

### *4.2.1 Thermal Energy Input: Polymer Softening and Drug Dissolution Mechanisms*

Thermal energy enables polymer transitions above its glass transition ($T_g$) or melting point ($T_m$), thereby creating the molecular mobility required for drug incorporation into the carrier matrix and the formation of homogeneous systems, particularly amorphous solid dispersions (ASDs) [10, 11]. Precise temperature control defines the processing window and ensures product stability, performance, and regulatory compliance. Key considerations include polymer softening and viscosity reduction, drug distribution within the molten matrix, crystalline-to-amorphous transformation, Drug-polymer miscibility, and Prevention of API thermal degradation [12].

### 4.2.2 *Mechanical Energy Contribution: Shear-Induced Mixing and Homogenization*

Mechanical energy generated by screw rotation provides essential distributive and dispersive mixing, ensuring uniform drug distribution and consistent product quality. Specific mechanical energy serves as a key process parameter for monitoring mixing efficiency and batch reproducibility in continuous manufacturing [13]. Critical functions include deagglomeration of API particles, enhancement of drug-polymer interfacial contact, promotion of amorphization, control of residence time and mixing intensity, and contribution to viscous dissipation heating [14].

### 4.2.3 *Pressure Development and Melt Conveying Behaviour*

Pressure generated along the screw channel enables continuous material conveying and controlled die shaping, ensuring dimensional and structural consistency of the extrudate [15]. Effective pressure control supports scalable, reproducible pharmaceutical manufacturing. Important process aspects include stable melt flow toward the die, control of extrudate morphology and surface quality, influence on strand or filament diameter uniformity, impact on downstream processing efficiency, and maintenance of process robustness under continuous operation [9].

## 4.3 Theoretical Considerations

The theoretical framework underpinning HME is critical for rational formulation design, prediction of physical stability, and process optimization. Solid-state transformations during HME are governed by thermodynamics (miscibility and phase behaviour), molecular mobility, and kinetic stabilization during cooling. A mechanistic understanding of these principles supports robust development of amorphous solid dispersions (ASDs) or solid dispersions (SDs) and advanced delivery systems.

### 4.3.1 Classification of Solid Dispersions (SDs)

Solid dispersions formed by HME vary based on drug-carrier miscibility and processing conditions, ranging from crystalline eutectic systems to fully amorphous dispersions (Fig. 4.2). Proper classification is essential to predict dissolution performance, stability, and overall formulation behaviour [11]. The classifications are as follows:

(i) *Eutectic mixtures*: Drug and carrier crystallize together at a specific composition; improved dissolution, but remain crystalline [16].
(ii) *SDs*: Drug molecularly dispersed in a crystalline or amorphous carrier [17].
(iii) *ASDs*: Drug molecularly dispersed within an amorphous polymer matrix; enhanced solubility but prone to recrystallization [18].
(iv) *Co-amorphous systems*: Amorphous drug-drug or drug-small molecule systems stabilized by intermolecular interactions [19].
(v) *Phase-separated systems*: Partially miscible dispersions with drug-rich and polymer-rich domains [20].

The resulting system depends on thermodynamic compatibility, drug loading, and processing conditions.

### 4.3.2 Drug–Polymer Miscibility

Drug–polymer miscibility determines ASD stability, dissolution behaviour, and manufacturability. While high processing temperatures during HME promote mixing, long-term stability depends on thermodynamic compatibility at storage conditions [21].

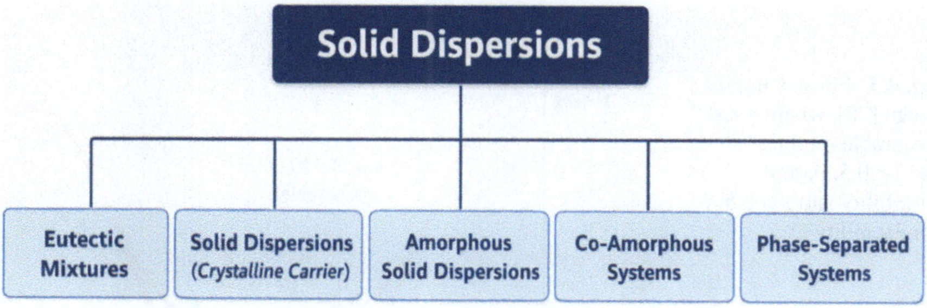


**Fig. 4.2** Classification of solid dispersions

#### 4.3.2.1 Hot Melt Extrusion Process Context

During the HME process, miscibility is promoted by polymer softening above $T_g$ or $T_m$, enabling drug dissolution in the molten matrix. Shear-enhanced mixing improves molecular distribution, while specific interactions (e.g., hydrogen or ionic bonding) support phase stability. Rapid cooling then kinetically stabilizes the amorphous state and limits phase separation [22].

#### 4.3.2.2 Predictive Approaches

Predictive thermodynamic models are widely applied to screen carrier candidates and estimate miscibility prior to experimental trials. These approaches reduce development risk and support Quality by Design (QbD) strategies [23]. The most common predictive tools are the following:

(i) *Flory–Huggins interaction parameter ($\chi$)*: The Flory–Huggins theory describes the thermodynamic interaction between a polymer and a low molecular weight compound (Fig. 4.3) using the interaction parameter [24].
(ii) *Hansen solubility parameters (HSP)*: Miscibility is likely when the Hansen distance ($R_a$) between the drug and polymer is small. HSP is widely applied in early-stage screening due to its practical accessibility, though experimental confirmation remains essential [25]. Hansen Solubility Parameters divide the total cohesive energy into: Dispersion ($\delta_D$), Polar ($\delta_P$), and Hydrogen-bonding ($\delta_H$).
(iii) *Glass transition temperature ($T_g$) modelling*: $T_g$ modelling predicts the physical stability of amorphous dispersions. The Gordon-Taylor equation (Eq. 4.1) estimates the mixture $T_g$ based on component $T_g$ values and composition [26]. A higher system $T_g$ generally correlates with reduced molecular mobility and improved long-term physical stability.

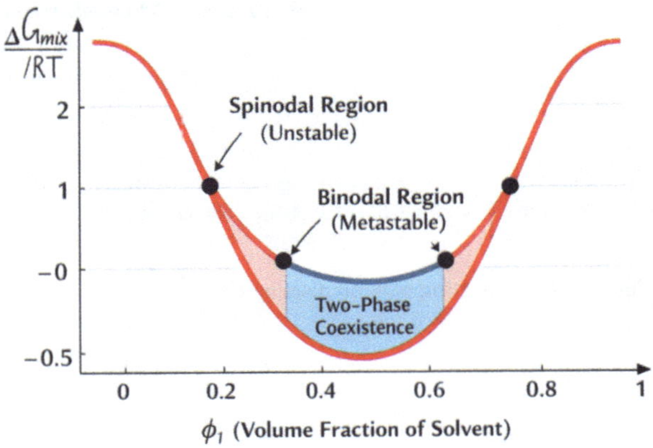


**Fig. 4.3** Flory–Huggins model [28], where $\chi < 0$: favourable mixing, $0 < \chi < 0.5$: partial miscibility and $\chi > 0.5$: immiscibility

$$T_g = \frac{w_1 T_{g_1} + k w_2 T_{g_2}}{w_1 + k w_2} \tag{4.1}$$

where $T_g$ is the glass transition temperature of the mixture (in K or °C), $T_{gi}$ are the glass transition temperatures of component 1 and component 2, respectively; $w_i$ are the weight fractions of component 1 and component 2, respectively, and $k$ is a fitting parameter (Gordon-Taylor constant).

(iv) *Phase diagram construction*: Phase diagrams map drug–polymer interactions across temperature and composition ranges. They help define miscibility limits and processing windows in HME [27]. Phase diagrams support rational formulation development and reduce empirical trial-and-error approaches.

## 4.4 Components of HME

Hot melt extrusion components integrate controlled feeding systems, a temperature-regulated barrel, modular screw assemblies (primarily co-rotating twin-screw), a shaping die, and downstream cooling units [1]. These components collectively ensure consistent mixing, stable processing, and scalable pharmaceutical manufacturing under continuous conditions.

### *4.4.1 Feeding Systems*

Accurate and consistent feeding is fundamental to achieving content uniformity, stable torque, and reproducible product quality in HME. Variations in feed rate directly affect residence time, specific mechanical energy, and melt pressure, impacting both process stability and final product performance [29].

#### 4.4.1.1 Gravimetric Feeding

Gravimetric feeders operate based on real-time mass flow measurement using load cells. They continuously adjust the feed rate to maintain a preset mass throughput. Gravimetric feeding is considered the standard in pharmaceutical-scale HME. The advantages of gravimetric feeding systems are high precision and reproducibility, essential for continuous manufacturing, minimize batch-to-batch variability, and support Quality by Design (QbD) strategies [14, 30].

#### 4.4.1.2 Volumetric Feeding

Volumetric feeders deliver material based on screw speed and calibrated volume without direct mass measurement. For robust commercial manufacturing, volumetric feeding is generally replaced by gravimetric systems. The characteristics of volumetric feeding are lower precision compared to gravimetric systems, sensitive to bulk density variation and powder flow properties, and commonly used in early-stage screening or feasibility studies [31].

### *4.4.2 Extruder Barrel*

The extruder barrel consists of multiple segmented, temperature-controlled zones designed to regulate material transformation along the process path. Independent temperature control across zones enables optimization of the thermal profile, reducing degradation risk while ensuring adequate mixing. Modular barrel segmentation enhances process flexibility and scalability [32]. The extruder barrel functions include controlled polymer melting or softening, progressive mixing, homogenization, and pressure development before the die.

### *4.4.3 Extruder Screws*

The screw configuration is the core determinant of mixing efficiency, shear exposure, and residence time distribution (Table 4.1) [29].

#### 4.4.3.1 Types of Screws

Co-rotating twin-screw extruders are the predominant choice in pharmaceutical hot-melt extrusion due to their excellent distributive and dispersive mixing efficiency, modular screw configurations, and precise process control, all of which facilitate consistent production of high-quality amorphous solid dispersions. In contrast, counter-rotating twin-screw systems generate higher pressure build-up and are advantageous for applications requiring substantial pumping capacity, but they offer comparatively weaker mixing performance and are therefore less commonly employed for complex drug–polymer matrices. Single-screw extruders, while characterized by mechanical simplicity and suitability for processing high-viscosity melts, provide limited mixing capability and are thus generally reserved for shaping or melt-forming operations rather than for achieving the molecular-level mixing required for pharmaceutical solid dispersion manufacture (Fig. 4.4).

**Table 4.1** Comparison of extruder types used in pharmaceutical hot melt extrusion

| Parameter | Single screw extruder | Co-rotating twin-screw extruder | Counter-rotating twin-screw extruder |
|---|---|---|---|
| Number of screws | One screw | Two screws rotating in the same direction | Two screws rotating in opposite directions |
| Mixing efficiency | Limited (mainly distributive) | Excellent distributive and dispersive mixing | Moderate mixing; less dispersive than co-rotating |
| Shear intensity | Low to moderate | High and adjustable via screw configuration | Moderate; more compressive than shear-dominant |
| Residence time control | Limited control | Good control via modular screw elements | Moderate control |
| Pressure generation | Moderate | Moderate | High-pressure build-up capability |
| Suitability for ASDs | Limited (less efficient amorphization) | Highly suitable; industry standard for ASDs | Suitable but less commonly used for ASDs |
| Screw configuration flexibility | Fixed/simple design | Highly modular (kneading blocks, mixing, reverse elements) | Less modular compared to co-rotating systems |
| Process control | Basic | Advanced; supports PAT integration | Moderate |
| Scalability and continuous manufacturing | Suitable but limited flexibility | Excellent; widely used in continuous pharma manufacturing | Suitable for specific applications |
| Typical Pharmaceutical Applications | Melt shaping, simple matrix systems | Amorphous solid dispersions, controlled release, 3D-printing filaments | High-viscosity formulations, pressure-sensitive systems |
| Industrial prevalence in pharma | Less common | Most widely used | Less common |

#### 4.4.3.2 Screw Elements

Modern co-rotating twin-screw extruders are inherently modular systems in which screw configuration can be rationally engineered to match formulation and process requirements. As highlighted by Crowley et al., the arrangement, length, and sequence of screw elements determine the balance between shear stress, residence time distribution, thermal input, and pressure generation [9]. This level of configurational flexibility allows formulators to tailor the extrusion environment for specific objectives, such as amorphous solid dispersion formation, polymer blending, or controlled degradation avoidance. For example, conveying elements primarily ensure forward material transport with minimal shear input, thereby stabilizing throughput and maintaining process continuity. In contrast, kneading blocks—available at stagger angles of 0°, 30°, 45°, 60°, and 90°—generate progressively higher shear forces, facilitating dispersive mixing and particle size reduction. The

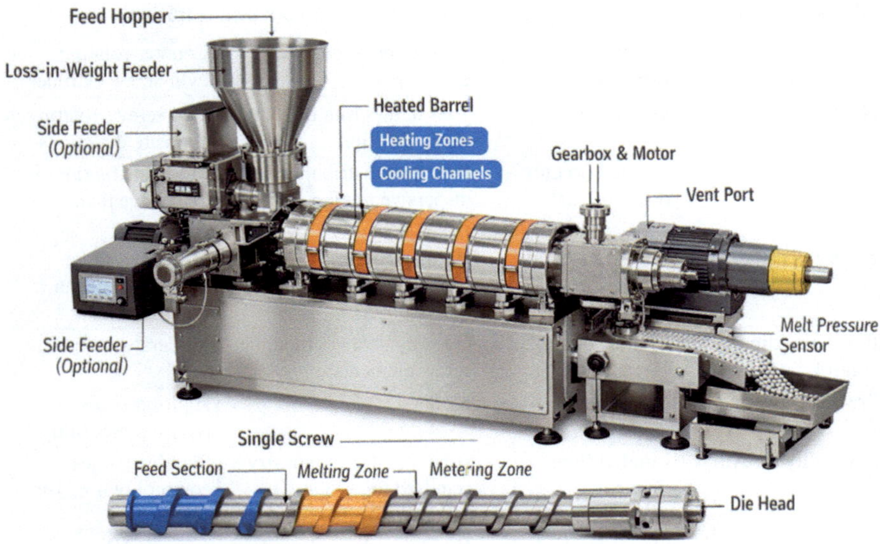


**Fig. 4.4** Single screw hot melt extruder

choice of stagger angle directly influences energy dissipation and mechanical work input, which are critical parameters for drug amorphization and uniform API distribution.

Mixing elements further enhances distributive mixing by promoting material redistribution without excessive shear, improving compositional homogeneity across the extrudate cross-section. Reverse or backward-conveying elements are strategically incorporated to increase residence time and localized pressure build-up, thereby improving melt consolidation and homogenization. However, excessive use of reverse elements may elevate torque and thermal load, potentially compromising thermolabile compounds. Consequently, screw design represents a critical process parameter in hot-melt extrusion, where optimized configurations enable precise control over microstructure development, drug-polymer interactions, and final product performance while maintaining process robustness and scalability.

### 4.4.4 *Dies*

The die is a critical component of the hot-melt extrusion (HME) system, as it governs not only the final geometry of the extrudate—such as strands, pellets, films, or filaments—but also influences flow behaviour, pressure development, and thermal history at the final stage of processing. The design parameters, including length-to-diameter (L/D) ratio, channel geometry, and land length, directly affect shear stress distribution and melt relaxation prior to exit. These factors are particularly important in pharmaceutical applications, where dimensional precision and surface quality can influence downstream operations such as pelletization, cutting, filament winding, or direct feeding into 3D printing systems. For instance, filament production for fused deposition modelling (FDM) requires tight diameter tolerances to ensure consistent feeding and dose accuracy, making die design and pressure stability especially critical.

Die pressure serves as a real-time indicator of melt viscosity, homogeneity, and overall process stability. Fluctuations in die pressure may signal variations in material feeding, incomplete melting, phase separation, or degradation phenomena. Maintaining stable die pressure ensures uniform extrudate dimensions, minimizes internal stresses, and enhances reproducibility across batches. Furthermore, controlled pressure profiles contribute to consistent residence time and shear exposure, thereby supporting uniform drug dispersion and solid-state stability. In continuous pharmaceutical manufacturing, die pressure monitoring is therefore considered a key process analytical parameter, facilitating robust scale-up and ensuring product quality through improved process control.

### *4.4.5 Downstream Components*

Downstream processing plays a decisive role in preserving the physicochemical attributes imparted during hot-melt extrusion (HME), particularly when manufacturing amorphous solid dispersions (ASDs). As highlighted by Grymonpré et al., controlled cooling and solidification are essential to stabilize the amorphous state and prevent phase separation or recrystallization [33]. The rate of heat removal directly influences molecular mobility, residual stresses, and free volume within the extrudate matrix. Rapid and uniform cooling reduces residual mobility of the drug-polymer system, thereby minimizing the risk of post-extrusion crystallization and ensuring long-term physical stability. Effective downstream design is therefore critical for maintaining product quality while supporting scalable continuous manufacturing.

Several downstream components are strategically selected depending on the intended final dosage form. Chill rolls provide rapid and uniform cooling for films and sheets, controlling thickness and surface morphology. Air-cooling systems facilitate gradual and controlled solidification of extruded strands, preventing deformation or internal stress accumulation. Pelletizers and strand cutters enable precise size reduction, ensuring narrow particle size distributions for capsule filling or further processing. Spooling systems are essential for filament production, maintaining a consistent diameter and tension for subsequent 3D printing applications. Milling units may be employed for granulation, preparing extrudates for tablet compression, or encapsulation. Collectively, optimized downstream operations ensure dimensional consistency, physical stability, and efficient integration into continuous pharmaceutical manufacturing platforms.

## 4.5 Process Parameters and Their Influence on Product Quality

HME is a thermo-mechanical process in which both thermal energy and shear are applied to transform a drug-excipient blend into a homogeneous extrudate. The performance of an ASD or any HME-based product depends strongly on the selection and control of critical process parameters (CPPs) [31]. These parameters govern the melt state, molecular mixing, drug dissolution within the polymer, and the final physical stability of the product (Table 4.2). The following subsections summarize the key parameters and their mechanistic impact on product quality attributes.

### *4.5.1 Temperature Profile*

The temperature profile along the barrel is often the most influential CPP in HME. It must provide sufficient thermal energy to soften or melt the polymer and enable drug incorporation without inducing thermal degradation [34]. A well-designed temperature profile ensures controlled viscosity, uniform melt flow, and consistent residence time distribution.

If the process is carried out below the polymer softening/melting temperature, then inadequate mixing with high torque takes place, which leads to incomplete amorphization and drug crystals in the final product. In addition, when operating above the optimal melt temperature, there is a great risk of polymer degradation and drug decomposition, where colour changes, and reduction in molecular weight (particularly for polyvinylpyrrolidone or polyethylene oxide) are observed.

Zonal control is a key element where early zones may be cooler to prevent premature melting near the feed throat, while mid-barrel zones provide the bulk of melting and mixing, and the final zone stabilizes the melt before die entry.

### *4.5.2 Screw Speed (RPM)*

Screw speed governs the balance between shear input and residence time. In most ASD processes, moderate to high screw speeds (100–400 RPM for laboratory equipment) facilitate drug dissolution in the polymer without overexposing the melt

**Table 4.2** Process parameters and their influence on product quality

| Process parameter | Role/effect |
|---|---|
| Temperature | Degradation/amorphization |
| Screw speed | Mixing/RTD |
| Feed rate | Pressure/homogeneity |
| Specific mechanical energy | Drug dispersion |
| Cooling | Recrystallization risk |

to heat. The higher screw speed results in lower residence time, higher shear rate and specific mechanical energy, improved distributive and dispersive mixing, and lower melt viscosity due to shear thinning [35]. The lower screw speed results in longer residence time, higher risk of degradation (if high temperatures are used), and less intensive mixing.

### *4.5.3 Feed Rate and Fill Level*

Feed rate affects barrel fill, pressure generation, and overall process stability. Maintaining an appropriate fill level (typically 40–80% for twin-screw systems) ensures efficient conveying, good mixing, and stable melt pressure at the die [1]. The low feed rate results in reduced torque, higher residence time, and potential for over-shearing or overwriting thermal exposure. The high feed rate results in increased torque and die pressure, risk of insufficient melting or incomplete mixing if the barrel becomes overfilled, and variability in strand shape or filament diameter (for 3D printing filaments).

### *4.5.4 Specific Mechanical Energy*

Specific mechanical energy (SME, kJ/kg) is a critical process parameter in hot-melt extrusion (HME), representing the mechanical work transferred from the rotating screws to the material per unit mass processed. SME (Eq. 4.2) integrates the effects of screw speed, torque, screw configuration, and mass throughput, thereby providing a quantitative measure of the mechanical intensity experienced by the formulation. Because it captures both shear stress and residence-related energy input, SME is frequently used as a scale-up parameter to align laboratory- and commercial-scale extrusion systems [29, 36]. Maintaining comparable SME values across scales can help ensure similar melt homogeneity, amorphization efficiency, and product performance, even when equipment geometry differs.

$$\mathrm{SME} = \frac{2 \times \pi \times n \times M}{\dot{m}} \left( \frac{\mathrm{kWh}}{\mathrm{Kg}} \right) \tag{4.2}$$

where $n$ (rpm) is the screw rotation speed, $M$ (Nm) is the net torque, and $\dot{\mathbf{m}}$ is the feed rate (kg/h).

Higher SME levels generally promote enhanced dispersive mixing, facilitating particle size reduction and improved molecular-level drug dispersion within the polymer matrix. This is particularly advantageous for the preparation of amorphous solid dispersions, where sufficient mechanical energy can compensate for lower barrel temperatures, reducing the reliance on thermal input and thereby mitigating the risk of thermally induced degradation. However, excessively high SME may increase localized shear heating and mechanical stress, potentially leading to polymer chain scission or oxidative degradation if not properly controlled. Conversely, low SME conditions may result in inadequate mixing, incomplete drug amorphization, and residual crystallinity within the extrudate. Therefore, SME must be carefully optimized to balance efficient dispersion and solid-state transformation with material stability, making it a central parameter in both process development and robust scale-up of pharmaceutical HME systems.

### *4.5.5 Residence Time Distribution (RTD)*

RTD defines how long material resides inside the extruder and is crucial for predicting thermal exposure and mixing efficiency. RTD is influenced by screw configuration (conveying vs kneading blocks), screw speed, feed rate, and barrel fill [32]. With the Narrowed RTD, predictable mixing and controlled amorphization, on the other hand, with the broad RTD leads to the risk of overexposure, which leads to variability in crystallinity and degradation.

### *4.5.6 Melt Viscosity and Rheology*

Viscosity influences torque, pressure, mixing, and die swell. Drug loading strongly affects melt viscosity: high drug loadings (>40% w/w) often reduce viscosity because low-molecular-weight APIs act as plasticizers [37]. Formulators often target a melt viscosity window of 50–500 pascal-seconds (Pa·s) for optimal processing, depending on equipment and formulation design. Monitoring rheology helps optimize the mixing efficiency, dispersive capabilities, filament quality for HME-based 3D printing, and the feasibility of high-drug-load systems.

### *4.5.7 Screw Configuration*

Screw configuration is a fundamental determinant of process performance in hot-melt extrusion (HME), as it governs the balance between material conveying, melting, mixing, and pressure generation. By strategically arranging modular screw elements, formulators can tailor shear exposure, residence time distribution, and melt homogeneity to meet the specific requirements of amorphous solid dispersion (ASD) development. For ASD systems, screw designs typically incorporate multiple kneading blocks staggered at angles between 30° and 60°, which provide sufficient dispersive shear to promote drug dissolution into the polymer matrix while maintaining process stability (Fig. 4.5). The number, position, and stagger angle of

**Fig. 4.5** Screw types that are assembled on high torque splined shafts (left) and flanged barrels, electrically heated and liquid cooled (right)

these kneading elements directly influence mechanical energy input and mixing efficiency.

Conveying elements are primarily responsible for the forward transport of the material and contribute to progressive melting through controlled shear and heat transfer from the barrel. Kneading blocks, in contrast, generate elevated shear stresses that enhance dispersive mixing, reduce particle size, and facilitate molecular-level drug distribution within the polymer. Reverse elements or specialized mixing zones are often introduced to increase local residence time and pressure build-up, improving melt consolidation and homogeneity. However, excessive back-mixing can elevate torque and thermal load, potentially compromising thermally sensitive formulations. Therefore, optimized screw design represents a critical process parameter in HME, enabling precise control over solid-state transformation, microstructure formation, and final product performance.

### 4.5.8 Die Geometry and Pressure

Die geometry plays a critical role in determining extrudate shape, surface characteristics, and initial cooling behaviour. It directly influences melt flow, pressure development, and dimensional consistency in hot melt extrusion. Die pressure serves as a sensitive indicator of process stability, reflecting changes in viscosity, feed consistency, and temperature. Continuous monitoring of die pressure can be integrated into Process Analytical Technology (PAT) frameworks to support real-time quality control and ensure robust pharmaceutical manufacturing [38].

Narrow dies generate higher back pressure, which can enhance mixing and densification; however, excessive pressure may increase torque, thermal load, and risk of melt fracture. Wide or slit dies are typically employed for film and sheet extrusion, enabling uniform thickness and surface finish. Circular dies are preferred for producing strands, pellets, and filaments used in downstream processing, including 3D printing applications.

### 4.5.9 Cooling and Downstream Processing

Cooling is a critical step in hot melt extrusion, as it determines whether the amorphous dispersion is kinetically stabilized or undergoes phase separation and recrystallization. The rate of temperature reduction directly influences molecular mobility and the ability of the system to retain a single-phase structure. Rapid cooling (air-cooling units, cooling plates, or chill rolls) quickly reduces molecular mobility, effectively locking the drug within the polymer matrix and minimizing the risk of recrystallization. Slow cooling allows continued molecular movement, increasing the likelihood of phase separation, crystal growth, or structural heterogeneity [38].

Following solidification, downstream operations such as palletization, milling, cutting, or filament spooling must be carefully controlled. These steps should minimize additional heat generation, mechanical stress, and exposure to humidity, all of which can compromise amorphous stability and product performance. Properly designed downstream processing is essential to preserve the physicochemical integrity and long-term stability of HME formulations.

## 4.6 Excipients for ASDs

Excipients are central to the successful development of ASDs prepared by hot melt extrusion. Among them, polymers function as the primary structural matrix, providing both molecular dispersion capability and long-term stabilization against recrystallization. The selection of excipients directly influences drug-polymer miscibility, processing feasibility, dissolution enhancement, and physical stability throughout shelf life [39]. Unlike conventional solid dosage forms, ASDs rely heavily on excipient functionality to maintain the drug in a high-energy amorphous state while ensuring manufacturability under thermal and shear stress conditions. Therefore, excipient choice must consider thermodynamic compatibility, glass transition temperature ($T_g$), hygroscopicity, melt rheology, and regulatory acceptability.

### 4.6.1 Functional Roles of Excipients

Polymers constitute the structural and functional backbone of amorphous solid dispersions (ASDs) prepared by hot-melt extrusion. Frequently employed pharmaceutical polymers include PVP/VA (polyvinylpyrrolidone-vinyl acetate copolymer), HPMC, HPMCAS, PEO, Eudragit® grades, Soluplus®, and various cellulose derivatives such as HPC, HPMC, EC, and CMC [40]. Selection of a suitable carrier is governed by critical attributes including high glass transition temperature ($T_g$), thermal stability under processing conditions, appropriate melt viscosity, and physicochemical compatibility with the API. These properties ensure adequate processability while maintaining the drug in a stabilized amorphous state.

Functionally, polymers enhance apparent solubility by sustaining the drug in a molecularly dispersed form and promoting favourable thermodynamic interactions, such as hydrogen bonding or ionic interactions, that improve miscibility. They suppress recrystallization by reducing molecular mobility and inhibiting nucleation and crystal growth during storage and dissolution. Furthermore, polymers support supersaturation maintenance through the "spring and parachute" mechanism, thereby prolonging drug exposure in solution. Beyond stabilization, the polymer matrix provides the mechanical integrity necessary for downstream shaping into tablets, films, or filaments.

Excipients such as plasticizers, surfactants, stabilizers, and anti-plasticizers are often incorporated to fine-tune performance. Plasticizers reduce melt viscosity and lower processing temperatures by decreasing polymer $T_g$, thereby improving flow and reducing torque—particularly important for thermally labile APIs. However, excessive plasticization can increase molecular mobility and compromise long-term physical stability. Surfactants enhance wettability and dissolution by lowering interfacial tension, facilitating micellar solubilization, and improving supersaturation retention, provided they remain thermally stable and polymer-compatible during extrusion. Stabilizers mitigate chemical degradation pathways, including oxidation and moisture-induced hydrolysis, thereby extending shelf life. Conversely, anti-plasticizers elevate system $T_g$, reduce molecular mobility, and strengthen mechanical properties, which is particularly advantageous in high drug-loading or hygroscopic systems where stability margins are reduced. Collectively, rational polymer and excipient selection is central to achieving robust processing and durable ASD performance.

### 4.6.2 Strategic Considerations in Excipient Selection

A rational, mechanism-based excipient strategy is essential for developing robust, high-performance amorphous solid dispersions suitable for commercial pharmaceutical manufacturing. Excipient selection for ASDs must balance thermodynamic compatibility, processability under HME conditions, long-term physical and chemical stability, dissolution enhancement capability, and regulatory acceptance and scalability. Some of the polymers/carriers used in HME-based ASDs and their key functions are listed in Table 4.3.

## 4.7 Formation of ASDs

The formation of ASDs (Fig. 4.6) via hot melt extrusion is a thermally and mechanically driven process in which the crystalline drug is transformed into a molecularly dispersed amorphous state within a polymer matrix [48]. Successful ASD formation depends on controlled melting, efficient mixing, favourable intermolecular interactions, and kinetic stabilization during cooling.

**Table 4.3** Polymers/carriers used in HME-based ASDs and their key functions

| Polymer/carrier | Key functional attributes in ASDs (HME) |
|---|---|
| PVP/VA (polyvinylpyrrolidone-vinyl acetate copolymer) [41] | Strong hydrogen-bonding capability; enhances drug-polymer miscibility and solubilization; widely used for ASD formation |
| HPMC (hydroxypropyl methylcellulose) [42] | Provides physical stability; high $T_g$; suitable for sustained/controlled release applications |
| HPMCAS (hydroxypropyl methylcellulose acetate succinate) [43] | Enteric polymer with pH-dependent solubility; effective precipitation inhibitor; maintains supersaturation |
| PEO (polyethylene oxide) [44] | Suitable for controlled release systems; high chain mobility; useful in high viscosity melt systems |
| Eudragit grades [45] | Methacrylate-based polymers with pH-dependent solubility enable targeted and modified release profiles |
| Soluplus [46] | Amphiphilic graft copolymer specifically developed for ASD formation; enhances solubility and dissolution of poorly soluble drugs |
| Cellulose derivatives [47] | Provide mechanical strength, stability, and modulation of drug release behaviour in extruded systems |

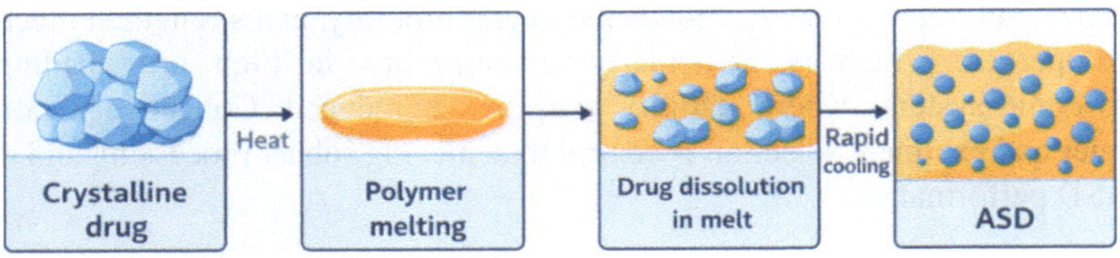


**Fig. 4.6** Formation of amorphous solid dispersions

## *4.7.1 Mechanisms for the Formation of ASDs*

Amorphous solid dispersion (ASD) formation during hot-melt extrusion (HME) is governed by a series of interrelated thermal and mechanical phenomena that collectively drive the transformation of a crystalline or partially crystalline drug into a molecularly dispersed amorphous phase. Initially, as the polymer is heated above its glass transition temperature ($T_g$) or melting temperature ($T_m$), it transitions into a viscoelastic or molten state capable of solubilizing the active pharmaceutical ingredient (API). Drug dissolution into the polymer melt is a critical step, as it determines whether a true single-phase system can be achieved. The extent of dissolution depends on drug-polymer miscibility, processing temperature, residence time, and the thermodynamic driving force for mixing.

For crystalline APIs with melting points within or near the extrusion temperature range, partial or complete melting may occur, facilitating molecular dispersion within the polymer matrix. In cases where the API melting point exceeds practical processing temperatures, dissolution may instead proceed through solid-state solubilization driven by polymer plasticization and shear-induced disruption of crystal lattices. Shear-assisted mixing plays a central role in this context; the mechanical energy imparted by kneading and mixing elements promotes deagglomeration, reduces particle size, and enhances interfacial contact between drug and polymer. This mechanical contribution accelerates dissolution kinetics and supports the formation of a homogeneous, single-phase amorphous system.

Following molecular dispersion, rapid cooling at the die exit is essential to kinetically stabilize the amorphous structure. The abrupt temperature reduction decreases molecular mobility and effectively "freezes" the drug within the polymer matrix before nucleation and crystal growth can occur. The cooling rate, final extrudate temperature, and downstream handling conditions, therefore, critically influence the long-term physical stability of the ASD. Together, controlled thermal input, optimized shear exposure, and efficient quenching define the mechanistic pathway of ASD formation during HME and determine the quality and durability of the resulting dispersion.

### 4.7.2 Intermolecular Interactions

The long-term stability of ASDs depends largely on the strength and nature of drug-polymer interactions. These interactions reduce molecular mobility and inhibit nucleation and crystal growth [49]. Key stabilizing mechanisms include hydrogen bonding between drug functional groups and polymer chains, ionic interactions in systems containing ionizable components, dipole-dipole interactions that enhance miscibility, polymer chain entanglement that restricts molecular diffusion, and high $T_g$ matrices that reduce molecular mobility and slow recrystallization kinetics [50].

### 4.7.3 Factors Involved

The formation and long-term stability of amorphous solid dispersions (ASDs) are governed by a complex interplay of formulation composition and process conditions. Among these variables, drug loading is one of the most critical determinants of system performance. Increasing drug concentration enhances the thermodynamic driving force for phase separation, particularly when the solubility limit of the drug in the polymer matrix is exceeded. At high loadings, the system may transition from a single-phase molecular dispersion to a biphasic structure containing amorphous drug-rich domains, which significantly increases the likelihood of nucleation and recrystallization during storage. Therefore, identification of the drug-polymer miscibility limit through phase diagrams or solubility parameter analysis is essential during formulation development.

Polymer properties, particularly glass transition temperature ($T_g$) and molecular weight, strongly influence both processing behaviour and physical stability.

Polymers with high $T_g$ values reduce molecular mobility within the matrix, thereby suppressing nucleation and crystal growth according to the Williams–Landel–Ferry (WLF) relationship. Higher molecular weight polymers can further restrict chain mobility and enhance kinetic stabilization; however, they also increase melt viscosity, which may compromise processability, elevate torque, and require higher processing temperatures. Consequently, a balance must be achieved between stability enhancement and manufacturability, often necessitating the use of plasticizers or optimized screw configurations to maintain manageable rheological properties during extrusion.

Processing conditions, particularly cooling rate, play a decisive role in kinetically trapping the drug in its amorphous state. Rapid quenching at the die exit reduces molecular mobility before phase separation or crystallization can occur, effectively preserving the single-phase structure formed in the melt. Inadequate or slow cooling may allow structural relaxation and facilitate nucleation, especially in marginally stable systems. Downstream handling conditions, therefore, become integral to ASD stabilization, particularly in continuous manufacturing settings.

Environmental factors, most notably humidity exposure, critically impact ASD stability. Moisture acts as a plasticizer, lowering the system $T_g$ and increasing molecular mobility, which accelerates crystallization kinetics. Hygroscopic polymers are especially susceptible to this effect, and even modest increases in relative humidity can significantly reduce the physical stability window. Appropriate packaging, moisture barriers, and desiccants are therefore essential for long-term storage.

Finally, the intrinsic crystallization tendency of the API itself remains a fundamental constraint. Drugs characterized by high lattice energy, strong intermolecular interactions, or inherently high molecular mobility exhibit a greater propensity for recrystallization. Such compounds may require lower drug loading, higher $T_g$ polymers, or stronger intermolecular drug-polymer interactions to achieve durable stabilization. Collectively, rational optimization of composition and process parameters is essential to ensure robust ASD formation and sustained amorphous stability throughout the product lifecycle.

## 4.8 Stability of ASDs

Although ASDs enhance apparent solubility and dissolution performance, their thermodynamic instability presents significant formulation challenges [51]. The high-energy amorphous state tends to revert to a lower-energy crystalline form over time. Key stability concerns (Table 4.4) include physical stability (recrystallization), phase separation, and chemical degradation (chemical stability) during processing and storage [52].

**Table 4.4** Stability factors influencing HME-based amorphous solid dispersions

| Factor | Mechanism | Mitigation strategy | CQA (impact on product quality) |
|---|---|---|---|
| Low glass transition temperature ($T_g$) | Increased molecular mobility enhances nucleation and crystal growth | Use high-$T_g$ polymers; reduce drug loading; incorporate anti-plasticizers | Physical stability; crystallinity level; dissolution profile |
| High drug loading | Exceeds drug solubility in polymer matrix → phase separation and recrystallization | Optimize drug-to-polymer ratio; perform miscibility screening; apply $T_g$ modelling | Crystallinity; dissolution rate; bioavailability |
| Moisture exposure (humidity) | Moisture plasticizes matrix → lowers $T_g$ → increases recrystallization risk | Use moisture-protective packaging; include desiccants; select less hygroscopic polymers | Physical stability; dissolution performance; shelf life |
| Poor drug–polymer miscibility | Limited thermodynamic compatibility → formation of drug-rich domains | Hansen solubility screening; calculate Flory-Huggins $\chi$; select interactive polymers | Phase homogeneity; dissolution; long-term stability |
| Thermal degradation | Elevated temperature induces API or polymer breakdown | Optimize barrel temperature; reduce residence time; add antioxidants | Impurity levels; assay; regulatory compliance |
| Oxidative degradation | Oxygen exposure causes API oxidation | Oxygen-barrier packaging; nitrogen purging; antioxidants | Impurity profile; assay potency; shelf life |
| Hydrolytic degradation | Moisture-induced chemical degradation (e.g., ester/amide hydrolysis) | Control humidity; moisture-barrier packaging; stable polymer selection | Impurity levels, chemical stability, safety |
| Polymer molecular weight reduction | Thermal/mechanical stress causes chain scission → altered matrix properties | Control SME; optimize temperature profile | Release profile; mechanical strength; content uniformity |
| Phase separation during storage | Drug-rich and polymer-rich domains form due to thermodynamic instability | Improve miscibility; reduce drug loading; rapid cooling | Dissolution consistency; bioavailability; physical stability |
| Inadequate cooling rate | Slow cooling permits molecular rearrangement and crystallization | Use rapid cooling (chill rolls/air-cooling); minimize thermal exposure | Residual crystallinity; dissolution rate; product performance |

### 4.8.1 *Physical and Chemical Stability*

Physical instability in amorphous solid dispersions (ASDs) is fundamentally governed by molecular mobility within the amorphous matrix. In contrast to crystalline systems, the amorphous state is thermodynamically metastable and characterized by higher free energy. As molecular mobility increases, the system gains the kinetic freedom necessary to undergo structural relaxation, nucleation, and ultimately crystal growth [53]. Therefore, long-term stabilization of ASDs is primarily a kinetic challenge: maintaining sufficiently low molecular mobility throughout storage is essential to suppress recrystallization and preserve performance attributes such as enhanced solubility and dissolution rate.

Glass transition temperature ($T_g$) is one of the most critical determinants of molecular mobility. Below $T_g$, the amorphous matrix exists in a glassy state with restricted segmental motion; above $T_g$, it transitions into a rubbery state where molecular diffusion and rearrangement become significantly more rapid. For robust stability, the $T_g$ of the ASD should remain meaningfully above the intended storage temperature—commonly by at least 20–50 °C—to provide a sufficient kinetic barrier against nucleation. Systems with marginal $T_g$-storage temperature separation are particularly vulnerable to physical aging and crystallization over time.

Humidity exposure represents a major destabilizing factor because water acts as a potent plasticizer. Moisture absorption lowers the effective $T_g$ of the system by increasing free volume and enhancing chain mobility. Even limited water uptake can shift the matrix closer to or below storage temperature, dramatically accelerating crystallization kinetics. Hygroscopic polymers are especially susceptible, making environmental control, protective packaging, and moisture-resistant formulations critical components of stability strategy.

Thermodynamic miscibility between drug and polymer also plays a central role. Limited compatibility promotes phase separation, leading to the formation of drug-rich amorphous domains with locally elevated mobility and reduced $T_g$. These domains act as preferential sites for nucleation. Similarly, higher drug loading increases the probability of exceeding the solubility limit of the API within the polymer matrix, thereby increasing the thermodynamic driving force for crystallization. Consequently, optimization of drug loading, polymer selection, and environmental protection collectively determines the ability of an ASD to maintain its amorphous state throughout its shelf life.

In addition to physical instability, ASDs may undergo chemical degradation, particularly under elevated temperature and humidity. Chemical instability can compromise product efficacy, safety, and regulatory compliance [54]. Potential degradation pathways include thermal degradation during extrusion or storage, oxidation of susceptible functional groups, hydrolysis, especially in hygroscopic systems, and polymer molecular weight reduction, which may alter mechanical and release properties.

### 4.8.2 Stabilization Strategies

To enhance the stability of amorphous solid dispersions (ASDs), a multifaceted approach that addresses both physical and chemical instability is critical for the development of robust and commercially viable formulations [51]. Physical stabilization focuses on limiting molecular mobility, which is a primary driver of recrystallization. This is typically achieved through the use of high-glass-transition-temperature (high-$T_g$) polymers, which increase the overall $T_g$ of the system and reduce molecular diffusion. Additionally, careful control of the thermal history during processes such as hot-melt extrusion, particularly through controlled cooling rates, can kinetically trap the drug in an amorphous state, further mitigating the risk of crystallization. Formulation strategies may also include the use of polymer blends or copolymers to optimize the drug-polymer interactions and improve physical stability over the product's shelf life.

Chemical stabilization, on the other hand, targets potential degradation pathways such as oxidation, hydrolysis, or other chemical reactions. Incorporating antioxidants or other stabilizing excipients into the ASD formulation can mitigate these risks, while moisture-protective packaging prevents plasticization caused by environmental humidity, which can accelerate both chemical and physical instability. Together, these strategies create a synergistic effect, where physical stabilization reduces the likelihood of recrystallization and chemical stabilization limits molecular degradation. By combining careful polymer selection, excipient optimization, process control, and protective packaging, formulators can develop ASD products with enhanced shelf-life and reproducibility, ultimately supporting the translation of lab-scale formulations into commercially viable dosage forms.

## 4.9 Comprehensive Analytical Characterization of Amorphous Solid Dispersions

A rigorous analytical framework is essential for understanding the structural, physicochemical, and chemical attributes of amorphous solid dispersions (ASDs). Because the amorphous state is inherently metastable and susceptible to crystallization, molecular mobility changes, and chemical degradation, a multifaceted characterization strategy is required to assess formulation performance, establish stability, and ensure regulatory compliance. An integrated analytical approach typically includes thermal, structural, spectroscopic, and microscopic methods in combination with chromatographic and mass-based chemical analyses (Fig. 4.7). Together, these techniques provide a robust assessment of amorphous structure, drug-polymer interactions, phase behaviour, crystallization risk, and chemical integrity—attributes that critically determine the efficacy and stability of ASD-based drug products.

### *4.9.1 Thermal and Structural Characterization*

Robust evaluation of thermal behaviour and solid-state structure is central to confirming amorphous conversion, miscibility, and physical stability (Fig. 4.7).

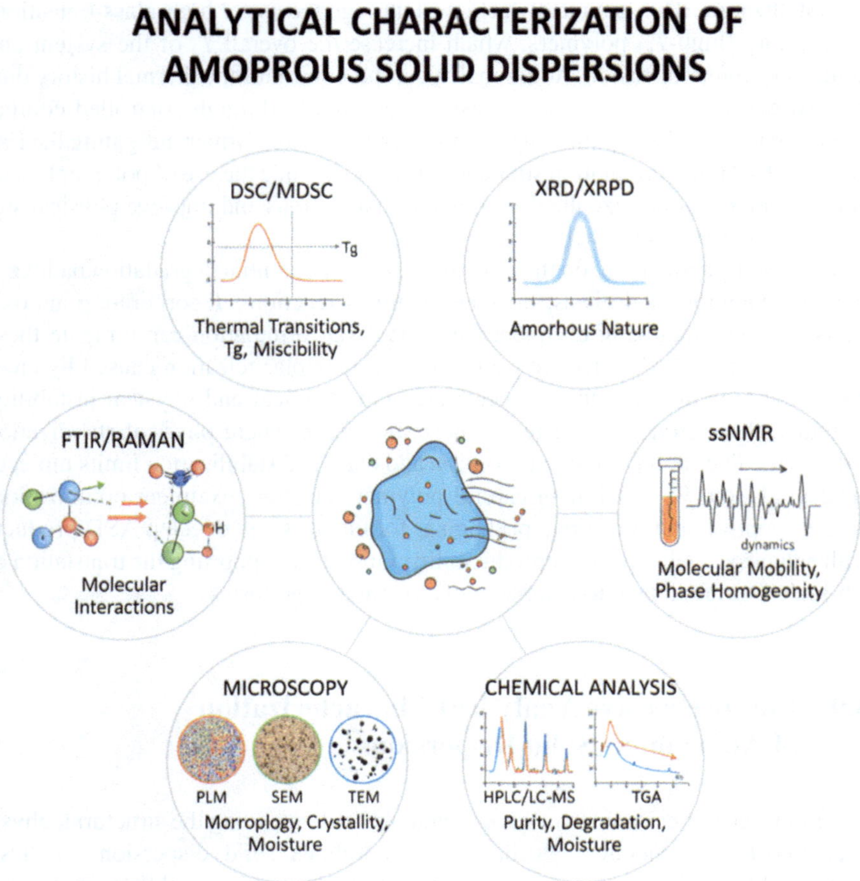


**Fig. 4.7** Summary of analytical characterization techniques for amorphous solid dispersions

#### 4.9.1.1 Differential Scanning Calorimetry (DSC) and Modulated DSC (mDSC)

DSC and mDSC represent cornerstone techniques for assessing the thermal properties of ASDs. Determination of the glass-transition temperature ($T_g$) allows evaluation of molecular mobility and the likelihood of physical instability. A single $T_g$ is widely interpreted as evidence of a molecularly homogeneous and miscible drug-polymer system, whereas multiple $T_g$s suggest partial phase separation or incomplete dispersion [10, 55].

mDSC further enhances interpretation by deconvoluting heat-capacity-related (reversing) transitions from kinetic events such as enthalpic relaxation or recrystallization (non-reversing signals), enabling greater sensitivity to low levels of residual crystallinity and subtle relaxation processes. These insights are critical during formulation screening and accelerated stability studies.

#### 4.9.1.2 Thermogravimetric Analysis (TGA)

TGA evaluates mass loss as a function of temperature, providing critical information on moisture content and thermal degradation behaviour. Moisture has profound effects on amorphous systems, including plasticization of the matrix, reduction of $T_g$, and enhancement of molecular mobility, ultimately increasing the risk of recrystallization. TGA also identifies degradation onset temperatures, helping define safe processing windows and storage requirements. When interpreted alongside DSC and XRPD data, TGA contributes significantly to a holistic understanding of ASD stability.

Amorphous solid dispersions require a comprehensive suite of analytical tools to fully characterize their physical, chemical, and structural attributes. The combination of thermal analysis, diffraction methods, vibrational spectroscopy, ssNMR, microscopy, and chromatographic techniques enables a multidimensional understanding of ASD behaviour. Such a holistic analytical approach not only ensures the scientific robustness of formulation development but also supports regulatory compliance, enhances predictive stability modelling, and ultimately contributes to the successful translation of ASDs into clinically effective drug products.

Overall, the solid-state characterization confirms phase behaviour, miscibility, and structural integrity of ASDs (Table 4.5). The techniques mentioned below collectively validate successful amorphization and long-term structural stability [56].

#### 4.9.1.3 X-Ray Diffraction (XRD/XRPD)

X-ray powder diffraction is the most widely used technique for verifying the amorphous state of ASDs. The absence of sharp Bragg peaks and the presence of broad diffuse halos indicate loss of long-range molecular order, a hallmark of amorphous materials. XRPD provides high specificity for detecting even trace crystalline fractions, supporting identification of recrystallization phenomena during storage or thermal stress. Its non-destructive nature and high reproducibility make XRPD indispensable both in early-stage formulation development and in routine quality control [57].

### 4.9.2 *Molecular Interaction and Microstructural Analysis*

Understanding interactions at the molecular level is essential for predicting the stability and performance of ASDs, as drug-polymer interactions often govern amorphous stabilization.

**Table 4.5** Solid-state characterization techniques used for HME-based amorphous solid dispersions

| Technique | Purpose | Detects | Limitation |
|---|---|---|---|
| Differential scanning calorimetry (DSC ()) | Determine thermal transitions and miscibility | Glass transition temperature ($T_g$), melting point ($T_m$), recrystallization events, and residual crystallinity | Limited sensitivity for low levels of crystallinity (<1–2%); overlapping thermal events may complicate interpretation |
| Modulated DSC (mDSC) | Improve resolution of overlapping thermal events | Accurate $T_g$ measurement, separation of reversible/non-reversible transitions | More complex data analysis; longer experimental time |
| X-ray powder diffraction (XRPD) | Confirm amorphous or crystalline state | Sharp diffraction peaks (crystalline) vs halo pattern (amorphous); polymorphic transitions | Limited detection threshold (≈1–5% crystallinity); may miss nanocrystals |
| FTIR spectroscopy | Evaluate drug-polymer intermolecular interactions | Hydrogen bonding, ionic interactions, functional group shifts | Interpretation can be qualitative; overlapping peaks are possible |
| Raman spectroscopy | Monitor crystallinity and molecular interactions; suitable for PAT | Phase transitions, polymorphs, and crystallinity changes | Fluorescence interference requires chemometric modelling for quantitative analysis |
| Solid-state NMR (ssNMR) | Assess molecular mobility and phase homogeneity | Molecular-level dispersion, phase separation, and interaction strength | Expensive instrumentation; complex data interpretation |
| Polarized light microscopy (PLM) | Visual identification of crystalline domains | Birefringence indicating crystal presence | Low sensitivity for very small or low-level crystals |
| Scanning electron microscopy (SEM) | Evaluate surface morphology | Surface structure, phase separation, particle distribution | Does not directly confirm amorphous state |
| Transmission electron microscopy (TEM) | High-resolution morphological analysis | Nano-scale phase separation, nanocrystals | Time-consuming sample preparation; limited bulk representativity |
| Thermogravimetric analysis (TGA) | Evaluate thermal stability and moisture content | Weight loss due to degradation or moisture | Does not directly measure crystallinity or miscibility |

#### 4.9.2.1 FTIR and Raman Spectroscopy

Vibrational spectroscopies such as FTIR and Raman provide rich detail about functional-group environments and intermolecular interactions. Shifts in characteristic vibrational bands can indicate hydrogen bonding, π–π interactions, or dipole-driven associations that stabilize the amorphous drug within the polymer matrix [58]. Raman spectroscopy additionally enables spatial mapping for detecting localized crystallization or phase heterogeneity, an important advantage for evaluating large-scale manufacturing processes. FTIR, particularly via ATR, supports rapid screening of interaction strength and provides indirect evidence of disruption of the crystalline lattice upon amorphization.

#### 4.9.2.2 Solid-State NMR (ssNMR)

Solid-state NMR offers unparalleled insight into molecular mobility, chemical environments, and phase heterogeneity. Techniques such as magic-angle spinning (MAS) and relaxation-time measurements provide quantitative data on segmental mobility, which correlates strongly with physical stability and recrystallization tendencies [58]. ssNMR can distinguish miscible from partially phase-separated systems, characterize drug-polymer interactions at the atomic level, and detect subtle structural domains undetectable by DSC or XRPD. Although resource-intensive, ssNMR remains one of the most powerful tools for mechanistic understanding of ASD stability.

#### 4.9.2.3 Microscopy (PLM, SEM, TEM)

Microscopy techniques complement thermal and spectroscopic methods by providing direct visualization of morphology and crystallinity. Polarized light microscopy (PLM) allows rapid detection of birefringent crystalline domains, useful for assessing amorphous conversion efficiency. Scanning electron microscopy (SEM) reveals surface morphology and particle architecture, offering insights into the influence of processing methods such as spray drying or hot-melt extrusion on microstructure. Transmission electron microscopy (TEM) further enables nanoscale imaging to detect submicron crystalline domains or phase-separated regions, providing an additional layer of sensitivity for comprehensive structural assessment.

### *4.9.3 Chemical Stability and Degradation Analysis*

Beyond physical stability, chemical integrity of the API is a critical determinant of both efficacy and safety. Chemical characterization ensures that degradation pathways are identified early and controlled throughout product development.

#### 4.9.3.1 Chemical Integrity and Regulatory Importance

Chemical stability evaluation ensures that the API remains intact within the ASD matrix and does not undergo degradation due to processing conditions, interactions with excipients, or prolonged storage. This aspect is central for regulatory approval, as chemical purity and well-defined impurity profiles are required for establishing shelf life and ensuring patient safety [59]. The amorphous state may increase reactivity due to higher molecular mobility, necessitating vigilant monitoring during formulation design and stability studies.

#### 4.9.3.2 HPLC and LC–MS

High-performance liquid chromatography (HPLC) is the primary method for quantifying API content and assessing content uniformity in ASDs. It enables routine monitoring of purity and detection of degradants that may arise during storage or high-temperature manufacturing processes. Liquid chromatography–mass spectrometry (LC–MS) further enhances analytical specificity by combining chromatographic separation with mass-based identification. LC–MS can detect low-level impurities, characterize degradation pathways, and identify interaction products arising from drug–polymer reactions. Because ASDs often involve processing conditions that expose the API to heat or shear, these methods are indispensable for ensuring chemical robustness of the final product.

### *4.9.4 Process Analytical Technologies (PAT)*

Analytical characterization is essential for confirming amorphous structure, understanding drug-polymer interactions, and ensuring product stability in HME-based systems. Integration of process analytical technology (PAT) further enables real-time monitoring and control under continuous manufacturing frameworks [60].

PAT tools enable real-time monitoring and control of HME processes, supporting Quality by Design (QbD) and continuous manufacturing. Integration of advanced analytical tools with data-driven modelling enhances process robustness, reduces variability, and supports real-time release testing in pharmaceutical extrusion [56].

In advanced HME operations, process analytical technology (PAT) tools play an essential role in real-time monitoring and quality assurance, enabling continuous assessment of critical material attributes and process parameters. In-line near-infrared (NIR) spectroscopy is extensively used to evaluate blend uniformity, drug distribution, and melt state, providing immediate feedback on whether the drug is fully dispersed within the polymer matrix. Likewise, in-line Raman spectroscopy offers high chemical specificity and is particularly effective for detecting crystallinity, monitoring phase transitions, and identifying any residual or newly formed crystalline domains during extrusion. These spectroscopic tools, integrated directly into the extrusion barrel or die, support proactive process adjustments and ensure consistent production of amorphous solid dispersions.

Complementing spectroscopic monitoring, mechanical sensors such as torque and pressure transducers provide valuable indirect indicators of melt viscosity, feed consistency, and overall process stability. Changes in torque or die pressure can signal variations in material properties, the onset of degradation, or deviations in feed rate, enabling operators to intervene before product quality is compromised. To fully leverage the diverse data streams generated by PAT tools, chemometric approaches such as principal component analysis (PCA) and partial least squares (PLS) modelling are routinely employed. These multivariate methods facilitate the interpretation of complex datasets, support the development of predictive control models, and enhance real-time decision-making in continuous manufacturing environments. Together, the integration of spectroscopic monitoring, mechanical sensing, and chemometrics significantly strengthens process understanding and control within HME-based pharmaceutical production.

## 4.10 Dosage Form Development Using HME

HME is a versatile platform technology that enables the development of diverse pharmaceutical dosage forms. By producing homogeneous drug-polymer matrices, particularly ASDs, HME supports improved solubility, immediate and controlled release, and personalized medicine applications [4]. Its compatibility with continuous manufacturing further enhances scalability and reproducibility.

### *4.10.1 Oral Solid Dosage Forms*

HME-derived intermediates can be processed into conventional oral solid dosage forms, including compressed tablets, capsule-filled extrudates, or milled granules, and pellets for multi-particulate systems. ASDs prepared via HME significantly enhance dissolution rate and apparent solubility of poorly water-soluble drugs (BCS Class II and IV), leading to improved bioavailability [14, 30]. Extruded materials typically exhibit excellent content uniformity and compressibility, supporting robust downstream processing.

### *4.10.2 Immediate-Release Systems*

HME-derived extrudates are milled and processed into conventional immediate-release formulations such as compressed tablets, capsules, granules, and pellets. By converting crystalline APIs into amorphous dispersions within hydrophilic polymers, HME significantly enhances dissolution rate and apparent solubility, enabling rapid drug release in gastrointestinal fluids. Polymer selection (hydrophilic carriers), optimized drug loading, and appropriate downstream processing are critical to achieving fast release profiles without compromising stability [61]. Thus, HME provides a scalable and solvent-free approach for developing high-performance immediate-release oral dosage forms.

Beyond its efficiency as a manufacturing method, immediate-release systems developed using HME offer significant clinical and industrial advantages. These systems facilitate rapid disintegration and dissolution, which directly translates to improved oral bioavailability for poorly soluble BCS Class II and IV drugs. Furthermore, the process ensures uniform drug distribution and high content consistency, effectively reducing dose variability. From a production standpoint, HME also provides robust manufacturability, allowing these high-performance formulations to be produced reliably under continuous processing conditions.

### *4.10.3 Controlled-Release Systems*

HME enables the fabrication of modified and controlled-release systems through precise polymer matrix engineering. By adjusting polymer type, drug loading, and matrix density, release kinetics can be tailored [62]. The ability to manipulate polymer composition and geometry allows predictable and reproducible release profiles.

These engineering capabilities support a wide range of specialized applications, including the development of matrix tablets designed for sustained drug release and osmotic systems that utilize controlled drug diffusion. Furthermore, HME is instrumental in creating multi-particulate systems, such as pellets that allow for flexible dosing, as well as extended-release implants intended for long-term therapy.

### *4.10.4 Films and Transdermal Systems*

HME facilitates the production of uniform, drug-loaded polymeric films with controlled thickness and mechanical properties [63]. These systems are suitable for oral thin films, buccal delivery, and transdermal patches. Film-based systems benefit from improved dosing accuracy and rapid or sustained drug delivery depending on polymer selection.

The adoption of HME for film production offers several distinct industrial and clinical advantages over traditional casting methods. Most notably, it allows for completely solvent-free processing, which eliminates the need for costly and time-consuming drying steps. This mechanical process also ensures a highly uniform drug distribution and precise thickness control across the entire film. From a clinical perspective, these features contribute to enhanced patient compliance, as the resulting films are often more portable and easier to administer than traditional oral or injectable dosage forms.

## 4.11 Applications of HME

HME has evolved from a formulation technique to a multifunctional platform technology in modern pharmaceutics. Its ability to produce homogeneous drug–polymer matrices under continuous processing conditions has enabled applications ranging from solubility enhancement to advanced and personalized drug delivery systems [1, 64].

### *4.11.1 Solubility and Bioavailability Enhancement*

One of the most established applications of HME is the preparation of ASDs to address poor aqueous solubility. Many new chemical entities fall under BCS Class II and IV, where dissolution is the rate-limiting step for absorption [65].

HME-based ASDs significantly enhance the performance of poorly water-soluble drugs by transforming them from their stable crystalline form into a high-energy amorphous state with superior dissolution characteristics. The intimate mixing of drug and polymer within the extruder promotes molecular-level dispersion, thereby improving apparent solubility and dissolution rates once the dosage form encounters gastrointestinal fluids. Furthermore, polymers selected for ASD formulations often exhibit solubilizing or precipitation-inhibiting properties, which help sustain drug supersaturation for extended periods and minimize the likelihood of recrystallization during dissolution. This maintenance of supersaturation is critical for ensuring that poorly soluble drugs achieve adequate concentration gradients for absorption across the intestinal epithelium.

As a result of these mechanistic advantages, HME-based ASDs frequently demonstrate enhanced oral bioavailability and reduced inter- and intra-patient pharmacokinetic variability compared with their crystalline counterparts. Improved absorption efficiency can enable the use of lower doses while still achieving therapeutic plasma concentrations, reducing pill burden and potentially minimizing dose-related side effects. These benefits collectively contribute to improved therapeutic performance and a more consistent patient response, making HME-based ASDs an attractive strategy for modern drug development, particularly for molecules falling within the low-solubility categories of the Biopharmaceutics Classification System (BCS).

### *4.11.2 Advanced Oral Delivery*

HME supports the development of multifunctional oral dosage forms through matrix engineering and polymer selection [64]. The key applications include:

(i) *Taste-masking*: Encapsulation of bitter APIs within polymer matrices to improve palatability, particularly for paediatric formulations.
(ii) *Abuse-deterrent formulations*: Development of mechanically robust or gel-forming matrices to prevent dose dumping or tampering.
(iii) *Fixed-dose combinations (FDCs)*: Uniform dispersion of multiple APIs within a single matrix to improve adherence and simplify therapy.
(iv) *Gastro-retentive systems*: Floating or swellable matrices that prolong gastric residence time and improve absorption of drugs with narrow absorption windows.

### 4.11.3 3D-Printing Filaments

HME is the primary method for producing drug-loaded filaments used in fused deposition modelling (FDM) 3D printing. These filaments serve as feedstock for fabricating personalized dosage forms [66]. The integration of HME with 3D printing represents a major advancement toward patient-specific medicine and decentralized pharmaceutical manufacturing.

Recent advances in hot-melt extrusion (HME) have increasingly focused on producing drug-loaded filaments for fused-deposition modelling (FDM) 3D printing, enabling personalized dosage forms with tuneable geometry and drug release profiles. One of the most notable developments has been the use of HME to generate filaments containing high drug loads while maintaining printability and mechanical integrity. Goyanes et al. demonstrated one of the first successful integrations of HME and FDM by producing paracetamol-loaded filaments capable of being printed into tablets with individualized shapes and doses [67]. This work established the feasibility of combining continuous manufacturing with additive manufacturing to produce patient-specific oral dosage forms. Building upon this foundation, Melocchi et al. expanded the applicability of HME for filament fabrication by developing polymeric matrices with enhanced thermal stability and flexibility, enabling the extrusion of filaments incorporating thermolabile drugs and complex release-modifying excipients [68].

More recent studies have moved beyond conventional polymers to explore innovative carrier systems designed specifically for HME-based filament production. Solanki et al. introduced amphiphilic polymer blends formulated to promote drug solubilization while preserving the mechanical characteristics required for FDM [69]. Their work revealed that tailored polymer combinations could improve filament homogeneity, drug distribution, and print reliability, even at elevated drug loadings. Another significant innovation was provided by Awad et al., who utilized HME to produce filaments incorporating nanostructured drug carriers, enabling multi-scale control over drug release kinetics [70]. This hybrid strategy combined the benefits of amorphous solid dispersion formation with nanoscale engineering, resulting in filaments with enhanced dissolution rates and customizable release profiles.

As a result, these novel studies highlight the growing applications of HME-based filament manufacturing, demonstrating its potential to support fully personalized medicine, expand the innovative space for poorly soluble drugs, and accelerate the development of on-demand pharmaceuticals. As polymer science, extrusion engineering, and additive manufacturing continue to converge, HME-produced filaments represent one of the most promising frontiers in advanced drug delivery.

## 4.12 Quality by Design (QbD) and Design of Experiments (DoE)

HME development benefits significantly from a structured Quality by Design (QbD) framework. QbD promotes a science- and risk-based approach to formulation and process optimization, ensuring consistent product quality, regulatory compliance, and scalable manufacturing (Table 4.6) [71, 72]. By systematically identifying critical attributes (Quality and Material attributes) and process parameters, manufacturers can establish a robust control strategy and define an operational design space.

### *4.12.1 Critical Quality Attributes (CQAs)*

Critical Quality Attributes (CQAs), which are measurable properties that directly impact product safety, efficacy, and performance. For HME-based ASDs, key CQAs include Glass transition temperature ($T_g$) (indicator of molecular mobility and physical stability), crystallinity level (confirms successful amorphization), dissolution profile (determines bioavailability and therapeutic performance), and impurity levels (Reflect chemical stability and regulatory compliance) [73].

### *4.12.2 Critical Material Attributes (CMAs)*

Critical Material Attributes (CMAs), such as polymer grade, molecular weight, drug particle size, hygroscopicity, and drug loading, also significantly influence product quality. Early identification of CQAs and CMAs enables risk assessment and focused experimental design [74].

**Table 4.6** Critical process parameters (CPPs) and their impact in hot melt extrusion

| CPP | Increase effect | Decrease effect | Associated risk |
|---|---|---|---|
| Barrel temperature | → Improved polymer softening and drug dissolution<br>→ Reduced melt viscosity<br>→ Enhanced amorphization | → Incomplete melting<br>→Poor mixing<br>→ Residual crystallinity | → API degradation<br>→Polymer degradation<br>→ Colour change or impurity formation |
| Screw speed (RPM) | → Higher shear rate<br>→ Improved distributive/ dispersive mixing<br>→ Reduced residence time<br>→ Lower apparent viscosity (shear thinning) | → Longer residence time<br>→ Lower shear input<br>→ Potential incomplete mixing | → Overheating from excessive shear<br>→ Drug degradation (low RPM + high temperature)<br>→ Poor homogeneity |
| Feed rate | → Higher throughput<br>→ Increased barrel fill and die pressure | → Lower torque<br>→ Higher residence time | → Overfilling → incomplete melting<br>→ Underfilling → overexposure to heat/shear<br>→ Variability in extrudate dimensions |
| Specific mechanical energy (SME) | → Enhanced drug dispersion<br>→ Improved amorphization • Better homogeneity | → Insufficient mixing<br>→ Higher residual crystallinity | → Polymer chain scission (high SME)<br>→ Thermal/mechanical degradation |
| Residence time | → Greater mixing and dissolution (if optimized) | → Insufficient melting or dispersion | → Overexposure → degradation<br>→ Underexposure → incomplete amorphization |
| Barrel fill level | → Improved pressure generation<br>→ Better mixing efficiency | → Reduced melt pressure<br>→ Lower mixing efficiency | → Instability in torque and pressure<br>→ Poor content uniformity |
| Die pressure | → Improved densification<br>→ Better dimensional uniformity | → Irregular extrudate shape | → Melt fracture<br>→ Equipment strain<br>→ Process instability |
| Cooling rate | → Rapid kinetic stabilization<br>→ Reduced recrystallization | → Increased molecular mobility | → Phase separation<br>→ Post-extrusion recrystallization |

### *4.12.3 Identification of Critical Process Parameters (CPPs)*

Critical Process Parameters (CPPs) are operational variables that impact CQAs. Through process mapping and risk analysis, key CPPs in HME are identified. Typical CPPs include barrel temperature profile, screw speed, feed rate, specific mechanical energy, residence time, and die pressure [75]. Understanding how these parameters influence melting behaviour, mixing efficiency, and amorphization is essential for achieving reproducible product performance.

### 4.12.4 *Design of Experiments (DoE) and Design Space Development*

Design of Experiments (DoE) applies statistical methodologies to systematically evaluate the relationships between CQAs, CMAs, and CPPs. Multivariate models identify significant factors, interaction effects, and optimal processing conditions. Benefits of DoE in HME include quantitative understanding of parameter interactions, optimization of drug loading and processing temperature, minimization of variability, and establishment of a regulatory-compliant design space [75, 76].

The design space defines the multidimensional space of combinations of material attributes and process parameters that consistently produce acceptable product quality. Operating within this approved space provides manufacturing flexibility while maintaining regulatory assurance. Together, QbD and DoE strengthen process robustness, support continuous manufacturing strategies, and facilitate efficient regulatory submissions for HME-based pharmaceutical products [33].

## 4.13 Scale-Up and Continuous Manufacturing

Hot melt extrusion (HME) is inherently suited for continuous manufacturing; however, successful scale-up from laboratory to commercial production requires careful control of process dynamics, material behaviour, and regulatory compliance [77]. Maintaining product quality during scale transition depends on preserving key thermomechanical conditions and process consistency.

### 4.13.1 *Similarity Principles*

Successful scale-up of HME operations relies on maintaining critical process similarities between laboratory-scale and commercial-scale equipment to ensure that the final product exhibits consistent solid-state and performance characteristics. Because scale transitions inherently modify heat-transfer efficiency, torque behaviour, and pressure profiles within the extruder, preserving dynamic similarity is essential to avoid unintended recrystallization, degradation, or variability in drug-polymer mixing. Central to this process is the alignment of specific mechanical energy (SME), a parameter that reflects the shear and mechanical work imparted to the material. Matching SME across scales ensures that the degree of amorphization, dispersive and distributive mixing, and drug-polymer interactions remain comparable, thereby reducing the risk of scale-induced deviations in critical quality attributes [13, 78].

Equally important is the maintenance of a similar residence time distribution (RTD), which governs the duration of material exposure to thermal and mechanical energy. Discrepancies in RTD can result in insufficient amorphization, variable dissolution performance, or thermal degradation, especially for heat-sensitive drugs or polymers. Strategies to preserve RTD typically include matching screw speed, feed rate ratios, and screw element arrangements while adjusting barrel temperature profiles to account for scale-dependent heat transfer differences. Geometric similarity—including consistent screw configurations, length-to-diameter ratios, and the spatial arrangement of kneading blocks—further supports predictable mixing behaviour and enables more reliable translation from bench-top to commercial production [9, 79]. Together, these scale-up principles help ensure robust process performance and product uniformity across manufacturing scales.

### 4.13.2 Pilot to Commercial Transition

Transitioning from pilot-scale to commercial-scale extruders often requires parameter adjustments due to differences in barrel diameter, throughput capacity, and thermal mass. A structured scale-up strategy minimizes development risk and supports regulatory approval [48]. Critical considerations include optimization of screw design to maintain mixing efficiency, adjustment of feed rate and screw speed to balance specific mechanical energy and RTD, refinement of barrel temperature profiles to prevent overheating or incomplete melting, and monitoring torque and die pressure to ensure process stability [77].

### 4.13.3 Cleaning and GMP Considerations

For commercial HME operations, compliance with Good Manufacturing Practice (GMP) standards is essential. Effective GMP implementation ensures product safety, regulatory compliance, and sustainable large-scale manufacturing of HME-based formulations. Important aspects are cleaning validation to remove polymeric residues and prevent carryover, cross-contamination, particularly for potent or highly active compounds, integration of extrusion units into a continuous manufacturing line, including feeders, blenders, and PAT systems; documentation, and process control aligned with regulatory expectations [80].

### 4.13.4 Regulatory Considerations

Regulatory expectations for HME-based products emphasize a science and risk-based approach aligned with Quality by Design (QbD) principles and continuous manufacturing guidelines. Agencies require a comprehensive understanding and control of both material attributes and process parameters to ensure consistent product performance and long-term stability. HME-based products are increasingly well accepted by regulatory authorities [77]. Several commercially approved amorphous solid dispersions are manufactured using HME, reflecting regulatory confidence in the technology when supported by robust scientific justification, validated control strategies, and comprehensive stability data [81].

Key regulatory focus areas include:

(i) *Control of crystallinity*: Demonstration of complete or controlled amorphization using validated analytical methods (DSC, XRPD), with justification of acceptable limits.
(ii) *Impurity and degradation profiling*: Assessment of thermal degradation, oxidative pathways, and polymer-related impurities using stability-indicating methods.
(iii) *Design space justification*: Statistical validation of operating ranges (CPPs and CMAs) that ensure CQAs remain within specifications.
(iv) *PAT-enabled Real-Time Release Testing (RTRT)*: Integration of in-line analytical tools (NIR, Raman probes) for continuous monitoring of critical attributes, supporting reduced end-product testing.
(v) *Stability Testing (ICH Q1A)*: Long-term and accelerated stability evaluation to confirm physical and chemical integrity throughout shelf life.

## 4.14 Case Studies and Marketed Products

The translation of hot HME from development laboratories to commercial manufacturing underscores its maturity as a pharmaceutical platform technology. Marketed products (Table 4.7) demonstrate HME's versatility in addressing solubility limitations, enabling controlled release, improving stability, and enhancing patient compliance [48]. These real-world examples provide practical insight into formulation design, process optimization, and regulatory acceptance [59].

### 4.14.1 Commercially Marketed Products

*Kaletra*: One of the most cited examples of HME-based commercialization, Kaletra® utilizes hot melt extrusion to produce a solid dispersion of lopinavir and ritonavir within a polymeric matrix (Meltrex). The formulation improved solubility and

**Table 4.7** Commercial pharmaceutical products manufactured using hot melt extrusion

| Product | API(s) | Polymer/ matrix system | Indication | Release type | Approval year |
|---|---|---|---|---|---|
| Kaletra® | Lopinavir + Ritonavir | Copovidone-based solid dispersion (Meltrex® technology) | HIV-1 infection | Immediate release (enhanced solubility ASD) | 2000 (US tablet reformulation 2005) |
| Isoptin® SR | Verapamil HCl | HPMC/HPC/ alginate matrix | Hypertension, angina, arrhythmia | Sustained release | 1980s |
| Nurofen® Meltlets | Ibuprofen | Polymer-based extruded amorphous granules | Pain, inflammation, fever | Rapid disintegration/ immediate release | ~2010 |
| NuvaRing® | Etonogestrel + Ethinyl oestradiol | Poly (ethylene-vinyl acetate) (EVA) elastomer | Contraception | Controlled-release vaginal ring | 2001 |
| Estring® | Estradiol | EVA membrane system | Menopausal vaginal atrophy | Controlled-release vaginal ring | 1996 |
| Femring® | Estradiol acetate | EVA-based vaginal ring | Menopausal symptoms | Controlled-release vaginal ring | 2003 |
| Implanon®/ Nexplanon® | Etonogestrel | EVA subdermal implant | Contraception | Long-acting implant (3 years) | 2006 (Implanon US), 2011 (Nexplanon US) |

oral bioavailability while eliminating the refrigeration requirement associated with earlier soft-gel formulations. This product demonstrated how HME can improve stability, simplify supply chains, and enhance patient adherence [9].

*Isoptin SR*: An example of HME-enabled sustained-release matrix design using verapamil hydrochloride within a polymer matrix (hydroxypropyl cellulose/ hydroxypropyl methyl cellulose/sodium alginate) [82]. Isoptin SR® uses polymer-controlled diffusion to regulate drug release over extended periods. The extrusion process ensures uniform drug distribution within the matrix, enabling predictable and reproducible release kinetics.

*Nurofen meltlets orally disintegrating tablet (ODT)*: HME-enabled heating and mixing of the active ingredient (ibuprofen) with excipients, amorphous, high-surface-area granules that dissolve more quickly in the mouth. This applied to rapid disintegration dosage forms, enhancing patient compliance through ease of administration and improved palatability [83].

*Generic HME-based products*: Numerous generic formulations now utilize HME for producing amorphous solid dispersions or modified-release matrices. The adoption by generics manufacturers reflects confidence in the scalability, cost-efficiency, and regulatory familiarity of extrusion technology [84].

### 4.14.2 Formulation and Processing Insights

Commercial HME products demonstrate that successful development relies on selecting compatible high-$T_g$ polymers, optimizing drug loading, and carefully controlling processing parameters such as temperature, screw configuration, and specific mechanical energy to ensure complete amorphization without degradation. Maintaining consistent mixing and residence time during scale-up is essential to preserve stability and product performance. Analysis of commercial products reveals several consistent development strategies.

Regarding polymer selection and stability, high-$T_g$ polymers with strong drug-polymer interactions are selected to ensure physical stability and minimize recrystallization risk. Miscibility studies and $T_g$ modelling guide drug loading limits. In terms of thermal and shear optimization, processing conditions are carefully optimized to achieve complete amorphization without inducing degradation. Barrel temperature profiles and screw configurations are tailored to balance dissolution enhancement with chemical stability. For scale-up robustness, commercialization requires maintaining specific mechanical energy and residence time distribution during scale transition. Pilot studies are used to validate geometric and thermodynamic similarity. As for regulatory strategy, successful products demonstrate strong integration of Quality by Design (QbD), validated analytical methods for crystallinity control, and comprehensive ICH stability data.

### 4.14.3 Market and Industrial Impact

These case studies collectively demonstrate that HME is not merely a formulation enhancement tool but a scalable, industrially proven manufacturing technology capable of delivering stable, high-performance pharmaceutical products across multiple therapeutic categories. The commercial success of HME products confirms several broader advantages.

One primary benefit is the elimination of organic solvents compared to spray drying. This shift reduces environmental impact and removes the need for complex solvent recovery systems in the manufacturing facility. Another advantage is the inherent compatibility with continuous manufacturing. This alignment allows for real-time monitoring and a more streamlined production flow than traditional batch processing. Furthermore, the technology leads to improved supply chain efficiency. By simplifying the production steps and reducing the footprint of the machinery required, manufacturers can respond more quickly to market demands. The process also results in reduced batch variability. Precise control over mechanical and thermal inputs ensures that each unit of the product meets exact specifications for amorphization and stability. Finally, there is strong regulatory acceptance for this method. Because it integrates well with Quality by Design principles, regulatory agencies often find the data packages for these products to be robust and predictable.

## 4.15 Challenges and Future Perspectives

Although HME is a mature and commercially validated technology, several scientific and industrial challenges limit its universal application. As drug molecules become increasingly complex, highly lipophilic, thermally sensitive, or high-dose, greater innovation in formulation science and process control is required. Future advancements will likely arise from interdisciplinary integration of polymer chemistry, digital modelling, and continuous manufacturing frameworks.

Despite its advantages, HME faces several technical challenges. Processing thermolabile APIs is difficult due to the elevated temperatures required, which may lead to degradation. High drug loading in amorphous solid dispersions can reduce miscibility and increase the risk of recrystallization. In addition, accurately predicting long-term physical stability, particularly under humid storage conditions, remains a significant limitation requiring improved mechanistic and predictive approaches.

Many modern drug candidates, including peptides and highly functionalized small molecules, exhibit thermal sensitivity. HME typically requires elevated temperatures to soften polymers and enable molecular mixing. Even short exposure to high temperatures can lead to degradation, impurity formation, or loss of potency. Although strategies such as plasticizer incorporation, reduced residence time, and optimized barrel temperature zoning are employed, they may compromise long-term stability. Therefore, expanding HME applicability to thermosensitive compounds remains a key technical frontier.

Commercial feasibility often demands high drug loading to reduce pill burden and excipient mass. However, increasing drug concentration narrows the miscibility window and increases the risk of phase separation, recrystallization, and reduced $T_g$. High drug loading can also increase melt viscosity, complicating processing and scale-up. Developing predictive tools and advanced polymer systems capable of stabilizing high drug fractions remains a major research priority.

While thermodynamic models, F–H theory (Flory–Huggins), and $T_g$-based predictions provide valuable insight, they do not fully capture real-time aging behaviour, humidity effects, and molecular mobility under storage conditions [85]. Moisture-induced plasticization and micro-phase separation can occur even when initial characterization suggests homogeneity. More accurate predictive modelling and accelerated stability frameworks are needed to minimize late-stage product failure.

### *4.15.1 Emerging Technologies*

#### 4.15.1.1 Machine Learning and Data-Driven Modelling

Artificial intelligence (AI) and machine learning (ML) are increasingly applied to predict drug–polymer miscibility, optimal processing parameters, and stability outcomes. By analysing historical formulation data, ML models can reduce experimental burden and accelerate development timelines [86]. Integration with PAT data streams may enable adaptive process control in real time.

#### 4.15.1.2 Hybrid HME Spray Drying Approaches

Combining HME with spray drying or other solid-state technologies may allow better control over particle engineering, drug loading, and dissolution performance [87]. Hybrid systems can leverage the scalability of extrusion with the particle-level control of spray processes.

#### 4.15.1.3 $CO_2$-Assisted Extrusion

The incorporation of pressurized or supercritical $CO_2$ acts as a temporary plasticizer, lowering melt viscosity and processing temperature [88]. This approach may expand HME applicability to thermolabile APIs and enable porous extrudate formation for rapid dissolution systems.

#### 4.15.1.4 Custom-Designed Polymers

Next-generation polymers specifically engineered for HME are being developed with optimized $T_g$, improved miscibility profiles, and tailored rheological behaviour. Amphiphilic and multifunctional polymers may further enhance supersaturation maintenance and long-term stability [89].

### *4.15.2 Future Outlook*

The future of HME lies in its integration with continuous manufacturing, PAT-driven real-time release testing, digital twins, and personalized medicine platforms. As regulatory frameworks increasingly support advanced manufacturing technologies, HME is expected to play a central role in decentralized production, 3D printing applications, and next-generation drug delivery systems.

Addressing current technical gaps through materials innovation, predictive modelling, and process digitalization will further solidify HME as a cornerstone technology in modern pharmaceutics.

## 4.16 Conclusion

Hot melt extrusion (HME) has emerged as a versatile and industrially validated platform technology in modern pharmaceutics. Its ability to produce homogeneous amorphous solid dispersions in a solvent-free and continuous manner makes it highly effective for enhancing the solubility and bioavailability of poorly water-soluble drugs. As a green manufacturing approach, HME reduces solvent use, waste generation, and environmental impact, supporting sustainable pharmaceutical production.

From an industrial standpoint, HME aligns with Quality by Design (QbD), continuous manufacturing, and PAT-enabled process control, enabling robust scale-up and regulatory flexibility. Its applicability across immediate-release, controlled-release, taste-masked, implantable, and 3D-printed dosage forms highlights its broad formulation versatility.

While challenges such as thermolabile drug processing and long-term stability prediction remain, advances in polymer science and digital modelling are expanding its potential. Overall, HME represents a scalable, sustainable, and future-ready technology for efficient and patient-centric pharmaceutical manufacturing.

## References


1. Patil H, Tiwari RV, Repka MA. Hot-melt extrusion: from theory to application in pharmaceutical formulation. AAPS PharmSciTech. 2016;17(1):20–42.
2. Tambe S, Jain D, Agarwal Y, Amin P. Hot-melt extrusion: highlighting recent advances in pharmaceutical applications. J Drug Deliv Sci Technol. 2021;63:102452.
3. Li M, Gogos CG, Ioannidis N. Improving the API dissolution rate during pharmaceutical hot-melt extrusion I: effect of the API particle size, and the co-rotating, twin-screw extruder screw configuration on the API dissolution rate. Int J Pharm. 2015;478(1):103–12.
4. Repka MA, Majumdar S, Kumar Battu S, Srirangam R, Upadhye SB. Applications of hot-melt extrusion for drug delivery. Expert Opin Drug Deliv. 2008;5(12):1357–76.
5. Zhao J, Tian G, Qu H. Pharmaceutical application of process understanding and optimization techniques: a review on the continuous twin-screw wet granulation. Biomedicines [Internet]. 2023; 11(7):[1923 p.].
6. Andrews GP, Jones DS, Diak OA, Margetson DN, McAllister MS. Hot-melt extrusion: an emerging drug delivery technology. Pharm Technol Eur. 2009;21(1):18–24.
7. da Silva Júnior WF, de Oliveira Pinheiro JG, Moreira CDLFA, de Souza FJJ, de Lima ÁAN. Chapter 15—Alternative technologies to improve solubility and stability of poorly water-soluble drugs. In: Grumezescu AM, editor. Multifunctional systems for combined delivery, biosensing and diagnostics. Elsevier; 2017. p. 281–305. Radarweg 29, PO Box 211, 1000 AE Amsterdam, Netherlands.
8. Leister D, Geilen T, Geissler T. Twin-screw extruders for pharmaceutical hot-melt extrusion: technology, techniques and practices. In: Hot-melt extrusion: pharmaceutical applications. 2012. p. 23–42. John Wiley & Sons Ltd, The Atrium, Southern Gate, Chichester, West Sussex, PO19 8SQ, United Kingdom
9. Crowley MM, Zhang F, Repka MA, Thumma S, Upadhye SB, Battu SK, et al. Pharmaceutical applications of hot-melt extrusion: part I. Drug Dev Ind Pharm. 2007;33(9):909–26.
10. Hancock BC, Zografi G. Characteristics and significance of the amorphous state in pharmaceutical systems. J Pharm Sci. 1997;86(1):1–12.
11. Leuner C, Dressman J. Improving drug solubility for oral delivery using solid dispersions. Eur J Pharm Biopharm. 2000;50(1):47–60.
12. Baghel S, Cathcart H, O'Reilly NJ. Polymeric amorphous solid dispersions: a review of amorphization, crystallization, stabilization, solid-state characterization, and aqueous solubilization of biopharmaceutical classification system class II drugs. J Pharm Sci. 2016;105(9):2527–44.

13. Djuris J, Nikolakakis I, Ibric S, Djuric Z, Kachrimanis K. Preparation of carbamazepine-Soluplus solid dispersions by hot-melt extrusion, and prediction of drug-polymer miscibility by thermodynamic model fitting. Eur J Pharm Biopharm. 2013;84(1):228–37.
14. Maniruzzaman M, Boateng J, Douroumis D, Snowden MJ. Hot-melt extrusion (HME): from process to pharmaceutical applications. In: Sezer AD, editor. Recent advances in novel drug carrier systems. London: IntechOpen; 2012.
15. Martin C. Twin screw extrusion for pharmaceutical processes. In: Repka MA, Langley N, DiNunzio J, editors. Melt extrusion: materials, technology and drug product design. New York: Springer New York; 2013. p. 47–79.
16. Chakraborty S, Chormale JH, Bansal AK. Deep eutectic systems: an overview of fundamental aspects, current understanding and drug delivery applications. Int J Pharm. 2021;610:121203.
17. Vo CL, Park C, Lee BJ. Current trends and future perspectives of solid dispersions containing poorly water-soluble drugs. Eur J Pharm Biopharm. 2013;85(3 Pt B):799–813.
18. Shi Q, Chen H, Wang Y, Wang R, Xu J, Zhang C. Amorphous solid dispersions: role of the polymer and its importance in physical stability and in vitro performance. Pharmaceutics. 2022;14(8). p 1747
19. Silva JFC, Rosado MTS, Maria TMR, Pereira Silva PS, Silva MR, Eusébio MES. Introduction to pharmaceutical co-amorphous systems using a green co-milling technique. J Chem Educ. 2023;100(4):1627–32.
20. Olsson M, Storm R, Björn L, Lilja V, Krupnik L, Chen Y, et al. Phase-separated polymer blends for controlled drug delivery by tuning morphology. Commun Mater. 2024;5(1):231.
21. Bhatta HP, Han HK, Maharjan R, Jeong SH. Recent techniques to improve amorphous dispersion performance with quality design, physicochemical monitoring, molecular simulation, and machine learning, Pharmaceutics 2025;17(10). p 1249
22. Lang B, McGinity JW, Williams RO 3rd. Hot-melt extrusion--basic principles and pharmaceutical applications. Drug Dev Ind Pharm. 2014;40(9):1133–55.
23. Wang N, Zhong H, Xiong P, Zhu J, Ouyang D. Quality by computational design: harnessing AI to advance rational drug development. Adv Drug Deliv Rev. 2026;229:115764.
24. Klueppelberg J, Handge UA, Thommes M, Winck J. Composition dependency of the Flory–Huggins interaction parameter in drug–polymer phase behavior. Pharmaceutics [Internet]. 2023; 15(12):[2650 p.].
25. Milliman HW, Boris D, Schiraldi DA. Experimental determination of Hansen solubility parameters for select POSS and polymer compounds as a guide to POSS–polymer interaction potentials. Macromolecules. 2012;45(4):1931–6.
26. Červinka C, Fulem M. Structure and glass transition temperature of amorphous dispersions of model pharmaceuticals with nucleobases from molecular dynamics. Pharmaceutics. 2021;13(8). p 1253
27. Tian Y, Booth J, Meehan E, Jones DS, Li S, Andrews GP. Construction of drug–polymer thermodynamic phase diagrams using Flory–Huggins interaction theory: identifying the relevance of temperature and drug weight fraction to phase separation within solid dispersions. Mol Pharm. 2013;10(1):236–48.
28. Qian D, Michaels TCT, Knowles TPJ. Analytical solution to the Flory-Huggins model. J Phys Chem Lett. 2022;13(33):7853–60.
29. Thompson SA, Williams RO. Specific mechanical energy—an essential parameter in the processing of amorphous solid dispersions. Adv Drug Deliv Rev. 2021;173:374–93.
30. Maniruzzaman M, Boateng JS, Snowden MJ, Douroumis D. A review of hot-melt extrusion: process technology to pharmaceutical products. ISRN Pharm. 2012;2012:436763.
31. Alshetaili A, Alshahrani SM, Almutairy BK, Repka MA. Hot melt extrusion processing parameters optimization. Processes [Internet]. 2020; 8(11):[1516 p.].
32. Wesholowski J, Berghaus A, Thommes M. Investigations concerning the residence time distribution of twin-screw-extrusion processes as indicator for inherent mixing. Pharmaceutics. 2018;10(4). p 207

33. Grymonpré W, Bostijn N, Herck SV, Verstraete G, Vanhoorne V, Nuhn L, et al. Downstream processing from hot-melt extrusion towards tablets: a quality by design approach. Int J Pharm. 2017;531(1):235–45.
34. Winck J, Gottschalk T, Thommes M. Predicting residence time and melt temperature in pharmaceutical hot melt extrusion. Pharmaceutics. 2023;15(5):1417.
35. Luker K. Single-screw extrusion: principles. In: Hot-melt extrusion: pharmaceutical applications. 2012. John Wiley & Sons Ltd, The Atrium, Southern Gate, Chichester, West Sussex, PO19 8SQ, United Kingdom
36. Domenech T, Peuvrel-Disdier E, Vergnes B. The importance of specific mechanical energy during twin screw extrusion of organoclay based polypropylene nanocomposites. Compos Sci Technol. 2013;75:7–14.
37. Kimmel V, Ercolin E, Zimmer R, Yörük M, Winck J, Thommes M. Measuring and modeling of melt viscosity for drug polymer mixtures. Pharmaceutics. 2024;16(3). p 301
38. Censi R, Gigliobianco MR, Casadidio C, Di Martino P. Hot melt extrusion: highlighting physicochemical factors to be investigated while designing and optimizing a hot melt extrusion process. Pharmaceutics. 2018;10(3). p 89
39. Alzahrani A, Nyavanandi D, Mandati P, Youssef AAA, Narala S, Bandari S, et al. A systematic and robust assessment of hot-melt extrusion-based amorphous solid dispersions: theoretical prediction to practical implementation. Int J Pharm. 2022;624:121951.
40. Kachrimanis K, Nikolakakis I. Polymers as formulation excipients for hot-melt extrusion processing of pharmaceuticals. In: Handbook of polymers for pharmaceutical technologies: processing and applications, vol. 2. 2015. pp. 121–49. John Wiley & Sons, Inc. Hoboken, New Jersey, and Scrivener Publishing LLC, Salem, Massachusetts
41. Danilova A, Gusev K, Arkhipova N, Danilov L, Vishnyakov E, Maimistov D, et al. Assessment of polyvinylpyrrolidone vinyl acetate effect on quercetin properties in binary solid dispersion prepared by hot melt extrussion. Drug Dev Regist. 2024. Vol. 14, p 127–137
42. Karandikar H, Ambardekar R, Kelly A, Gough T, Paradkar A. Systematic identification of thermal degradation products of HPMCP during hot melt extrusion process. Int J Pharm. 2015;486(1–2):252–8.
43. Butreddy A, Sarabu S, Almutairi M, Ajjarapu S, Kolimi P, Bandari S, et al. Hot-melt extruded hydroxypropyl methylcellulose acetate succinate based amorphous solid dispersions: impact of polymeric combinations on supersaturation kinetics and dissolution performance. Int J Pharm. 2022;615:121471.
44. Cantin O, Siepmann F, Danede F, Willart JF, Karrout Y, Siepmann J. PEO hot melt extrudates for controlled drug delivery: importance of the molecular weight. J Drug Deliv Sci Technol. 2016;36:130–40.
45. Dos Santos J, da Silva GS, Velho MC, Beck RCR. Eudragit(®): a versatile family of polymers for hot melt extrusion and 3D printing processes in pharmaceutics. Pharmaceutics. 2021;13(9).1424. p 1424
46. Li S, Zhang Z, Gu W, Gallas M, Jones D, Boulet P, et al. Hot melt extruded high-dose amorphous solid dispersions containing lumefantrine and Soluplus. Int J Pharm. 2024;665:124676.
47. Chatterjee T, O'Donnell KP, Rickard MA, Nickless B, Li Y, Ginzburg VV, et al. Rheology of cellulose ether excipients designed for hot melt extrusion. Biomacromolecules. 2018;19(11):4430–41.
48. Agrawal AM, Dudhedia MS, Zimny E. Hot melt extrusion: development of an amorphous solid dispersion for an insoluble drug from mini-scale to clinical scale. AAPS PharmSciTech. 2016;17(1):133–47.
49. Schittny A, Huwyler J, Puchkov M. Mechanisms of increased bioavailability through amorphous solid dispersions: a review. Drug Deliv. 2020;27(1):110–27.
50. He Y, Ho C. Amorphous solid dispersions: utilization and challenges in drug discovery and development. J Pharm Sci. 2015;104(10):3237–58.

51. Al-Japairai K, Hamed Almurisi S, Mahmood S, Madheswaran T, Chatterjee B, Sri P, et al. Strategies to improve the stability of amorphous solid dispersions in view of the hot melt extrusion (HME) method. Int J Pharm. 2023;647:123536.
52. Martynek D, Ridvan L, Sivén M, Šoóš M. Stability and recrystallization of amorphous solid dispersions prepared by hot-melt extrusion and spray drying. Int J Pharm. 2025;672:125331.
53. Corrie L, Ajjarapu S, Banda S, Parvathaneni M, Bolla PK, Kommineni N. HPMCAS-based amorphous solid dispersions in clinic: a review on manufacturing techniques (hot melt extrusion and spray drying), marketed products and patents. Materials (Basel). 2023;16(20):6616.
54. Sarode AL, Sandhu H, Shah N, Malick W, Zia H. Hot melt extrusion for amorphous solid dispersions: temperature and moisture activated drug-polymer interactions for enhanced stability. Mol Pharm. 2013;10(10):3665–75.
55. Qian F, Huang J, Hussain MA. Drug-polymer solubility and miscibility: stability consideration and practical challenges in amorphous solid dispersion development. J Pharm Sci. 2010;99(7):2941–7.
56. Netchacovitch L, Dumont E, Cailletaud J, Thiry J, De Bleye C, Sacré PY, et al. Development of an analytical method for crystalline content determination in amorphous solid dispersions produced by hot-melt extrusion using transmission Raman spectroscopy: a feasibility study. Int J Pharm. 2017;530(1):249–55.
57. Bates S, Zografi G, Engers D, Morris K, Crowley K, Newman A. Analysis of amorphous and nanocrystalline solids from their X-ray diffraction patterns. Pharm Res. 2006;23(10):2333–49.
58. Paudel A, Geppi M, Mooter GV. Structural and dynamic properties of amorphous solid dispersions: the role of solid-state nuclear magnetic resonance spectroscopy and relaxometry. J Pharm Sci. 2014;103(9):2635–62.
59. Matić J, Alva C, Witschnigg A, Eder S, Reusch K, Paudel A, et al. Towards predicting the product quality in hot-melt extrusion: small scale extrusion. Int J Pharm: X. 2020;2:100062.
60. Hitzer P, Bäuerle T, Drieschner T, Ostertag E, Paulsen K, van Lishaut H, et al. Process analytical techniques for hot-melt extrusion and their application to amorphous solid dispersions. Anal Bioanal Chem. 2017;409(18):4321–33.
61. AlSuwais AK, Alqudah M, Alshayeb M, Banat F, Abu Alhaija A, Alawamleh E, et al. Investigating the use of hot-melt extrusion as a surrogate to conventional techniques as the principal manufacturing technique for the preparation of lornoxicam tablets. J Pharm Innov. 2025;20(6):253.
62. Jain DD, Tambe SM, Amin PD. Oral controlled drug delivery by hot-melt extrusion technology. In: Verma S, Verma SK, editors. Advancements in controlled drug delivery systems. Hershey: IGI Global Scientific Publishing; 2022. p. 205–36.
63. Karki S, Kim H, Na S-J, Shin D, Jo K, Lee J. Thin films as an emerging platform for drug delivery. Asian J Pharm Sci. 2016;11(5):559–74.
64. Douroumis D. Practical approaches of taste masking technologies in oral solid forms. Expert Opin Drug Deliv. 2007;4(4):417–26.
65. Sahu G, Gupta C. Exploration of solubilization strategies: enhancing bioavailability for low solubility drugs. Int J Newgen Res Pharm Healthc. 2023;1:96–115.
66. Douroumis D. 3D printing of pharmaceutical and medical applications: a new era. Pharm Res. 2019;36(3):42.
67. Goyanes A, Buanz AB, Basit AW, Gaisford S. Fused-filament 3D printing (3DP) for fabrication of tablets. Int J Pharm. 2014;476(1–2):88–92.
68. Melocchi A, Parietti F, Maroni A, Foppoli A, Gazzaniga A, Zema L. Hot-melt extruded filaments based on pharmaceutical grade polymers for 3D printing by fused deposition modeling. Int J Pharm. 2016;509(1–2):255–63.
69. Solanki NG, Tahsin M, Shah AV, Serajuddin ATM. Formulation of 3D printed tablet for rapid drug release by fused deposition Modeling: screening polymers for drug release, drug-polymer miscibility and printability. J Pharm Sci. 2018;107(1):390–401.
70. Awad A, Trenfield SJ, Gaisford S, Basit AW. 3D printed medicines: a new branch of digital healthcare. Int J Pharm. 2018;548(1):586–96.

71. Butreddy A, Bandari S, Repka MA. Quality-by-design in hot melt extrusion based amorphous solid dispersions: an industrial perspective on product development. Eur J Pharm Sci. 2021;158:105655.
72. Yu LX, Amidon G, Khan MA, Hoag SW, Polli J, Raju GK, et al. Understanding pharmaceutical quality by design. AAPS J. 2014;16(4):771–83.
73. Moseson DE, Taylor LS. Crystallinity: a complex critical quality attribute of amorphous solid dispersions. Mol Pharm. 2023;20(10):4802–25.
74. Yu LX. Pharmaceutical quality by design: product and process development, understanding, and control. Pharm Res. 2008;25(4):781–91.
75. Maniruzzaman M, Ross SA, Dey T, Nair A, Snowden MJ, Douroumis D. A quality by design (QbD) twin-screw extrusion wet granulation approach for processing water insoluble drugs. Int J Pharm. 2017;526(1–2):496–505.
76. Stocker E, Becker K, Hate S, Hohl R, Schiemenz W, Sacher S, et al. Application of ICH Q9 quality risk management tools for advanced development of hot melt coated multiparticulate systems. J Pharm Sci. 2017;106(1):278–90.
77. Wesholowski J, Hoppe K, Nickel K, Muehlenfeld C, Thommes M. Scale-up of pharmaceutical hot-melt-extrusion: process optimization and transfer. Eur J Pharm Biopharm. 2019;142:396–404.
78. Van Renterghem J, Van de Steene S, Digkas T, Richter M, Vervaet C, De Beer T. Assessment of volumetric scale-up law for processing of a sustained release formulation on co-rotating hot-melt extruders. Int J Pharm. 2019;569:118587.
79. Keleb EI, Vermeire A, Vervaet C, Remon JP. Continuous twin screw extrusion for the wet granulation of lactose. Int J Pharm. 2002;239(1–2):69–80.
80. Bhujbal SV, Mitra B, Jain U, Gong Y, Agrawal A, Karki S, et al. Pharmaceutical amorphous solid dispersion: a review of manufacturing strategies. Acta Pharm Sin B. 2021;11(8):2505–36.
81. Amidon KS, Langguth P, Lennernäs H, Yu L, Amidon GL. Bioequivalence of oral products and the biopharmaceutics classification system: science, regulation, and public policy. Clin Pharmacol Ther. 2011;90(3):467–70.
82. Vajna B, Pataki H, Nagy Z, Farkas I, Marosi G. Characterization of melt extruded and conventional Isoptin formulations using Raman chemical imaging and chemometrics. Int J Pharm. 2011;419(1):107–13.
83. Gryczke A, Schminke S, Maniruzzaman M, Beck J, Douroumis D. Development and evaluation of orally disintegrating tablets (ODTs) containing Ibuprofen granules prepared by hot melt extrusion. Colloids Surf B Biointerfaces. 2011;86(2):275–84.
84. Li S, Liu D, Jones D, Andrews G. Hot-melt extrusion: an emerging manufacturing technology for drug products. Emerging drug delivery and biomedical engineering technologies: CRC Press; 2023. p. 15–29. Boca Raton, FL 33487, United States
85. Moffat JG, Qi S, Craig DQ. Spatial characterization of hot melt extruded dispersion systems using thermal atomic force microscopy methods: the effects of processing parameters on phase separation. Pharm Res. 2014;31(7):1744–52.
86. Jiang J, Lu A, Ma X, Ouyang D, Williams RO 3rd. The applications of machine learning to predict the forming of chemically stable amorphous solid dispersions prepared by hot-melt extrusion. Int J Pharm X. 2023;5:100164.
87. Poozesh S, Bilgili E. Scale-up of pharmaceutical spray drying using scale-up rules: a review. Int J Pharm. 2019;562:271–92.
88. Almutairi M, Srinivasan P, Zhang P, Austin F, Butreddy A, Alharbi M, et al. Hot-melt extrusion coupled with pressurized carbon dioxide for enhanced processability of pharmaceutical polymers and drug delivery applications—an integrated review. Int J Pharm. 2022;629:122291.
89. Pereira GG, Figueiredo S, Fernandes AI, Pinto JF. Polymer selection for hot-melt extrusion coupled to fused deposition modelling in pharmaceutics. Pharmaceutics. 2020;12(9):795.

# Chapter 5
# Solvent Evaporation Approaches for the Preparation of Amorphous Solid Dispersions

**Seth P. Nowaseb, Edwin Chun Yin Chong, and Hisham Al-Obaidi**

**Abstract** Recent studies have shown clear benefits to using various solvent and polymer manipulation techniques to formulate poorly water-soluble oral pharmaceuticals. This is primarily because many of the active pharmaceutical ingredients (APIs) currently on the market or in the pipeline are poorly water-soluble compounds, which results in limited or poor oral bioavailability. One of the methods used to overcome the challenge of poor water solubility of orally delivered drugs has been the use of amorphous solid dispersions with polymers and other excipients. Amorphous solid dispersions (ASDs) involve a mixture of an API, a polymer, and possibly an additional excipient to form a solid dispersion. The aim of this chapter is to provide an overview of the significance of amorphous oral dispersions produced using solvent evaporation methods to achieve higher aqueous solubility and bioavailability while maintaining relative stability during production and upon storage after manufacture. The roles of solvents, polymers, and excipients in stabilizing amorphous systems and preventing recrystallisation are critically examined, alongside an overview of the advantages and limitations inherent to these formulations. This chapter focuses on solvent evaporation approaches, particularly spray drying and freeze drying, as key techniques in the preparation of ASDs. Recent developments in polymer technologies, nozzle design, and process analytical tools have greatly increased the robustness and application of solvent-based ASD methods, despite issues with stability, residual solvent removal, and large-scale manufacture. The challenges of using solvent evaporation methods are discussed, and any new developments under consideration will be elucidated to conclude the chapter.

S. P. Nowaseb
The School of Pharmacy, University of Reading, Reading, UK

School of Pharmacy, University of Namibia, Windhoek, Namibia

E. C. Y. Chong · H. Al-Obaidi (✉)
The School of Pharmacy, University of Reading, Reading, UK
e-mail: h.-al-obaidi@reading.ac.uk

 
I. Nikolakakis, I. Partheniadis (eds.), *Amorphous Solid Dispersions*, AAPS Introductions in the Pharmaceutical Sciences 5,
https://doi.org/10.1007/978-3-032-24883-1_5

**Keywords** Spray drying · Solvent evaporation · Bioavailability · Polymers · Supersaturation · Biofilms · Pulmonary delivery

## 5.1 Introduction

The formulation of drugs for oral application has been a challenge because most of these compounds have poor aqueous solubility, which results in poor bioavailability [1]. Studies of both marketed drugs and those in research and development indicate that this issue is widespread. Approximately 40% of drugs currently on the market, and 90% of those under development, present similarly poor aqueous solubility, resulting in low bioavailability [2, 3].

Various methods are used to improve the aqueous solubility of poorly water-soluble drugs. These methods include particle size reduction, salt formation, co-crystallization, solid dispersions, cyclodextrin complexation, and more [2]. In this chapter, the focus is on the use of solid dispersions in general and amorphous solid dispersions (ASDs) in particular, to address issues of poor water solubility for various drugs [4].

In many cases, only a small number of novel compounds can readily form salts; thus, the use of solid dispersions is often preferred to enhance solubility [5]. Examples of drugs with improved solubility when formulated as solid dispersions include carbamazepine, indomethacin, and griseofulvin, which exhibit solubility increases of 45%, 50%, and 35%, respectively [6]. Building on evidence that solid dispersions can enhance solubility, advanced formulation strategies now utilize ASDs to overcome the limitations posed by strong crystalline lattice structures [7]. Many studies demonstrated that ASDs could significantly enhance solubility for many compounds [8].

For instance, the solubility of itraconazole, ketoconazole, and nifedipine increased to a large extent upon ASD formulation [9–11]. This chapter will examine specific manufacturing approaches, advantages, and disadvantages of ASDs, solvents, and polymers used in manufacturing, and recent research findings in drug formulation and development.

## 5.2 Amorphous Solid Dispersions

Formulating poorly water-soluble drugs as part of solid dispersions is one of the most common strategies to enhance their aqueous solubility and bioavailability [4]. A solid dispersion typically involves dispersing a poorly water-soluble drug in a hydrophilic carrier matrix [6] (Fig. 5.1). In some cases, a third component, such as a surfactant or stabilizer, is included to further improve the system's performance

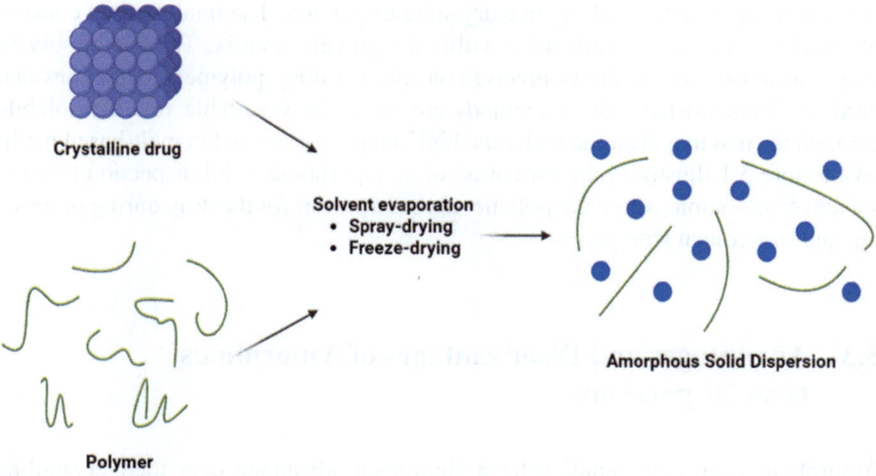


**Fig. 5.1** Crystalline drug co-formulation with an amorphous polymer to form ASD

[8]. First-generation solid dispersions generally use crystalline carriers like urea or mannitol, while second-generation solid dispersions utilize amorphous polymers such as polyvinylpyrrolidone (PVP) or hydroxypropyl methylcellulose (HPMC) to form ASDs [12]. Both crystalline and amorphous drugs have been used in solid dispersions, including examples like indomethacin, itraconazole, and nifedipine [12]. However, crystalline systems often suffer from poor solubility due to their strong lattice energies and intermolecular forces [7]. To overcome this limitation, recent studies have focused on formulating ASDs using amorphous forms of the drug instead of their crystalline counterparts [10]. The improved solubility in ASDs is attributed to the lack of a crystalline lattice, leading to higher free energy, weaker intermolecular interactions, and improved wettability [13]. Many studies have shown that ASDs of drugs such as ketoconazole, fenofibrate, and naproxen result in significantly enhanced solubility and bioavailability [9]. ASDs are a widely used strategy to enhance the solubility and bioavailability of poorly water-soluble drugs, particularly those in BCS Class II and IV [4, 6, 14]. In ASDs, the drug is dispersed in a hydrophilic polymer matrix, resulting in improved dissolution due to the absence of a crystalline lattice and reduced intermolecular forces [13]. Commonly used carriers include PVP, HPMC, and Soluplus®, which help maintain the amorphous state and inhibit recrystallisation[7].

Solid dispersions of poorly water-soluble drugs are commonly manufactured using a range of techniques, primarily based on solvent evaporation or mechanical processing. Methods such as spray drying, freeze drying, rotary evaporation, and hot-melt extrusion are widely employed, each offering specific advantages depending on the physicochemical properties of the drug and polymer. The focus of this chapter is on solvent-based approaches, where the drug and polymer are co-dissolved or dispersed in a volatile solvent to form a homogeneous solution. This solution is then subjected to controlled solvent removal, typically via spray drying or rotary evaporation, resulting in a dry solid dispersion. The final product consists of the drug molecularly embedded within the polymer matrix. The use of solvent evaporation techniques allows precise control over drug–polymer interactions and particle characteristics. These methods are particularly suitable for thermolabile compounds that may degrade under the high temperatures used in melt-based methods. Figure 5.1 illustrates the formation of an amorphous solid dispersion through solvent evaporation, where the polymer acts as a carrier for the drug during processing and subsequent storage.

## 5.3 Advantages and Disadvantages of Amorphous Oral Dispersions

Amorphous drug compounds offer a significant advantage over their crystalline counterparts by exhibiting higher apparent solubility and faster dissolution rates [15, 16]. This improvement is largely due to the absence of a crystalline lattice and the higher thermodynamic energy state of amorphous materials, which promotes rapid molecular dispersion upon contact with dissolution media [13]. Additionally, the manufacturing of ASDs often results in reduced particle size and increased surface area, both of which contribute to enhanced dissolution kinetics and improved oral bioavailability [9, 10]. These properties are particularly beneficial for BCS Class II drugs, as demonstrated by formulations of ritonavir, ketoconazole, and naproxen, where ASD technologies have led to measurable improvements in clinical performance [4, 9, 10]. Despite their benefits, ASDs present several critical formulation and manufacturing challenges, the most significant being physical and chemical instability. The amorphous state is metastable, and without adequate stabilization, drug molecules may recrystallize over time, particularly during storage or under environmental stress, potentially reversing solubility and bioavailability gains [13, 16]. In addition, hygroscopicity is a common issue in amorphous systems. Due to their high surface energy and small particle size, ASD powders are prone to moisture uptake, which can induce deliquescence, hydrolysis, or further promote recrystallisation, leading to chemical degradation [17]. From a processing standpoint, amorphous powders often exhibit poor flowability and compressibility due to particle aggregation and cohesive intermolecular forces, complicating downstream operations such as tabletting and encapsulation [18]. Furthermore, the complexity of maintaining physical stability while preserving solubility makes the manufacturing of ASDs more challenging than that of crystalline drug products. While these limitations must be carefully addressed during formulation development, the therapeutic and bioavailability benefits offered by ASDs continue to outweigh their challenges in many pharmaceutical applications [4].

## 5.4 Solvents, Excipients, and Polymers Used in the Manufacture of Amorphous Solid Dispersions

A variety of pharmaceutical solvents, polymers, and excipients are employed in the manufacture of ASDs, with the primary objective of enhancing solubility, bioavailability, and stability. This section outlines the role of these formulation components, beginning with solvents.

### *5.4.1 Pharmaceutical Solvents*

The primary role of solvents in the manufacture of ASDs is to dissolve both the API and the polymer carrier, enabling uniform dispersion before solvent removal traps the drug in an amorphous state. Solvent selection significantly influences drug-polymer miscibility, drying kinetics, process safety, particle morphology, and ultimately the performance of the final product. The impact of solvent properties can be observed in several marketed pharmaceutical products. For instance, Vascepa® (icosapent ethyl) uses ethanol to improve solubility and bioavailability, while Tavneos® (tepotinib) utilizes N-methyl-2-pyrrolidone (NMP). Nexium® (esomeprazole magnesium) is formulated using acetone, Zyflo® (zileuton) employs dimethyl sulfoxide (DMSO) [19], and Xeljanz® (tofacitinib) incorporates polyethylene glycol (PEG) as a solvent [20]. An ideal solvent must be capable of dissolving both the drug and polymer, evaporate rapidly to prevent crystallization, and comply with regulatory and safety standards during manufacturing [21]. Solvent polarity plays a central role in selection; polar solvents like water are suitable for polar solutes, while non-polar solvents such as dichloromethane (DCM) or ethyl acetate are preferable for non-polar compounds [31]. When solubility mismatches exist between hydrophobic APIs and hydrophilic polymers, mixed solvent systems are often required to bridge the polarity gap. Volatility, determined primarily by boiling point, governs drying behaviour. High-boiling solvents like DMSO or water require additional energy input and may necessitate secondary drying, whereas low-boiling solvents such as acetone and dichloromethane enable rapid evaporation and efficient drug entrapment in the amorphous matrix with reduced thermal stress [32].

From a regulatory standpoint, solvent choice is guided by ICH Q3C classifications based on residual toxicity [22, 23]. Class 1 solvents (e.g., benzene, carbon tetrachloride) are considered unacceptable due to their high toxicity. Class 2 solvents (e.g., methanol, acetonitrile, DCM, and tetrahydrofuran) may be used under stringent residual limits, while Class 3 solvents (e.g., ethanol, acetone, ethyl acetate) are generally favoured in pharmaceutical applications due to their lower toxicity profiles and ease of removal. The intended dosage form further affects solvent suitability. For example, topical formulations may benefit from solvents like ethanol or isopropyl alcohol (IPA), which also aid in skin penetration. Injectable formulations require minimal residual solvents and exclude toxic options altogether. Inhalation products permit only Class 3 solvents due to pulmonary sensitivity, while oral ASDs can tolerate a broader range of solvents as long as residual levels remain compliant with regulatory guidelines [22, 23]. Environmental impact is another growing consideration. Safer, biodegradable solvents such as water and ethanol are generally preferred over halogenated or toxic species for sustainability and reduced environmental persistence. Solvent selection also influences the microstructure and crystallinity of the final ASD. Differences in solvent polarity can result in changes to product architecture—for example, amorphous lactose is more likely to form from water, while crystalline lactose may result from ethanol processing [24]. Additionally, fast-evaporating solvents can lead to surface enrichment or phase separation, both of which can affect long-term stability and performance.

When spray drying ASDs, the choice of solvent is crucial since it affects not only the immediate ability to produce a homogenous drug-polymer solution but also the end product's safety, quality, and compliance with regulations. Table 5.1 lists most of the solvents used for this purpose and includes some properties of each solvent. Class 3 solvents-such as water, ethanol, acetone, ethyl acetate, etc. are typically chosen for pharmaceutical formulations due to their relatively low toxicity and safety profiles [25]. To meet a variety of formulation requirements, they provide a range of polarity alternatives, with water at one extreme and acetone/ethyl acetate at the moderately polar/nonpolar end. Class 2 solvents, such as methanol, DCM, and others, can still be utilized in rare situations or in early development, but their usage requires rigorous elimination down to ppm levels and good justification [26]. Solvents such as NMP and chloroform have faced regulatory scrutiny and bans in certain jurisdictions due to concerns about toxicity and environmental impact. The European Union has particularly stringent regulations regarding the use of harmful solvents in pharmaceuticals. The industry is moving towards reducing or doing away with Class 2 solvents, frequently by adopting mixed-solvent techniques or "green" solvents. Unconventional solvents, such as DMSO or novel deep eutectic solvent (DES), are occasionally investigated to address difficult solubility issues. Nevertheless, these present process difficulties and are rarely used in production. In the future, solvent consumption may be minimized or abolished thanks to innovative methods like supercritical fluid spray drying, which may completely avoid some solvent-related problems. The solvent must always provide an amorphous product with a suitable residual solvent concentration and performance, and it must be compatible with the API and excipients. Formulators can choose a solvent solution that meets the practical requirements of manufacturing while producing a stable, effective amorphous dispersion by carefully balancing safety/regulatory limitations with physical attributes (polarity, volatility). As a guide to the range of solvents available for spray-dried ASDs across various application domains, the thorough comparison above emphasizes that the "best" solvent depends on the specific context, including the chemistry of the API and excipients and the final route safety requirements. The ongoing development of new formulations and processes increases the scope of what is possible while upholding the crucial values of patient safety and product quality [23].

**Table 5.1** Examples of solvents used in the formulation and manufacturing of ASDs [18, 20, 25, 26]

| Solvent | Boiling point (°C) | Advantages | Disadvantages | Common use | ICH Q3C class | Concentration limit (ppm) |
|---|---|---|---|---|---|---|
| Water | 100 | Non-toxic | High boiling point | Inhalation and oral ASDs | Not classified | Safe |
| Ethanol | 78.5 | Low toxicity | May not dissolve extremely nonpolar API | Inhalation and oral ASDs | Class 3 | No limit |
| Isopropanol | 82.5 | Fully miscible with water | Slightly less effective for very polar solutes | Alternative to ethanol in oral ASDs | Class 3 | No limit |
| Methanol | 64.6 | Very volatile (Dries quickly) | Must limit residual due to toxicity | Often used with DCM or others as a cosolvent in research | Class 2 | 3000 |
| Acetone | 56.2 | Low boiling point (dries quickly) | Highly flammable | Oral ASDs | Class 3 | No limit |
| Methyl ethyl ketone (MEK) | 79.6 | Fast drying | Partial water miscibility | Oral ASDs: Alternative to acetone | Class 3 | No limit |
| Dichloromethane (DCM) | 39.8 | Excellent for lipophilic drugs; Ultra-volatile (dry at low temp) | Toxic and regulated; environmental hazard | Use in oral ASDs (lab and some older products) | Class 2 | 600 |
| Chloroform | 61.7 | Dissolves very nonpolar drugs and some polymers | Highly toxic (carcinogen) | Banned for pharma ASDs; academic research; avoided in manufacturing. | Class 2 | 60 |
| Tetrahydrofuran (THF) | 66 | Fairly volatile (dries easier than DMSO) | Toxic requiring limits | Limited to R&D trials for ASDs | Class 2 | 720 |

(continued)

**Table 5.1** (continued)

| Solvent | Boiling point (°C) | Advantages | Disadvantages | Common use | ICH Q3C class | Concentration limit (ppm) |
|---|---|---|---|---|---|---|
| Acetonitrile | 82 | Good polarity (dissolves both polar and nonpolar aspects) | Toxic; relatively expensive | Rare in ASD prep; not common in production | Class 2 | 410 |
| Ethyl acetate | 77 | Green solvent | Not as polar (Won't dissolve very hydrophilic polymers) | Oral ASDs and injectable particle manufacturing | Class 3 | No limit |
| DMSO | 189 | Non-volatile at room T (easy handling) | Extremely high boiling point (hard to dry) | Used in early formulation screening | Class 3 | No limit |
| NMP | 202 | Dissolves tough polymers | Very high boiling; Toxic | Only as lab solvent for screening | Class 2 | 530 |

## 5.4.2 Pharmaceutical Polymers

It has already been indicated that the formulation of ASDs involves combining an API, a solvent, a polymer, and sometimes an extra excipient [4]. From that, it can be deduced that the API is an active ingredient that provides the therapeutic effect, whereas the other two or three ingredients are extra functional entities that provide a process and functional medium [24]. The solvent is used to dissolve and mix the otherwise insoluble API [5]. The polymer is used as a dispersant and can improve solubility and stability [2]. As a result, it contributes to the improvement in the bioavailability of the formulation [27]. Some of the examples of pharmaceutical products that are formulated using various polymers as part of the amorphous solid dispersion are griseofulvin, itraconazole, and fenofibrate [6].

Only a small number of polymers are often utilized in FDA-approved ASD medications (2012–2023), according to a recent survey (Table 5.2). Notably, copovidone (PVP-VA) is included in around 49% of products, and hypromellose acetate succinate (HPMC-AS) is present in about 30% [28]. Like most ASD carriers, these polymers have official monographs that represent their approved use in pharmaceutical goods and are generally recognized as safe (GRAS) pharmaceutical excipients [29]. The chemical classes of pharmacopeial polymer excipients frequently used in ASDs include cellulosic polymers (such as various hypromellose derivatives), vinyl polymers (such as polyvinylpyrrolidone and its copolymers), acrylate/methacrylate copolymers, and specific polyethylene glycols or surfactant copolymers [29]. The glass transition temperature ($T_g$), hygroscopicity, aqueous solubility profile, and other physicochemical characteristics of each polymer determine how well it performs as an ASD carrier [30]. The optimal ASD polymer generally possesses a low crystalline content, a relatively high $T_g$ to maintain the drug–polymer matrix glassy at room temperature, the capacity to form particular intermolecular interactions like hydrogen bonds with the drug, and appropriate solubility in the intended delivery environment for quick drug release [31]. Crucially, the polymer needs to be safe from a chemical and toxicological standpoint. In fact, many ASDs based polymers are inert, non-toxic, and not absorbed from the gastrointestinal tract [29].

Strong supersaturation "springs" are created when Vinyl lactams, like PVP and PVP-VA, dissolve quickly in stomach fluids and stabilize APIs through hydrogen bonding; nevertheless, their hygroscopicity and reduced crystallization inhibition sometimes call for high loadings or ternary systems [30]. Cellulose derivatives (HPMC, HPMC-AS) combine strong nucleation suppression with lower hygroscopicity and greater $Tg$. Widely utilized in authorized ASDs, HPMC-AS has pH-dependent solubility and works especially well for weak bases that need intestinal release. High $Tg$ is provided by older enterics (HPMCP, CAP), however they are fragile and prone to hydrolysis [30, 33]. Eudragit® methacrylate copolymers increase functionality, cationic E allows for gastric release and taste masking, but precipitates at neutral pH, while anionic grades (L/S) offer pH-triggered intestinal release. Instead of improving bioavailability, neutral RL/RS play specialized roles in sustained-release matrices [30]. By plasticizing high-$Tg$ polymers and improving

**Table 5.2** Commonly used polymers in the preparation of amorphous solid dispersions [28, 30–32]

| Polymer (trade names) | $T_g$ (°C) | Solubility profile | Typical uses and routes |
|---|---|---|---|
| Polyvinylpyrrolidone (PVP) (Kollidon®, Plasdone®) | ≈150–160<br>≈177 | Water-soluble at all pH; dissolves rapidly in GI fluid | Oral tablets/capsules; solution binder; some topical films. Limited use inhalation |
| Copovidone (PVP-VA) (Kollidon® VA64) | ≈105 | Water-soluble (forms colloidal solutions); pH independent | Oral ASD tablets/ capsules (most common ASD polymer in products). Also used melt extrusion |
| PVCap-PVAc-PEG graft copolymer (Soluplus®) | ≈72 | Water-soluble (self-assembling amphiphile). pH independent | Oral ASDs via HME or spray-drying |
| Hydroxypropyl methylcellulose (HPMC) (Methocel®, etc.) | ≈120–150 (varies by grade)<br>≈139 (for low Mw) | Water-soluble (swells to gel); pH independent | Oral ASDs; often combined with enteric polymer. Used in ER matrix tablets, ocular lubricants |
| HPMC acetate succinate (HPMC–AS) (AQOAT®, Affinisol®) | ≈120–130 | Enteric: insoluble pH <5, dissolves pH ≥5–6 | Oral ASDs - prevalent in modern ASD drugs |
| HPMC phthalate (HPMCP) (HP-50, HP-55) | ≈143–147 (HP50/55) | Enteric: dissolves pH ≥5.0–5.5 | Oral enteric coatings; used in some ASDs |
| Cellulose acetate phthalate (CAP) | ≈175 | Enteric: dissolves pH >6.0 | Enteric coating for tablets/capsules |
| Amino methacrylate (Eudragit® E) | ≈50–55 | Soluble at pH <5 | Oral: taste-masked granules, chewables; ASD if gastric targeting needed |
| Methacrylic acid–methyl methacrylate Copolymers (Eudragit® L100–55 / L100 / S100) | L100–55: ≈111<br>L100: ≈195<br>S100: ≈173 | Enteric: dissolve at pH ≥5.5 (L100–55), ≥6.0 (L100), ≥7.0 (S100) | Oral: enteric-coated dosage forms; sometimes as ASD matrix alternative to HPMC-AS. S100 is mainly for colon targeting |
| Ammonio methacrylate copolymers (Eudragit® RL/RS) | ≈65–75 | Insoluble, pH-independent; swells and is semi-permeable | Oral sustained-release matrix or coatings. Not used for bioavailability enhancement |
| Polyethylene glycol (PEG)<br>Polyethylene oxide (PEO) | ≈55–65 (crystalline)<br>≈20 (amorphous) | PEG: very water soluble; PEO: water soluble but high viscosity; both dissolve at all pH | As a carrier in early ASDs (fusion method). Now mainly as a minor component in ASDs (plasticizer). PEO in OROS®/matrix tablets (not for ASD effect but ER). |

(continued)

**Table 5.2** (continued)

| Polymer (trade names) | $T_g$ (°C) | Solubility profile | Typical uses and routes |
|---|---|---|---|
| Poloxamers (Poloxamer 188, 407) (Pluronic F68/F127) | ≈52–57 (soft wax) | Amphiphilic, self-assembling. Freely soluble in water; pH independent | As a minor excipient in many ASDs to improve wetting. Primary matrix in a few melt dispersions. Used in injectable gels (407) |
| PLGA (poly(lactide-co-glycolide)) (Resomer®, Lactel®) | ≈45–55 (50:50) | Water insoluble. Slowly hydrolyzes; soluble in organics | Long-acting injectable microspheres, implants. Not used orally (polymer does not dissolve) |

wetting, adjuvant excipients like PEG and poloxamers facilitate processing and dissolution; however, their low *Tg* and crystallinity restrict their use on their own. Long-acting injectable methods for parenteral ASDs are made possible by PLGA, which stabilizes the medication in a degradable matrix but presents problems such as burst release and acidic microenvironments [19].

Everything from feed preparation to atomization/evaporation and, finally, the performance of the drug formulation is governed by the choice of polymeric carriers for spray-dried ASDs [19]. Amorphous drug kinetic stabilization through specific (H-bond) and nonspecific (van der Waals) interactions, carrier dissolution behaviour, and its effect on the "spring-parachute" profile in relevant biorelevant media, and regulatory constraints are the three interlocking criteria that formulators balance in practice [19]. The goal of data-driven polymer selection and blending, as well as next-generation amphiphilic polymers, is to increase drug loading while preserving physical stability and supersaturation. This is frequently accomplished by using hydrophobic segments and customized H-bond donors/acceptors to match API chemotypes. High-throughput miscibility screens and machine-learning-guided solvent/polymer design are increasingly supporting these tactics, which are often used as ternary systems, speeding up the transition from pre-formulation to reliable, scalable spray-drying processes [33, 34].

## 5.5 Solvent Evaporation Techniques in Drug Manufacture

### *5.5.1 Spray Drying*

Spray drying involves dissolving the drug and polymer in a common solvent and then atomizing the solution in a heated chamber [35]. The solvent evaporates, leaving behind a solid dispersion [19]. The particles of the powders from the dissolution are generated by being forced through the atomizer, whose orifice determines the size and shape of the particles [16]. The powder is atomized and forced through the orifice while hot gas is applied, and this leads to the evaporation of the solvent, leaving behind the drug-polymer solid particles, which are amorphous in nature [11]. The reason why it is possible for the drugs formulated by spray drying to result in amorphous powders is that the process of atomization and drying is so fast that the process does not allow gradual particle formation, which would normally lead to crystallization [36]. In addition, the use of a supersaturated feed solution favours the formation of amorphous powders [37]. The process starts with the preparation of a stock or feed solution in the chamber [35]. This solution is then atomized into droplets using different atomizers, such as rotary atomizers, pressure nozzles, or two-fluid nozzles [19]. Once the droplets are created through the atomizer or nozzles, they are introduced to a dry chamber where they are dried by gas [16]. While the powder particles are being dried, they are collected using a cyclone, a separator bag, or an electrostatic separator [11]. As the gas is used to dry the powder particles, it also absorbs the liquid and pumps it out of the chamber through the exhaust system [36]. The remaining product is a dry solid powder that consists of a drug-polymer or drug-polymer-excipient mix [37].

Some examples include treatments for cystic fibrosis such as Orkambi (Lumacaftor/Ivacaftor) and Kalydeco (Ivacaftor); HIV treatments such as Intelence (Etravirine) and Reyataz (Atazanavir); and antifungal products such as Sporanox (Itraconazole) and Onmel (Itraconazole) [11]. Some of the obvious advantages of the use of spray drying in the manufacture of amorphous solid dispersions are the formation of uniform particles and improved dissolution rates, while the disadvantages include environmental challenges related to the solvents used in the process and the removal of residual solvents from the particles [35].

Several marketed pharmaceutical products have successfully employed spray-dried ASDs technology to enhance the bioavailability of poorly soluble drugs. These formulations typically utilize hydrophilic or enteric polymers—such as PVP-VA, HPMCAS, or HPMC—as carriers to stabilize the amorphous form and prevent recrystallization. While solvent systems such as acetone–water or ethanol–water are commonly used, specific solvent choices for some products remain undisclosed in the public domain. Nevertheless, all entries are substantiated by regulatory documentation, peer-reviewed literature, or patent filings, offering a comprehensive overview of the current use of spray-dried ASDs in commercial medicines.

#### 5.5.1.1 Mechanisms of Drying in Spray Drying

Spray drying is a very advanced technology tool that converts liquids containing dissolved or suspended solids into dried powders. The liquid spray solution is delivered to the spray dryer, where it is atomized into fine droplets, enhancing heat and mass transfer, as well as providing control of the final particle size. Drying takes place instantly as the heated drying gas concurrently enters the drying chamber and encounters the droplets, causing rapid evaporation and solidification of dried particles. Dried particles are then separated from the drying gas using a cyclone collector. The drying gas is then usually sent through a condenser to remove any excess solvent, enabling the reuse of the drying gas and reducing gas consumption at a large scale. Figure 5.2 provides a general overview of the spray drying process [38].

Spray drying is a valuable and adaptable method with a wide variety of industrial applications. This method can produce powders with regulated particle characteristics, such as size and shape. In the food industry, spray drying is used to increase the shelf life as well as the easy handling of products [39]. Furthermore, spray drying has been a vital contributor to the manufacturing of dried powdered milk in the dairy industry [40]. The dairy sector uses spray-drying to make a variety of products, such as coffee powder, milk powder, and much more on a big industrial scale whilst considering the nutritional value [40].

In addition, spray drying is crucial in the pharmaceutical industry. This method is used to produce controlled release of vaccine, production of ultra-fine powders for pulmonary drug delivery, injectables for intravenous and intramuscular drug delivery, as well as lipid dry particles or nanoparticle-loaded microspheres that may be utilized as drug encapsulation and delivery vehicles [41]. Compared to other drying techniques, spray drying fluid extracts provide a product with superior properties since it creates homogeneous powders, which is more favourable for the industry [42, 43].

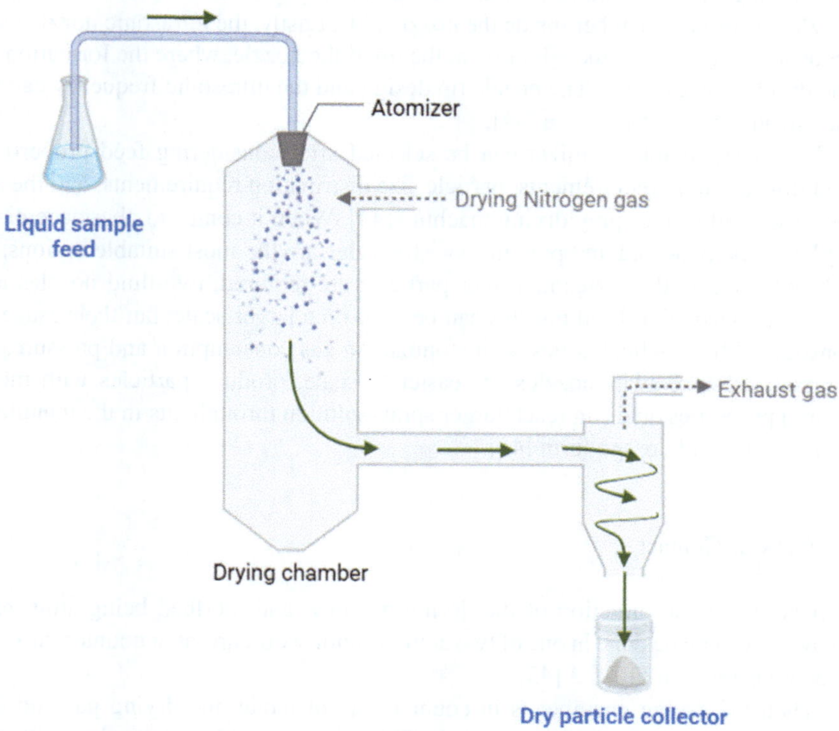


**Fig. 5.2** Simple diagram of spray drying process

The process of a spray dryer is divided into four main steps: Atomization; Droplet-gas contact; Droplet drying and particle formation; Separation of solid particles from the wet drying gas [44].

### Atomization

Atomization helps the formation of a powder or liquid suspension in a gas, or for a liquid, to the concurrent decrease in particle size. The dispersion is created within a drying gas during spray drying, which aids the formation of very large surface areas that are visible to the drying gas. This huge surface area allows heat transmission from the heated drying gas to the atomized droplets, resulting in solvent evaporation [43]. The atomization phase is the only method that allows the formation of particles with desired physicochemical and morphological characteristics.

There are different device types that can be used in this step, including: Rotary atomizers comprise a fast-spinning revolving wheel or disc that distributes the liquid into tiny droplets. They are widely employed in sectors like dairy processing and are very good at creating tiny droplets [43]; Hydraulic pressure nozzles use high forces to break the liquid into fine droplets, which allows for more control over the droplet size [43]; Pneumatic two-fluid nozzles use a gas stream to atomize the feed. The mixing of the liquid stream and the atomizing gas can be performed in the nozzle tips or in a chamber inside the nozzle [44]. Lastly, the ultrasonic nozzle uses acoustic energy to start the vibration at the tip of the nozzle, where the formation of fine droplets takes place. The nozzle tip design and the ultrasonic frequency can be used to adjust the droplet size [44].

The most suitable atomizer can be selected after considering feed properties, feed flow capacity requirements, particle size distribution requirements, and the air dispenser unit of the spray drying machine [44]. When it comes to pharmaceutical applications, two-fluid and pressure-swirl nozzles are the most suitable options. If reduced solution flow rates and finer particles are required, two-fluid nozzles are usually utilized. Two-fluid nozzles can be used on a larger scale, but their usage is constrained by practical issues with atomization gas consumption and pressure. In contrast, pressure-swirl nozzles are easier to scale, produce particles with more desired properties, and can reach larger spray-solution throughputs in the manufacturing of the oral dosage form [45].

### Droplet Gas Contact

Depending on the direction of the drying gas flow and the feed being atomized, spray drying can be done in one of two configurations: co-current or counter-current, as demonstrated in Fig. 5.3 [43].

When the drying chamber is in counter-current mode, the drying gas and the atomized feed come from different ends. This arrangement lengthens the residence time within the drying chamber. The pharmaceutical sector uses co-current setups

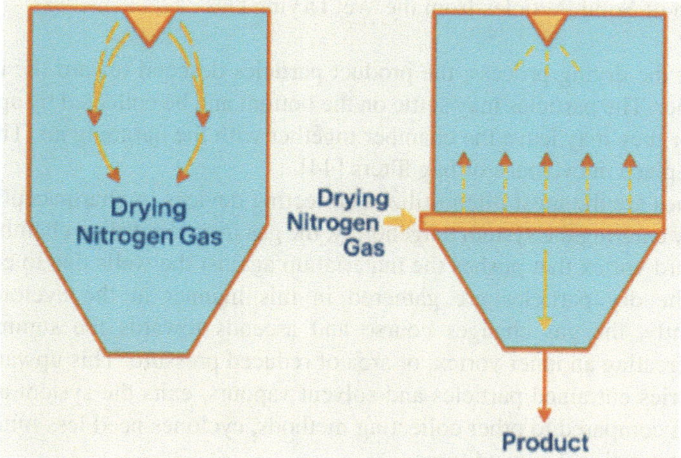


**Fig. 5.3** A diagram showing the co-current (left) and the counter-current (right) flow of nitrogen gas, hence the direction of droplet contact

more frequently than the counter-current due to the shorter residence duration, which results in less heat stress on the particles [43].

Droplet Drying and Particle Formation

Gas dispersers are critical components of the spray-drying unit because they establish the airflow pattern within the drying chamber, particularly near the nozzle. They affect droplet/particle collisions, particle residence times, and drying efficiency. Gas dispersers are situated right before the drying gas enters the chamber. To ensure that the droplets dry uniformly, the drying gas flow inside the chamber must be distributed evenly. If not, wall deposits may form, resulting in reduced yields.

Advancements in computational fluid dynamics led to changes in gas disperser design, which allowed for significant advancements in this field. The spray dryer's arrangement and the atomization nozzle that will be placed determine which design is best. Given that co-current spray driers with pressure or pneumatic nozzles are the most often utilized in the pharmaceutical sector.

The dispensers can be divided into two groups depending on the flow pattern. In one arrangement, the atomization nozzle is surrounded by a rotating pattern. While the solvent evaporation capacity is increased by this rotation, the powder at the top of the chamber is encouraged to circulate, which might result in product deposits on the ceiling. In contrast, another kind of disperser seeks to produce a streamlined flow that runs parallel to the vertical axis of the chamber. Because of the way this sort of dispersion flows, the chambers connected to it are often higher than those connected to rotating dispersers [43].

Separation of Solid Particles from the Wet Drying Gas

Following the drying process, the product particles descend toward the bottom of the chamber. The particles may settle on the bottom and be collected by appropriate devices, or they may leave the chamber together with the outgoing air. The separation takes place in cyclones or bag filters [44].

Cyclones are the most often utilized collecting devices in pharmaceutical spray drying. By entering the system tangentially, the gas from the drying chamber creates a downward vortex that pushes the material up against the walls due to centrifugal forces. The dry particles are gathered in this manner at the cyclone's base. Concurrently, the gas changes course and ascends towards the summit of the cyclone, creating an inner vortex, or area of reduced pressure. This upward stream, which carries entrained particles and solvent vapours, exits the system as exhaust gas. When compared to other collecting methods, cyclones need less initial investment and operational expenditures.

Usually installed downstream of the cyclone in spray-drying systems, filter bags remove entrained particles from the exhaust flow. This keeps the powder from escaping into the surroundings or polluting supporting machinery. Filter bags are made up of many bags that are inserted into a stiff container. Typically, a dry clean-in-place system is built to enable ongoing operation. This works by dislodging the powder that has become entrained in the bags with a pulsed jet of pressurized gas. The efficacy of these cleaning methods varies depending on the substance being spray-dried and the material used to make the bags. When compared to cyclones, filter bags need a larger initial investment and ongoing maintenance, but if they provide a noticeably higher result, the balance works in favour of expensive goods [44].

Lastly, electrostatic precipitators create an electric field. The spray-dried powders pass through this electric field, causing the particles to stick to oppositely charged plates. This is usually used to collect nanoparticles. This method is rarely used for large-scale industry production due to its cost. Other collection systems provide higher yields considering the cost (Table 5.3) [44].

Spray dryers are available in a variety of forms and combinations, each tailored to meet needs and requirements. One study has shown that agitated bed driers mix the powder bed and improve drying efficiency. An impeller distributes the powder bed within the drying vessel to provide even product contact with the heated surfaces of the chamber. Heat transfer in this system is therefore governed by conduction. This type of dryer offers the benefits of vacuum oven drying with the additional advantage of improved heat conduction [44]. They also suggested that rotary dryers

**Table 5.3** Collection rates for all collection systems [44]

| Particle size | 10 μ | 5 μ | 1 μ |
|---|---|---|---|
| High-efficiency cyclones | 95–100% | 90–95% | 10% |
| Filter bags | 100% | 100% | 99% |
| Electrostatic precipitators | 100% | 99% | 86% |

consist of a rotating heated drying chamber to constantly tumble the powder bed for enhanced heat transfer. The drum is typically equipped with baffles to evenly distribute the powder along the walls of the chamber and improve drying efficiency [44]. Fluidized spray drying (FSD), another method which utilizes fluid bed and spray drying in a single system. Following atomization, the droplets are added to a bed of fluidized particles, where they congregate and dry to produce bigger, more regulated particles. In addition, FSD decreases the fine contents of the spray-dried products, which improves powder flowability. Legako et al. pointed out that a two-fluid nozzle produces tiny droplets by using both liquid and gas streams. Exact control over droplet size and dispersion is made possible by this method [46].

#### 5.5.1.2 Three-Fluid Nozzle Spray Dryer

The three-fluid nozzle sprayer dryer is not commonly used in the pharmaceuticals industry compared to other types of spray dryers. A three-fluid nozzle (3FN) has two separate channels, which allow spray drying of materials in two incompatible solution systems Fig. 5.4. This nozzle has been used to produce amorphous solid dispersions (ASDs) and has been reported to deliver comparable drug dissolution performance relative to those prepared using different nozzle types [47]. ASDs prepared using a 3FN have been reported to deliver comparable drug dissolution performance relative to those prepared using a two-fluid nozzle (2FN). However, few studies have systematically examined the effect of a 3FN spray drying on the physical stability of ASDs upon storage [47]. This demonstrates the importance of using a 3FN in the pharmaceutical industry to improve the bioavailability of ASDs. Furthermore, when the two components were spray-dried using a 3FN, the individual solutions of drug and polymer remained stable during the process.

In addition, a three-fluid nozzle has been used to construct high-performance complex nanoparticle clusters (CNCs) consisting of different functional nanofillers for dental restoration. The application of a three-fluid nozzle can effectively avoid

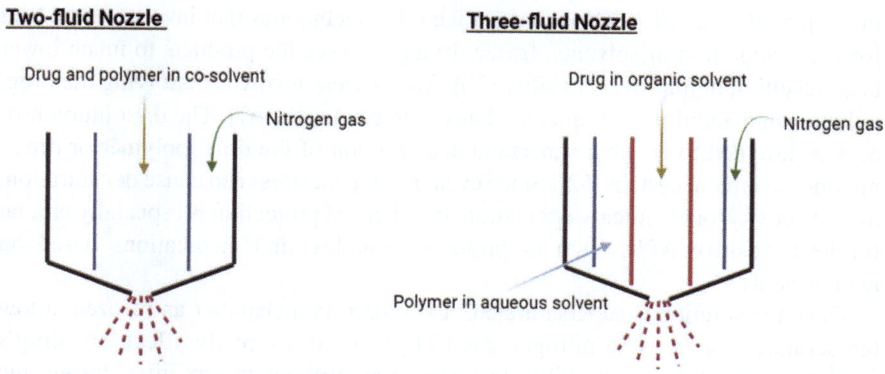


**Fig. 5.4** A simple zoomed-in schematic of the two types of nozzles (two-fluid and three-fluid) used in spray drying

the aggregation of different nanoparticles with opposite zeta potentials before the spray drying process to construct regularly shaped CNCs. Any CNCs that were produced by a three-fluid nozzle demonstrated to maintain exceptional mechanical properties [48]. In contrast, the use of a two-fluid nozzle displayed weak mechanical properties. The use of a three-fluid nozzle avoids the aggregation of incompatible particles before the spray drying process, which may occur with the use of a traditional two-fluid nozzle due to its single feed line and nozzle [48]. This is a clear example of the versatile use of a three-fluid nozzle and the effective use of this method across many industries.

A recent study explored the use of a 3FN in spray drying for microcapsule production [49]. The findings demonstrated that the stability of the emulsion fed into the inner channel of the 3FN was a critical factor in achieving lower surface oil and a more stable reconstituted emulsion. When a stable emulsion was used in the inner channel, increasing the concentration of wall material in the outer channel had only a limited effect on improving encapsulation efficiency.

The 3FN system was effective in reducing surface oil by introducing an additional layer of wall material through the outer channel. Powders produced using the 3FN exhibited smaller particle sizes compared to those produced with a conventional 2FN, while showing no significant differences in surface morphology or internal structure. Additionally, microcapsules produced with the 3FN demonstrated greater stability during 24-h refrigerated storage. Overall, the study highlights the 3FN as a promising microencapsulation technique with strong potential for encapsulating a variety of functional ingredients [49].

### 5.5.2 *Freeze Drying*

Freeze drying, which is also known as lyophilization, has been used in the formulation and manufacture of ASDs of drug-polymer or drug–polymer-excipient pharmaceutical systems. One of the primary reasons for using the freeze-drying technique in the manufacture of ASDs is that, unlike other techniques that involve direct heat for the evaporation of solvents, freeze drying exposes the products to much lower heat, resulting in improved stability [50]. The process involves dissolving the drug, polymer, and additional excipients, if any, into a solution [51]. The dissolution process is designed to create a supersaturated solution of the drug–polymer or drug–polymer-excipient system [52]. Since even minor heat stress can cause denaturation, loss of activity, or chemical degradation, this thermal protection is especially crucial for heat-sensitive APIs, such as proteins, peptides, and medications based on nucleic acids.

Once the solution is supersaturated, it is placed in a chamber and frozen at low temperatures, often using nitrogen gas [50]. Since this directly affects the drug's solubility, miscibility, and ability to form an amorphous matrix after drying, the crucial solvent system must be carefully selected. Rapid freezing by nitrogen gas leads to the formation of very small particles [53]. Rapid cooling locks the system in a high-energy, amorphous state by preventing the drug from recrystallizing and undergoing phase separation. The first stage of drying is performed by sublimation, where a vacuum is created while heat is gently applied to the chamber containing the frozen solution [51] (Fig. 5.5). This process changes the frozen solution into a solid without becoming liquid. The vacuum in the frozen chamber and gentle heating ensures the formation of small, amorphous particles, avoiding the formation of large crystals [54]. In order to prevent the collapse of the porous structure created during freeze sublimation, this step is the most energy intensive which requires careful management of the chamber pressure and shelf temperature.

The sublimation process, also known as the first drying stage, often leaves residual liquid. A second drying stage, called desorption, is then applied. During this process, the temperature is increased, and a vacuum is maintained to remove any remaining solvent particles. At this point, the amount of bound water molecules and organic solvent residues is reduced to levels that meet the International Council for Harmonization's (ICH) solvent residue requirements [55]. Because of the lower heat exposure and smaller particle sizes produced during freezing, freeze drying provides better product stability as compared to other solvent-evaporation techniques like spray drying [50]. The stability of freeze-dried products is due to the limited application of heat during processing, which reduces the likelihood of product degradation [53]. Furthermore, the freeze-dried powder's porous structure enhances dispersibility and solubility rates upon reconstitution, which is advantageous for medications that are not very soluble in water.

Common examples of freeze-dried ASDs on the market include medications for HIV such as Kaletra® (Lopinavir/Ritonavir), antifungal medications such as Vfend® (Voriconazole), oncology drugs such as Iressa® (Gefitinib), anti-inflammatory medications such as Celebrex® (Celecoxib), and antibiotics such as Zerbaxa® (Ceftolozane/Tazobactam) [54].

The most important advantage of using freeze drying, compared to other solvent evaporation techniques, is that it produces a more stable product because the lower heat and freezing process result in smaller, amorphous solid particles [50].

Despite these advantages, freeze-drying involves multiple steps and is considered relatively expensive to run [55]. Additionally, challenges exist in scaling the lyophilization process and achieving complete solvent removal. Furthermore, its industrial efficiency is limited by lengthy processing times, which can range from a

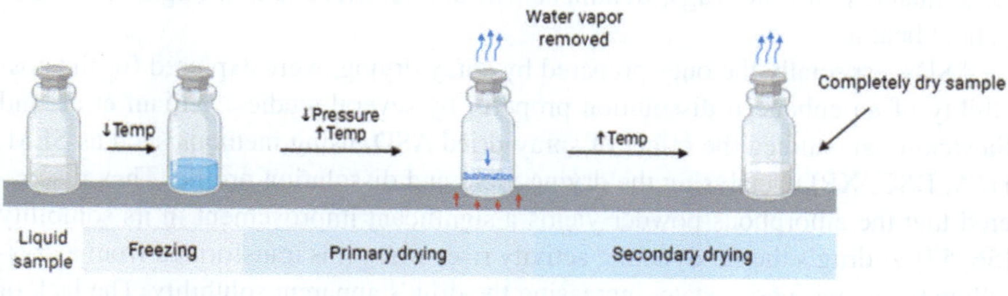


**Fig. 5.5** Diagrams showing typical steps of freeze-drying

few hours to several days, and the economic viability of medication classes is constrained by high manufacturing costs. By reducing cycle times and enhancing process uniformity, recent developments like controlled nucleation and microwave-assisted lyophilization technologies are being researched as potential solutions to these problems.

## 5.6 Dissolution and Dispersion Behaviour of Spray-Dried Amorphous Solid Dispersions

### *5.6.1 Oral and Pulmonary Delivery*

About 40% of commercialized oral drugs and 70–90% of drug candidates in development have poor water solubility. As suggested by Hancock et al. and discussed in the above sections, crystalline drugs often exhibit poor dissolution rate. Therefore, forming ASDs can be one of the best methods to improve water solubility and drug absorption, which can increase the bioavailability of these poorly soluble drugs [16]. In the pulmonary delivery route, faster dissolution rate, systemic uptake, and quicker onset are the benefits which inhaled drug will have from creating an ASDs structure. In terms of methods, spray-drying fits in perfectly as it can incorporate all the components of the drug (e.g., API, excipient, polymer, coating) with the benefit of the possibility of microencapsulation. It can also be engineered for the perfect aerodynamic particle size for the best flowability for pulmonary delivery.

#### 5.6.1.1 ASDs in Oral Drug Delivery

When a drug is taken by the oral route, it is exposed to GI fluid. In order to be absorbed into the bloodstream, it must first dissolve into the aqueous environment of the GI fluid. A poorly soluble drug will not be able to fully dissolve into the fluid, which results in less bioavailability and less therapeutic effect. To reach the desired potency, the formulator will often increase the content of each tablet or capsule, so even if the drug is not fully absorbed, the amount absorbed will still have an optimal therapeutic effect. This will cause wastage and is costly. If the patient cannot take a large quantity of oral drugs, treatment will not be effective and might jeopardize patient health.

ASDs, especially the ones prepared by spray drying, were explored for the possibility of an enhanced dissolution property by several studies. Wijiani et al. and Sawicki et al. studied the effect of spray-dried ASD, using methods such as SEM, DTA, DSC, XRD, analysing the drying yield and dissolution profile. They discovered that the amorphous powder yields a significant improvement in its solubility [56, 57]. A drug's thermodynamic activity rises when it is transformed from a crystalline to an amorphous state, increasing the drug's apparent solubility. The lack of a long-range lattice in the amorphous structure lowers the energy barrier to breakdown, leading to quicker initial kinetics of dissolution [4]. However, without a proper polymer matrix to stabilize it, this high-energy form is metastable and prone to recrystallization. Through drug-polymer interactions, the polymer not only prevents nucleation and crystal formation but also enhances wettability and dispersibility [30]. Moreover, the drug-polymer has improved miscibility and solubility with the addition of surfactant before spray-drying. The surfactant-containing ASDs exhibited a reduction in tendency for recrystallisation [30]. In order to achieve improved water solubility, a compilation of spray-dried drugs, polymers, and surfactants is crucial in ensuring the best results.

When an ASD is exposed to gastrointestinal fluids, it can produce a supersaturated solution, which is a state in which the concentration of the drug is higher than its crystalline solubility. This is because of the higher thermodynamic energy that ASDs have compared to their crystalline structure. This is known as the "spring" effect and its purpose is to promote absorption [58]. Nevertheless, supersaturation will collapse via precipitation in the absence of an agent of stabilization. As stabilizer, polymers like Soluplus®, hydroxypropyl methylcellulose acetate succinate (HPMC-AS), and polyvinylpyrrolidone (PVP) interact with drug molecules at the molecular level to postpone nucleation (also known as the "parachute" effect) [59]. Recent reviews describe how this balance between supersaturation and precipitation is governed by ionic interactions, hydrogen-bonding potential, and polymer viscosity [59].

A polymer is a crucial component when creating a stable and effective ASD. However, since the pH range of the GI tract is dynamic (about pH 1.5 in the stomach to pH 6.8–7.4 in the intestine), this implication could have a significant impact on ASDs performance [60]. Site-specific release is made possible and drug crystallization in the stomach is reduced by enteric polymers such as HPMC-AS (grades LG or MG), which dissolve at intestinal pH but remain insoluble at gastric pH. Research on weakly basic medications (such as delamanid and nintedanib) demonstrates that enteric polymer-based ASDs reduce surface crystallization and increase absorption by maintaining supersaturation at pH 6.8 more successfully than non-enteric matrices [61]. For accurate in vitro models to forecast in vivo performance of ASDs, the dynamic GI environment must be captured. The methods include: pH-shift dissolution (transition from acid to neutral), biphasic dissolution, which simulates absorption using both organic and aqueous solvents, and biorelevant media use (FaSSIF/FeSSIF) [62]. Monschke et al. suggested that the start and duration of supersaturation are significantly influenced by the polymer ratio, drug load, and particle size. Ketoconazole ASDs with HPMC-AS LG, for instance, showed fast release at intestinal pH but poor dissolution at gastric pH, with smaller particle sizes significantly enhancing dissolution kinetics [62]. This suggested the importance of considering the pH value when choosing the right polymer for the drug of choice when spray-drying ASDs.

The stability of ASDs is important as it is crucial to preserve the amorphous state while storing. There are factors that might influence stability, which are $T_g$, residual solvent content and moisture uptake [30]. Polymers are chosen to guarantee a $T_g$ far above storage temperatures, and spray-dried ASDs usually need to be packaged in low-humidity environments. To reproduce lab-scale morphology and dissolving performance, droplet drying kinetics must be carefully controlled when scaling up from lab to commercial spray dryers. In-line near-infrared spectroscopy is one example of an advanced process analytical technology (PAT) that enables consistent product quality and real-time monitoring [4].

To summarize, oral ASD formulations combat the dissolution-limited absorption of poorly water-soluble medications by utilizing the amorphous state, polymer-mediated supersaturation control, and precision spray-drying. ASDs have made the leap from bench to market, and are still being extended to novel chemical entities in the pharmaceutical pipeline through careful selection of polymers, process parameters, and in vitro model design.

#### 5.6.1.2 ASDs in Pulmonary Drug Delivery

There are many benefits to pulmonary drug delivery, including a large alveolar surface area that enables rapid absorption, reduced enzymatic breakdown, and avoidance of hepatic first-pass metabolism [63]. However, some obstacles must be overcome for effective pulmonary drug delivery, especially for drugs that are poorly soluble in water. Such drugs dissolve poorly in the lungs' limited fluid volume, reducing therapeutic effectiveness. Spray-drying is one of the most effective techniques for overcoming these solubility limitations, as it produces ASDs that greatly increase drug solubility, dissolution kinetics, and pulmonary bioavailability [64]. Drug particles must pass through physical barriers during inhalation in order to reach the intended absorption sites. Aerodynamic properties and size must meet the strict limitations set by the respiratory tract. If deep-lung deposition (alveolar area) is to be achieved, particles should preferably have aerodynamic sizes between 1 and 5 μm. While smaller particles (less than 1 μm) are expelled. Larger particles were to deposit prematurely in the upper airways via inertial impaction. Furthermore, in order to avoid clearance through mucociliary action in conducting airways or macrophage absorption within alveoli, particles of the ideal size that make it to deep lung regions must disintegrate rapidly in the tiny volume of alveolar fluid. Conventional crystalline formulations frequently dissolve excessively slowly, leading to lower bioavailability and sub-therapeutic doses. Therefore, novel formulations such as ASDs, which significantly speed up dissolution, are ideal for pulmonary medication administration [65].

As discussed in previous sections, amorphous materials, due to the absence of a crystalline lattice structure, have significantly higher thermodynamic energies than their crystalline counterpart. This elevated energy increases the initial rate of dissolution, reaching a state of supersaturation [59]. By creating a stronger concentration gradient at the alveolar epithelium, this temporary supersaturation greatly accelerates passive diffusion across biological membranes and improves systemic or local therapeutic effects. Additionally, the addition of polymers to ASDs prolongs the window for drug absorption by preserving supersaturation and reducing fast recrystallisation [64]. This mechanism is vital to pulmonary delivery as it can significantly enhance the bioavailability of a poorly water-soluble inhalation drug.

Using spray drying to precisely regulate the size, porosity, morphology, and surface properties of particles is a crucial formulation step for pulmonary drug administration. Particle characteristics are significantly impacted by factors such as atomization pressure, solvent selection, feed concentration, and inlet/outlet temperatures [19]. This adaptability makes it possible to create particles with ideal aerodynamic properties that are suited for effective lung deposition in the desired locations. Spray drying can ensure that particles remain amorphous and prevent crystallization by the quick evaporation of solvents. Additionally, by lowering interparticle cohesion, excipients like leucine or trehalose can be co-spray-dried with APIs to improve particle dispersibility [66]. This significantly improves aerosol performance by enabling dispersion in the airflow and encouraging uniform lung distribution. In terms of pulmonary delivery, polymers and excipients also have a significant impact on the dissolution and dispersion properties of ASDs. Hydrophilic polymers, such as copolymers like Soluplus®, hydroxypropyl methylcellulose (HPMC), hydroxypropyl methylcellulose acetate succinate (HPMC-AS), and polyvinylpyrrolidone (PVP), are widely used because of their capacity to maintain supersaturation, improve wettability, and stabilize the amorphous form. Furthermore, during spray drying, amino acids like L-leucine improve dispersion by creating a thin hydrophobic coating on particle surfaces, which greatly lowers particle agglomeration and improves aerosolization [67]. This type of surface engineering guarantees quick dissolving in alveolar fluid, effective deep-lung deposition, and ultimately increased bioavailability.

Physical stability and moisture sensitivity are important considerations not only in oral drug delivery but also in the pulmonary delivery route. The primary root causes are moisture-induced recrystallisation and particle agglomeration [63]. However, the enhanced dissolution and dispersion of ASDs come at the cost of reduced physical stability. The metastable amorphous state renders ASDs susceptible to ambient moisture uptake, which acts as a plasticizer, lowering the $T_g$ and speeding up recrystallisation [63]. Therefore, it is essential to regulate environmental moisture, control humidity during storage, and use protective packaging. Additionally, ASD stability can be significantly improved by using polymers with naturally high $T_g$ or low hygroscopicity, which preserves the best possible aerosol and dissolving performance during use and storage.

### 5.6.2 *Antimicrobial Delivery: Targeting Infections, Topical, or Systemic*

Many antibacterial, antifungal, and antiviral drugs (e.g., azithromycin, niclosamide, and itraconazole) are extremely lipophilic and have low water solubility. Rapid dissolution is necessary to reach the optimal therapeutic concentrations at the infection site before clearance or microbial multiplication takes place, whether that location is the thin lung lining fluid, an exudative skin lesion, or on the surface of biofilm. While the addition of a polymer matrix ("parachute") postpones recrystallisation, these spray-dried ASDs increase the drug's thermodynamic activity, creating an instantaneous "spring" of supersaturation upon contact with biological fluid—allowing sufficient time to penetrate biofilm and intracellular reservoirs. As a result, compared to crystalline or solution compounds, the concentration–time window for antibacterial action is longer and wider [28, 59].

#### 5.6.2.1 Pulmonary Delivery (Systemic and Localized)

With spray-drying, an antimicrobial–polymer solution can be transformed in a single step into respirable particles with mass–median aerodynamic diameters (MMAD) ranging from 1 to 5 μm, which is ideal for alveolar deposition. Co-excipient techniques, especially surface-migration of arginine or L-leucine, reduce cohesive forces and produce fine-particle fractions (FPF) greater than 60% in Next-Generation Impactor (NGI) testing [68]. They also speed up wetting and disintegration as soon as the particles touch down on the surface of the epithelium. NGI is a new cascade impactor that simulates the inside of a lung. This is particularly useful since it allows one to see and measure the distribution of particles throughout the respiratory tract, which is crucial data when testing ASDs for localized therapeutic effect.

Niclosamide was tested for its solubility and its interaction as an ASD. The solubility of crystalline niclosamide is less than 0.1 μg $mL^{-1}$, and its oral bioavailability is minimal. Jara et al. showed that over 25 μg $mL^{-1}$ of niclosamide was delivered into hamster lung fluid by a thin-film-frozen ASD within 5 min, resulting in a 3-log decrease in SARS-CoV-2 titres and total survival in a lethal challenge scenario [69]. Building on that achievement, a 2025 study by Su et al. also aimed to inhibit SARS-CoV-2 infection, co-sprayed niclosamide and ivermectin to create a stable co-amorphous powder that, after 6 months at 40 °C and 75% relative humidity, suppressed viral reproduction at <5 μM and maintained FPF ≈ 68% [70]. On that note, Chan et al. suggested that dry powder formulation can improve the bioavailability of azithromycin. The study showed enhanced in vitro deposition as in NGI stages 3–5, 55–70% of the metered dose was deposited by azithromycin ASD microparticles contained in a chitosan-PLGA matrix, which completely dissolved in simulated lung fluid (SLF) in less than 10 minv [71]. As the above studies have shown, formulating drugs as ASDs is often beneficial to bioavailability, particularly for localized respiratory infections.

Although alveolar epithelia lack first-pass hepatic metabolism, they are just as permeable as the mucosa in the small intestine. Near-instantaneous dissolution of the ASD produces a locally supersaturated drug concentration, which is advantageous for localized drug delivery. Furthermore, this effect also causes passive diffusion into pulmonary capillaries and produces systemic exposures that are comparable to those obtained through parenteral or oral routes at a much lower dosage. A 2023 study by Chan et al. demonstrated that all four azithromycin-loaded PLGA nanocomposite microparticle formulations successfully deposited drug in the deep lung regions in vitro, highlighting their potential for treating lower respiratory tract infections such as COVID-19 [71], suggesting that equivalent therapeutic effects can be achieved at substantially lower doses via inhalation. These findings demonstrate that ASDs not only improve targeted local delivery but also exploit the large surface area of the alveoli and provide effective systemic delivery.

#### 5.6.2.2 Topical Dermal and Mucosal Delivery

The majority of cutaneous and mucosal infection sites, in contrast to the lung or GI tract, are distinguished by high protein content, acidic to neutral pH, microliter-scale fluid volumes, and often have oxidative stress. Rarely does the thin layer of moisture in a vaginal fold or the exudate covering a diabetic ulcer exceed 20–50 μL $cm^{-2}$. If the medication remains crystalline, such limited aqueous media cannot dissolve a therapeutically effective dose of hydrophobic antifungals or antibiotics with intrinsic solubilities below 10 μg $mL^{-1}$ [28]. By presenting the API in a high-energy, easily wettable state that dissolves almost at the rate of hydration, spray-dried ASDs are able to overcome this limitation. The resulting transient supersaturation generates sharp concentration gradients that promote diffusion into keratinized tissue and disrupt biofilms. This is called the "supersaturation window," which is essential in fluid-restricted situations where clearance is slow but microorganisms are able to proliferate rapidly. Hence, with the addition of polymer, this window is extended from seconds to tens of minutes, which further impedes recrystallisation [59]. Therefore, the importance of choosing the ideal polymer is obvious. Polymers that produce a flexible, breathable matrix on the skin or mucosa, film-forming or mucoadhesive carriers (such as Soluplus®, Eudragit RS, and hydroxypropyl-β-cyclodextrin) are preferred because they extend residency and sustain drug release [72].

Gajera et al. conducted a study discussing the combination of clotrimazole-Soluplus® nanosuspension. It was spray-dried within the model framework of Quality by Design (QbD) driven by Design of Experiments (DOE) that methodically mapped aspiration, feed rate, and input temperature against key quality parameters such as yield, dispersibility, and residual solvent. According to data extracted from XRPD/DSC, the optimized powder was completely amorphous and dispersed to a median particle size of 230 nm in less than 30 . After in vitro dissolution testing, they found that the ASD form exhibited a higher dissolution rate than the physical mixture, attributable to its elevated Gibbs free energy and high surface-to-volume ratio [73]. Building that idea, the same API was used in a hot-melt-extrusion (HME) study by Sawant et al. to create mucoadhesive vaginal films. Results show the film exhibits tensile strengths >25 MPa, residence periods >6 h, and continuous release of 80% drug over 8 h were all demonstrated by the films on pig vaginal tissue. In vitro experiment uses a rat model of vulvovaginitis to demonstrate total fungal clearance in 3 days as opposed to 6 days for clotrimazole cream in vivo [74].

Regarding packaging requirements, since ASDs are highly moisture-sensitive, commercial packaging typically uses aluminium-foil blisters with desiccant, and formulations are required to maintain a minimum $T_g > 15$ °C above the highest ICH storage temperature, given the hygroscopic nature of amorphous APIs. According to Gajera et al.'s research, ASD-based vaginal films and nail lacquers reduce application frequency by half as compared to traditional creams, increasing adherence in practical situations [73].

#### 5.6.2.3 Oral Delivery

The foundation of systemic antibiotic therapy continues to be oral tablets and capsules, even with the pharmaceutical research and advancement in topical and pulmonary dosing forms. Unfortunately, many contemporary antibacterials (like oxazolidinones), antifungals (like triazoles), and repurposed antivirals are nearly insoluble in gastrointestinal (GI) fluids due to their lipophilic nature, which paradoxically also promotes membrane penetration. About 90% of anti-infective NCEs and 40% of marketed medications belong to BCS classes II or IV, where the main obstacle to absorption is dissolution rather than permeability [18]. When the drug comes into contact with GI media, spray-dried ASDs transform it into a high-energy amorphous form, increasing its thermodynamic activity and producing a burst of supersaturation. The polymer matrix then prevents precipitation long enough for absorption, a process known as the "spring-parachute" effect, which allows for enhanced dissolution and improved bioavailability. The rapid solvent evaporation inherent to spray-drying yields low-density, high-surface-area ASD particles with fast wetting and high dissolution rates in GI fluid.

In recent years, quality-by-design (QbD) toolkits are frequently used in conjunction with spray drying, thin-film freezing, and hot-melt extrusion to map the "design space" for drug loading, polymer ratio, residual solvent, and glass transition temperature. One important study by Xiang et al. utilized the polymer-overlap concentration ($c^*$) model, which predicts the maximum drug loading that preserves physical stability and single-phase miscibility [75]. Using $c^*$ as the governing metric, posaconazole tablets have been able to achieve a 50% drug load without compromising their six-month stability at room temperature. With ≤2 g of API, material-saving DoE screens may now deliver lead ASD tablets in less than 14 days, which is revolutionary for cytotoxic or rare anti-infective leads [75]. This provides further evidence of ongoing advances in this field, where predictive models are increasingly used to guide antimicrobial ASD design.

### *5.6.3 Biofilm Interaction: Disruption or Inhibition via Surface-Active ASD Formulations*

Biofilms are a primary cause of persistent infection wounds, lung infections, and different kinds of systemic infections. Bacterial and fungal biofilms produce extracellular polymeric substances (EPS) that delay diffusion and provide a 10- to

1000-fold increase in antibiotic tolerance [76]. They are surface-associated, cooperative communities embedded in a hydrated matrix of extracellular polymeric substances (EPS). These EPS usually consist of 10–15% macromolecules, primarily proteins, lipids, polysaccharides, and extracellular DNA (eDNA), and 85–90% water. In addition to providing mechanical cohesion and sorbing cations, the EPS matrix produces sharp pH, oxygen, and nutrition gradients that favour physiologically diverse subpopulations [77]. Due to this peculiar limitation, effective therapy consequently depends on how quickly the formulation can create a local supersaturated state that can penetrate and destabilize the extracellular matrix, as conventional crystalline medicines dissolve too slowly to get past this highly resistant barrier. More than 80% of chronic microbial infections, such as diabetic foot ulcers, otitis media, chronic rhinosinusitis, and cystic-fibrosis bronchopneumonia, are caused by biofilm formation. It is also the primary cause of failure for urinary stents, prosthetic joints, and central venous catheters. Long courses of combination antibiotics and implants are frequently required for device-associated biofilm infections, which increases morbidity and medical expenses [78]. Figure 5.6 illustrates the stages of biofilm maturation. At full maturity, the biofilm is at its most resistant and requires external intervention to disrupt its structure, whereas earlier stages are generally reversible. Overcoming the resistant mature biofilm through drug engineering is discussed in the following sections.

#### 5.6.3.1 Why Is Spray-Dried ASD Effective Against Biofilm

Considering this difficulty, using spray drying to create drug ASDs is often the most practical method to tackle this challenge. Drug molecules are trapped in a high-free-energy amorphous state after spray drying; when the powder comes into contact

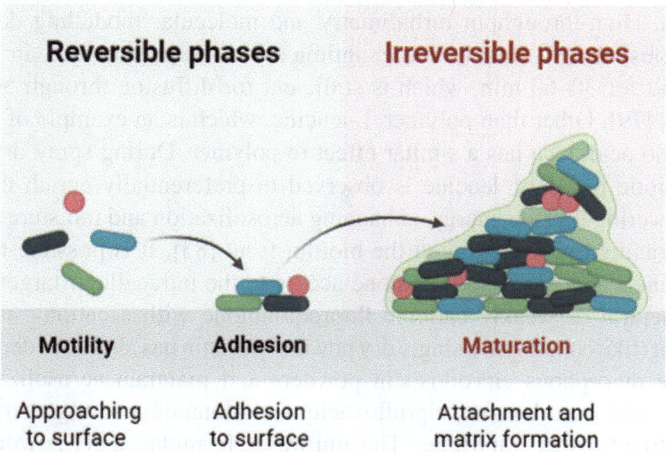


**Fig. 5.6** A diagram of the maturation of biofilm, from free-flowing bacteria to clusters of mature biofilms

with the thin layer of water that covers a biofilm, lattice-breaking dissolution produces drug concentrations that are many times higher than crystalline solubility in a matter of seconds (supersaturation state) [79]. It creates a driving force for passive diffusion through the polysaccharide mesh prior to active efflux or enzymatic breakdown. This mechanism is the foundation of the principle behind the action of ASDs against biofilms. Low surface energy molecules move towards the droplet-air interface during atomization and instant solvent drying process, encircling the amorphous drug core with a nanoscale shell. Chang et al. use leucine, which enhances powder dispersibility without compromising dissolving, reduces surface tension and moisture sorption, and crystallization at the particle surface [80].

Several materials can also be added to weaken the highly resistant EPS in the biofilm. For example, Zhang et al. suggested the disruption of biofilm EPS could be related to the electrostatic interaction between the positively charged $Ca^{2+}$ ion on the biofilm and the negatively charged alginate oligosaccharide (OligoG), creating a cross-linking phenomenon, breaking apart the biofilm. The study reported that co-spray-dried ciprofloxacin/OligoG microparticles (MMAD = 2.3 μm) were successfully produced and proven to reduce the mature biofilm by four logs in 4 h [81]. This novel inhalable formulation could potentially be a promising therapeutic treatment for biofilm-related infection. Building on this context, another study has been conducted by El-Naggar et al., which uses nano-to-micron chitosan-drug ASDs made by spray drying or ionic gelation. Incorporating an artificial intelligence-based optimization model, study results have shown that ASDs can inhibit metabolic activity and matrix protein content in *S. aureus* and *P. aeruginosa* biofilms. Chitosan contains protonated amines that bind anionic eDNA and lipopolysaccharide, which breaks the biofilm structure [82]. The amorphous core dissolves evenly throughout the biofilm surface thanks to these surface-active shells, which help improve wetting and weaken the EPS. Poly(vinylpyrrolidone) (PVP), hydroxypropyl methylcellulose (HPMC), and Soluplus® are examples of hydrophilic carriers that slow down nucleation and recrystallisation by forming hydrogen-bond networks with the drug in solution. High-throughput turbidimetry and molecular modelling demonstrate that Soluplus® keeps weakly basic antimicrobials supersaturated in simulated wound fluid for 30–60 min, which is sufficient for diffusion through 50–100 μm EPS layers [79]. Other than polymer, L-leucine, which is an example of an amphiphilic amino acid, also has a similar effect to polymer. During spray drying high-dose antibiotic powders, leucine is observed to preferentially enrich the particle surface, lowering surface energy, enhancing aerosolization and moisture resilience, and encouraging quick wetting at the biofilm front [83]. It is possible to increase water channels and speed up quinolone access to the intracellular targets by combining a neutral or weakly cationic fluoroquinolone with a cationic membrane-active agent (like colistin) in a single dry powder. Colistin has also been demonstrated to stabilize amorphous ciprofloxacin powders and maintain aerosolization [84]. Ross et al. have used leucine, ciprofloxacin, and glutamic acid together; the latter being a nutrient-dispersion trigger. The aim was to formulate a dry powder containing the combination for inhalation treatment of *Pseudomonas aeruginosa* bacterial biofilms. Leucine was shown to enhance aerosol performance (fine particle fraction

≈60%). The spray-dried powder containing glutamic acid triggers biofilm detachment/dispersion responses, which achieved lower MBEC against *P. aeruginosa*. biofilms compared to ciprofloxacin dry powder alone [85]. This work included thorough powder physiochemistry and quantified eradication dosage reductions, involving both physical and biochemical mechanisms.

#### 5.6.3.2 Methods to Quantify Biofilm Interactions

There are many methods used to demonstrate biofilm interaction. Ceri et al. measured the minimum biofilm inhibitory concentration (MBIC) and minimum biofilm eradication concentration (MBEC) in a controlled environment using the Calgary Biofilm Device (CBD). The CBD cultivates 96 comparable biofilms on pegged lids and offers a standardized procedure for comparing mature biofilms to various ASD formulations. Combined with quantitative microbiology and scanning electron microscopy, more comprehensive data can be obtained, which gives more insight into the effectiveness of each formulation. CBD remains a standard technique for comparing anti-biofilm drug efficacy and screening new antibiotic formulations; it also established the concept of MBEC [86]. Utilizing confocal laser scanning microscopy (CLSM) in conjunction with suitable fluorescent probes (such as polysaccharide/eDNA stains and viability dyes), Schlafer et al. were able to observe in real-time the deposition and dissemination of ASD particles across hydrated biofilm surfaces. They were also able to measure the extent of antimicrobial penetration and the patterns of spatial kill over time. This is because CLSM allows the collection of either qualitative or quantitative information. CLSM enables non-destructive evaluation of biofilm thickness, roughness, and matrix architecture in situ, which is an excellent tool for monitoring interfacial wetting and dissolving fronts produced by amorphous dispersions [87]. Quartz-crystal microbalance with dissipation (QCM-D) is a highly sensitive device that can monitor real-time mass uptake, and viscoelastic changes when EPS films (such as mucin layers or alginate/eDNA/protein composites) are exposed to ASD formulations. QCM-D determines if surface-active or chelating elements (such as alginate-oligosaccharides and leucine-enriched shells) break cross-links or soften the matrix, which is a mechanistic indicator that is consistent with improved penetration [88]. To quantify bacterial load (CFU), pulmonary histopathology, and inflammatory response, Christophersen et al. used chronic lung infection models that replicate biofilm characteristics. These models are often agar/alginate bead models of *P. aeruginosa* in mice. To assess whether surface-active ASDs lower bacterial load and change biofilm ecology in vivo, this study has improved bead size and host background (including CFTR-deficient animals) to replicate persistence and treatment response [89].

Alginate-focused adjuvants may not work the same against matrices that are rich in proteins or eDNA because different biofilms have different EPS chemistry. Therefore, it is important to optimize each formulation according to the target microorganism. High-load ASDs are nevertheless susceptible to moisture sensitivity, $T_g$ depression, and recrystallisation, which can reduce both antibiofilm performance and aerosolization. These create challenges when establishing stability mapping and desiccant packaging. Future directions include hybrid particles (such as ASD-within-lipid shells) that maintain the supersaturation advantage while adding mucus-penetrating stealth and model-informed ASD design that relates dissolution and diffusion reaction dynamics to MBEC outcomes. As a logical platform against established pulmonary biofilms, surface-active spray-dried ASDs can reduce MBECs, speed up killing, and stay inhalation-ready [85].

## 5.7 Conclusions

One of the most important issues of modern pharmaceutical science is the low aqueous solubility and bioavailability of many APIs. The creation of ASDs using solvent evaporation techniques has become a key component in tackling this issue. As discussed in this chapter, solvent-based techniques, in particular spray and freeze drying, provide a potent way to transform crystalline medications that are poorly soluble into high-energy amorphous systems with markedly improved dissolution rates and therapeutic potential. ASDs take advantage of the lack of a stiff crystalline lattice to promote better drug solubility, kinetics of dissolution, and, eventually, systemic bioavailability. By reducing the inherent challenge of recrystallisation. ASD formulations provide great stability in addition to increased solubility when paired with properly chosen polymers and excipients. The stability-performance balance has been further reinforced by the advent of sophisticated polymers like HPMC-AS, PVP-VA, and Soluplus®, which enable the regulated release of APIs via the oral, pulmonary, cutaneous, and mucosal routes. Furthermore, by adjusting variables like solvent selection, atomization technique, and drying arrangement, solvent evaporation techniques can yield particles with the best possible physicochemical and aerodynamic characteristics, increasing their use in specialized drug delivery systems. However, solvent evaporation techniques have drawbacks despite their potential. Since amorphous materials are naturally metastable and prone to recrystallisation, moisture absorption, and chemical deterioration, maintaining stability continues to be the fundamental obstacle. Thus, ongoing innovation is being done to enhance scalability, reproducibility, and commercial viability. Examples include the use of three-fluid nozzles for spray drying, polymer-overlap concentration models for stability prediction, and advanced PAT for real-time monitoring. Crucially, the use of ASDs has expanded beyond conventional oral dose forms, showing promising results in biofilm destruction, antimicrobial therapy, and pulmonary administration. ASDs improve therapeutic efficacy in both localized and systemic treatments by facilitating quick supersaturation and sustained drug release, frequently lowering dosage requirements and minimizing adverse effects. Their versatility in addressing a range of delivery issues, including topical penetration, pulmonary bioavailability, and gastrointestinal dissolution, highlights their extensive pharmacological potential. In summary, methods for manufacturing ASDs via solvent evaporation constitute an essential and developing technological platform that keeps revolutionizing drug development. Long-term stability, regulatory compliance, and large-scale production still present difficulties, but these are much outweighed by the benefits. The role of ASDs in the effective formulation of next-generation medications is anticipated to grow, as solvent systems, polymer technologies, and drying techniques continue to be improved. In the end, solvent-evaporated ASDs provide a link between clinically successful treatments and promising but poorly soluble APIs, guaranteeing that more patients can take advantage of the full potential of contemporary drug development.

## References


1. Waring MJ, Arrowsmith J, Leach AR, Leeson PD, Mandrell S, Owen RM, et al. An analysis of the attrition of drug candidates from four major pharmaceutical companies. Nat Rev Drug Discov. 2015;14(7):475–86.
2. Kalepu S, Nekkanti V. Insoluble drug delivery strategies: review of recent advances and business prospects. Acta Pharm Sin B. 2015;5(5):486–495.
3. Lipinski CA. Poor aqueous solubility–an industry wide problem in drug discovery. Am Pharm Rev. 2002;5(3):82–5.
4. Baghel S, Cathcart H, O'Reilly NJ. Polymeric amorphous solid dispersions: a review of amorphization, crystallization, stabilization, solid-state characterization, and aqueous solubilization of biopharmaceutical classification system class II drugs. J Pharm Sci. 2016;105(9).
5. Brough C, Williams RO, 3rd. Amorphous solid dispersions and nano-crystal technologies for poorly water-soluble drug delivery. Int. J. Pharm. 2013;453(1): 2527–2544.
6. Vasconcelos T, Sarmento B, Costa P. Solid dispersions as strategy to improve oral bioavailability of poor water soluble drugs. Drug Discov Today. 2007;12(23–24): 157–166.
7. Newman A, Knipp G, Zografi G. Assessing the performance of amorphous solid dispersions. J Pharm Sci. 2012;101(4) 1068–1075.
8. Vo CL-N, Park C, Lee B-J. Current trends and future perspectives of solid dispersions containing poorly water-soluble drugs. Eur J Pharm Biopharm. 2013;85(3) 1355–1377.
9. Huang Y, Dai W-G. Fundamental aspects of solid dispersion technology for poorly soluble drugs. Acta Pharm Sin B. 2014;4(1) 799–813.
10. Lin X, Hu Y, Liu L, Su L, Li N, Yu J, et al. Physical stability of amorphous solid dispersions: a physicochemical perspective with thermodynamic, kinetic and environmental aspects. Pharm Res. 2018;35(6) 18–25.
11. Williams HD, Trevaskis NL, Charman SA, Shanker RM, Charman WN, Pouton CW, et al. Strategies to address low drug solubility in discovery and development. Pharmacol Rev. 2013;65(1) 125.
12. Tekade AR, Yadav JN. A review on solid dispersion and carriers used therein for solubility enhancement of poorly water soluble drugs. Advanced Pharm Bull. 2020;10(3) 315–499.
13. Baird JA, Taylor LS. Evaluation of amorphous solid dispersion properties using thermal analysis techniques. Adv Drug Deliv Rev. 2012;64(5).
14. Leuner C, Dressman J. Improving drug solubility for oral delivery using solid dispersions. Eur J Pharm Biopharm. 2000;50(1).
15. Brough C, Williams RO 3rd. Amorphous solid dispersions and nano-crystal technologies for poorly water-soluble drug delivery. Int J Pharm. 2013;453(1):157–66.
16. Hancock BC, Zografi G. Characteristics and significance of the amorphous state in pharmaceutical systems. J Pharm Sci. 1997;86(1).

17. Yu L. Amorphous pharmaceutical solids: preparation, characterization and stabilization—PubMed. Adv Drug Deliv Rev. 2001;48(1):27–42, 1–12.
18. Bhujbal SV, Mitra B, Jain U, Gong Y, Agrawal A, Karki S, et al. Pharmaceutical amorphous solid dispersion: a review of manufacturing strategies. Acta Pharm Sin B. 2021;11(8).
19. Vehring R. Pharmaceutical particle engineering via spray drying. Pharm Res. 2008;25(5):999–1022.
20. Li N, Cape JL, Mankani BR, Zemlyanov DY, Shepard KB, Morgen MM, et al. Water-induced phase separation of spray-dried amorphous solid dispersions. Mol Pharm. 2020;17(10).
21. Zhang X, Rao Q, Qiu Z, Lin Y, Zhang L, Hu Q, et al. Using acetone/water binary solvent to enhance the stability and bioavailability of spray dried enzalutamide/HPMC-AS solid dispersions. J Pharm Sci. 2021;110(3) 4004–4017.
22. Chen H, Pui Y, Liu C, Chen Z, Su C-C, Hageman M, et al. Moisture-induced amorphous phase separation of amorphous solid dispersions: molecular mechanism, microstructure, and its impact on dissolution performance. J Pharm Sci. 2018;107(1).
23. Zheng W, Ke J, Liu K, Xue R, Chen F, Zheng W, et al. Aqueous solution spray drying preparations of binary amorphous solid dispersions. Crystals. 2025;15(4):323.
24. Jermain SV, Brough C, Williams RO. Amorphous solid dispersions and nanocrystal technologies for poorly water-soluble drug delivery—an update. Int J Pharm. 2018;535(1–2).
25. Patel BB, Patel JK, Chakraborty S, Shukla D. Revealing facts behind spray dried solid dispersion technology used for solubility enhancement. Saudi Pharm J. 2015;23(4).
26. U.S. Department of Health and Human Services Food and Drug Administration, Center for Drug Evaluation and Research (CDER), Center for Biologics Evaluation and Research (CBER). Q3C—tables and list guidance for industry. US 2017.
27. Savjani KT, Gajjar AK, Savjani JK. Drug solubility: importance and enhancement techniques—PubMed. ISRN Pharm. 2012;2012.
28. Moseson DE, Tran TB, Karunakaran B, Ambardekar R, Hiew TN. Trends in amorphous solid dispersion drug products approved by the U.S. Food and Drug Administration between 2012 and 2023. Int J Pharm: X. 2024;7.
29. Pandi P, Bulusu R, Kommineni N, Khan W, Singh M. Amorphous solid dispersions: an update for preparation, characterization, mechanism on bioavailability, stability, regulatory considerations and marketed products. Int J Pharm. 2020;586 100259.
30. Zhang J, Guo M, Luo M, Cai T. Advances in the development of amorphous solid dispersions: the role of polymeric carriers. Asian J Pharm Sci. 2023;18(4).
31. Keßler L, Mishra R, Hietala S, Lammens M, Peltonen L, Rades T, et al. Amorphous solid dispersions of amphiphilic polymer excipients and indomethacin prepared by hot melt extrusion. Eur J Pharm Sci. 2025;204 100834.
32. Adhikari BR, Das SC. Amorphicity and aerosolization of Soluplus-based inhalable spray dried powders. Pharmaceutics. 2022;14(12) 106960.
33. Shen Q, Du Y. A comprehensive review of advanced drug delivery systems for the treatment of rheumatoid arthritis. Int J Pharm. 2023;635.
34. He Y, Huang W, Zhang C, Chen L, Xu R, Li N, et al. Energy metabolism disorders and potential therapeutic drugs in heart failure. Acta Pharm Sin B. 2021;11(5).
35. Eedara BB, Alabsi W, Encinas-Basurto D, Polt R, Mansour HM, Eedara BB, et al. Spray-dried inhalable powder formulations of therapeutic proteins and peptides. AAPS PharmSciTech. 2021 22(5).
36. Van Drooge DJ, Hinrichs WLJ, Frijlink HW. Anomalous dissolution behavior of tablets prepared from sugar glass-based solid dispersions. J Pharm Sci. 2006;95(3):700–11.
37. Ferreira MPA, Martins JP, Hirvonen J, Santos HA. Spray-drying for the formulation of oral drug delivery systems. In: Nanotechnology for oral drug delivery. 2020. pp. 253–284.
38. Baumann JM, Adam MS, Wood JD, Baumann JM, Adam MS, Wood JD. Engineering advances in spray drying for pharmaceuticals. Annu Rev Chem Biomol Eng. 2021;12.
39. Verma A, Singh SV. Spray drying of fruit and vegetable juices—a review. Crit Rev Food Sci Nutr. 2015;55(5) 217–240.

40. O'Sullivan JJ, Norwood E-A, O'Mahony JA, Kelly AL. Atomisation technologies used in spray drying in the dairy industry: a review. J Food Eng. 2019;243 701–719.
41. Ré M-I. Formulating drug delivery systems by spray drying. Dry Technol. 2006;24(4) 57–69.
42. Sollohub K, Cal K. Spray drying technique: II. Current applications in pharmaceutical technology. J Pharm Sci. 2010;99(2) 433–446.
43. Cal K, Sollohub K. Spray drying technique. I: hardware and process parameters. J Pharm Sci. 2010;99(2) 587–597.
44. Miller DA, Gil M. Spray-drying technology. In: Formulating poorly water soluble drugs. New York: Springer; 2012 575–586.
45. Phapale BN, Patil AP. Application of spray drying technique for preparation amorphous solid dispersion. Int J Pharm Sci Res. 2017;8(4):1541–9.
46. Legako J, Dunford NT. Effect of spray nozzle design on fish oil–whey protein microcapsule properties. J Food Sci. 2010;75(6).
47. Bhujbal SV, Su Y, Pathak V, Zemlyanov DY, Cavallaro A-A, Munson EJ, et al. Effect of storage humidity on physical stability of spray-dried naproxen amorphous solid dispersions with polyvinylpyrrolidone: two fluid nozzle vs. three fluid nozzle. Pharmaceutics. 2021;13(7).
48. Yang D-L, Wang D, Niu H, Wang R-L, Liu M, Zhang F-M, et al. A general strategy for efficiently constructing multifunctional cluster fillers using a three-fluid nozzle spray drying technique for dental restoration. Engineering. 2022;8.
49. Shi X, Lee Y. Encapsulation of tributyrin with whey protein isolate (WPI) by spray-drying with a three-fluid nozzle. J Food Eng. 2020;281.
50. Abla KK, Mehanna MM. Freeze-drying: a flourishing strategy to fabricate stable pharmaceutical and biological products. Int J Pharm. 2022;628.
51. Chavan RB, Rathi S, Jyothi VGSS, Shastri NR. Cellulose based polymers in development of amorphous solid dispersions. Asian J Pharm Sci. 2019;14(3).
52. He Y, Ho C. Amorphous solid dispersions: utilization and challenges in drug discovery and development. J Pharm Sci. 2015;104(10) 248–264.
53. Pardeshi SR, Deshmukh NS, Telange DR, Nangare SN, Sonar YY, Lakade SH, et al. Process development and quality attributes for the freeze-drying process in pharmaceuticals, biopharmaceuticals and nanomedicine delivery: a state-of-the-art review. Future J Pharm Sci. 2023;9(1) 3237–3258.
54. Rogers TL, Johnston KP, Williams RO. Amorphous solid dispersions for solubility and dissolution enhancement: progress and challenges. Drug Dev Ind Pharm. 2018;34(8):757–65.
55. Hancock BC, Parks M. What is the true solubility advantage for amorphous pharmaceuticals? Pharm Dev Technol. 2016;5(2):115–22.
56. Wijiani N, Isadiartuti D, Rijal MAS, Yusuf H. Characterization and dissolution study of micellar curcumin-spray dried powder for oral delivery. Int J Nanomedicine. 2020;15.
57. Sawicki E, Beijnen JH, Schellens JHM, Nuijen B. Pharmaceutical development of an oral tablet formulation containing a spray dried amorphous solid dispersion of docetaxel or paclitaxel. Int J Pharm. 2016;511(2) 1787–1796.
58. Zhao P, Han W, Shu Y, Li M, Sun Y, Sui X, et al. Liquid–liquid phase separation drug aggregate: merit for oral delivery of amorphous solid dispersions. J Control Release. 2023;353 765–773.
59. Kong Y, Wang W, Wang C, Li L, Peng D, Tian B. Supersaturation and phase behavior during dissolution of amorphous solid dispersions. Int J Pharm. 2023;631.
60. Nguyen HT, Duong TV, Taylor LS. Impact of gastric pH variations on the release of amorphous solid dispersion formulations containing a weakly basic drug and enteric polymers. Mol Pharm. 2023;20(3) 122524.
61. Qin Y, Xiao C, Li X, Huang J, Si L, Sun M. Enteric polymer–based amorphous solid dispersions enhance oral absorption of the weakly basic drug nintedanib via stabilization of supersaturation. Pharmaceutics. 2022;14(9) 1681–1695.
62. Monschke M, Kayser K, Wagner KG, Monschke M, Kayser K, Wagner KG. Influence of particle size and drug load on amorphous solid dispersions containing pH-dependent soluble polymers and the weak base ketoconazole. AAPS PharmSciTech. 2021;22(1).

63. AboulFotouh K, Zhang Y, Maniruzzaman M, Williams RO, Cui Z. Amorphous solid dispersion dry powder for pulmonary drug delivery: advantages and challenges. Int J Pharm. 2020;587.
64. Chen L, Okuda T, Lu X-Y, Chan H-K. Amorphous powders for inhalation drug delivery. Adv Drug Deliv Rev. 2016;100 119711.
65. Labiris NR, Dolovich MB. BPS publications. Br J Clin Pharmacol. 2003;56(6) 102–115.
66. Zhao N, Wen X, Yang J, Li S, Wang Z. Modeling and prediction of viscosity of water-based nanofluids by radial basis function neural networks. Powder Technol. 2015;281 588–599.
67. Rattanupatam T, Srichana T. Budesonide dry powder for inhalation: effects of leucine and mannitol on the efficiency of delivery. Drug Deliv. 2014;21(6).
68. Lu W, Rades T, Rantanen J, Yang M. Inhalable co-amorphous budesonide-arginine dry powders prepared by spray drying. Int J Pharm. 2019;565.
69. Jara MO, Warnken ZN, Sahakijpijarn S, Moon C, Maier EY, Christensen DJ, et al. Niclosamide inhalation powder made by thin-film freezing: multi-dose tolerability and exposure in rats and pharmacokinetics in hamsters. Int J Pharm. 2021;603.
70. Su C, Saha T, Sinha S, Hird CP, Smith SXY, Quiñones-Mateu ME, et al. Inhalable spray-dried dry powders combining ivermectin and niclosamide to inhibit SARS-CoV-2 infection in vitro—PubMed. Int J Pharm. 2025;671.
71. Chan SHY, Sheikh K, Zariwala MG, Somavarapu S. Dry powder formulation of azithromycin for COVID-19 therapeutics—PubMed. J Microencapsul. 2023;40(4).
72. Guembe-Michel N, Nguewa P, González-Gaitano G. Soluplus®-based pharmaceutical formulations: recent advances in drug delivery and biomedical applications. Int J Mol Sci. 2025;26(4).
73. Gajera B, Shah H, Parekh B, Rathod V, Tilala M, Dave RH, et al. Design of experiments-driven optimization of spray drying for amorphous clotrimazole nanosuspension. AAPS PharmSciTech. 2024;25(6).
74. Sawant K, Elkanayati RM, Almotairy A, Repka MA, Almutairi M. Clotrimazole mucoadhesive films with extended-release properties for vaginal candidiasis—a hot-melt extrusion application. J Pharm Sci. 2025;114(2).
75. Tianyi Xiang, Sichen Song, A. Siegel R, Calvin Sun C, Tianyi Xiang, Sichen Song, et al. Efficient development of high drug loaded posaconazole tablets enabled by amorphous solid dispersion. RSC Pharm. 2025;2(1).
76. Sharma S, Mohler J, Mahajan SD, Schwartz SA, Bruggemann L, Aalinkeel R. Microbial biofilm: a review on formation, infection, antibiotic resistance, control measures, and innovative treatment. Microorganisms. 2023;11(6).
77. Martino PD. Extracellular polymeric substances, a key element in understanding biofilm phenotype. AIMS Microbiol. 2018;4(2).
78. Mishra A, Aggarwal A, Khan F. Medical device-associated infections caused by biofilm-forming microbial pathogens and controlling strategies. Antibiotics. 2024;13(7).
79. Sharma A, Arora K, Mohapatra H, Sindhu RK, Bulzan M, Cavalu S, et al. Supersaturation-based drug delivery systems: strategy for bioavailability enhancement of poorly water-soluble drugs. Molecules. 2022;27(9).
80. Chang RYK, Chan H-K. Advancements in particle engineering for inhalation delivery of small molecules and biotherapeutics. Pharm Res. 2022;39(12).
81. Zhang L, Bera H, Guo Y, Shi C, Ulrik Lind J, Radeke C, et al. Co-spray dried inhalable composite powders of ciprofloxacin and alginate oligosaccharide as anti-biofilm therapy—PubMed. Int J Pharm. 2024;654.
82. El-Naggar NE-A, Dalal SR, Zweil AM, Eltarahony M, El-Naggar NE-A, Dalal SR, et al. Artificial intelligence-based optimization for chitosan nanoparticles biosynthesis, characterization and in-vitro assessment of its anti-biofilm potentiality. Sci Rep. 2023;13(1).
83. Ordoubadi M, Shepard KB, Wang H, Wang Z, Pluntze AM, Ordoubadi M, et al. On the physical stability of leucine-containing spray-dried powders for respiratory drug delivery. Pharmaceutics. 2023;15(2).

84. Shetty N, Ahn P, Park H, Bhujbal S, Zemlyanov D, Cavallaro A, et al. Improved physical stability and aerosolization of inhalable amorphous ciprofloxacin powder formulations by incorporating synergistic colistin. Mol Pharm. 2018;15(9).
85. Ross SS, Gharse S, Sanchez L, Fiegel J. Dry powder aerosols to co-deliver antibiotics and nutrient dispersion compounds for enhanced bacterial biofilm eradication. Int J Pharm. 2017;531(1).
86. Ceri H, Olson ME, Stremick C, Read RR, Morck D, Buret A. The Calgary biofilm device: new technology for rapid determination of antibiotic susceptibilities of bacterial biofilms. J Clin Microbiol. 1999;37(6).
87. Schlafer S, Meyer RL. Confocal microscopy imaging of the biofilm matrix. J Microbiol Methods. 2017;138.
88. Latag GV, Nakamura T, Palai D, Mondarte EAQ, Hayashi T. Investigation of three-dimensional bacterial adhesion manner on model organic surfaces using quartz crystal microbalance with energy dissipation monitoring. ACS Appl Bio Mater. 2023;6(3).
89. Christophersen LJ, Trøstrup H, Damlund DSM, Bjarnsholt T, Thomsen K, Jensen PØ, et al. Bead-size directed distribution of Pseudomonas aeruginosa results in distinct inflammatory response in a mouse model of chronic lung infection. Clin Exp Immunol. 2012;170(2).
90. Di L, Fish PV, Mano T. Bridging solubility between drug discovery and development. Drug Discov Today. 2012;17(9–10).

# Chapter 6
# Mechanochemical Approaches for the Preparation of Amorphous Solid Dispersions

**Ioannis Partheniadis and Kyriakos Kachrimanis**

**Abstract** Poor aqueous solubility remains one of the principal challenges in the development of new chemical entities (NCEs) during drug discovery, driving continued interest in amorphous solid dispersions (ASDs) as an enabling formulation strategy. Mechanochemistry, i.e., the induction of physical and/or chemical reactions by mechanical processing, has emerged as a versatile, environment-friendly route for ASD preparation, offering an alternative to conventional thermal- and solvent-based methods, such as spray drying and hot-melt extrusion. By supplying mechanical energy through milling, extrusion, or acoustic mixing, crystalline materials can be transformed into disordered or molecularly mixed states, undergoing defect accumulation, lattice disruption, and stress-induced phase transitions. This chapter reviews the scientific foundations and practical implementation of mechanochemical approaches for ASD production. It outlines the theoretical principles of mechanochemistry and tribochemistry, describes major processing platforms and reactor designs, and explains how the mode of stress application, the energy input, temperature evolution, and reactor environment influence amorphization pathways. Special attention is given to drug–polymer systems, including the role of polymer properties, drug–polymer intermolecular interactions, and glass transition behavior in stabilizing mechanically generated amorphous phases. Key process parameters, such as milling intensity, media characteristics, temperature control, humidity, and formulation composition, are discussed in relation to amorphization efficiency, product stability, and reproducibility. Advantages, limitations, scale-up

I. Partheniadis
Department of Pharmacy, Faculty of Health and Medical Sciences, University of Copenhagen, Copenhagen, Denmark
e-mail: ioannis.partheniadis@sund.ku.dk

K. Kachrimanis (✉)
Laboratory of Pharmaceutical Technology, School of Pharmacy, Faculty of Health Sciences, Aristotle University of Thessaloniki, Thessaloniki, Greece

Center for Interdisciplinary Research & Innovation, Aristotle University of Thessaloniki, Thessaloniki, Greece
e-mail: kgk@pharm.auth.gr



I. Nikolakakis, I. Partheniadis (eds.), *Amorphous Solid Dispersions*, AAPS Introductions in the Pharmaceutical Sciences 5,
https://doi.org/10.1007/978-3-032-24883-1_6

considerations, and contamination risks are also evaluated, alongside emerging technologies such as reactive extrusion and resonant acoustic mixing. Collectively, the chapter positions mechanochemical processing as a powerful and increasingly scalable tool in the ASD formulation design and development toolbox, with strong relevance for green and sustainable pharmaceutical manufacturing.

**Keywords** Green Chemistry · Tribochemistry · Ball milling · Reactive extrusion · Resonant acoustic mixing · Polymer-assisted grinding

## 6.1 Introduction

The pharmaceutical industry currently faces a significant challenge where a vast majority of new chemical entities (NCEs) in the development pipeline exhibit poor water solubility, classifying them as Biopharmaceutics Classification System (BCS) Class II or IV drugs [1, 2]. To overcome the limitations of dissolution rate-limited absorption, the formulation of amorphous solid dispersions (ASDs) has emerged as a leading strategy [3]. By converting a crystalline active pharmaceutical ingredient (API) into its amorphous counterpart, stabilized within a carrier matrix, the long-range order of the crystal lattice is disrupted, resulting in a higher free energy state that offers enhanced apparent solubility and wettability [2, 4]. While conventional manufacturing methods for ASDs, such as hot melt extrusion (HME) and spray drying (SD), rely on thermal or solvent-based phase transformations, these approaches may be unsuitable for thermolabile drugs or environmentally hazardous due to the use of organic solvents. Consequently, mechanochemical approaches have gained prominence as a powerful, solvent-free alternative for inducing amorphization directly in the solid state [5, 6].

Mechanochemistry, historically developed in metallurgy and mineral science, utilizes mechanical energy to induce structural and microstructural changes in solids. When pharmaceutical solids are subjected to intensive mechanical stress, such as ball-milling, they accumulate crystal defects and lose long-range crystallographic periodicity, eventually reaching a disordered, glass-like state [5]. This process is governed by a complex competition between mechanically induced disordering (ballistic jumps) and thermally activated restoration (recrystallization), often described by the concept of "effective temperature." The efficiency of this solid-state amorphization is heavily influenced by processing conditions; for instance, milling performed below the glass transition temperature ($T_g$) of the material significantly favors the formation of the amorphous state, whereas processing above $T_g$ may induce polymorphic transformations or rapid recrystallization of the initial crystal form. Furthermore, mechanochemical processing aligns closely with the principles of Green Chemistry by eliminating or drastically reducing the need for hazardous solvents during the manufacturing of ASDs [2, 3, 5].

The successful preparation of ASDs via mechanical processing often requires the presence of polymeric carriers to catalyze amorphization and subsequently stabilize the metastable amorphous drug [7]. While neat grinding (NG) alone can often achieve amorphization, the high surface energy of the resulting particles causes physical instability [2, 8]. Therefore, techniques such as co-milling with polymeric additives, known as Polymer-Assisted Grinding (PoLAG) have become essential. In these systems, the mechanical energy facilitates the mixing of the drug and polymer at a molecular level, where specific and non-specific intermolecular interactions between the drug and polymer help stabilize the amorphous phase and prevent phase separation. Recent advancements have also introduced variations such as liquid-assisted grinding (LAG), which can accelerate cocrystal or amorphous phase formation, and novel technologies like acoustic mixing, expanding the toolkit available to formulation scientists beyond traditional ball-milling [2, 8].

Therefore, in the present chapter, we aim to provide a comprehensive overview of mechanochemical approaches for the preparation of ASDs, with emphasis on the fundamental theoretical principles, processing methods, and formulation variables that govern mechanically induced amorphization and drug–polymer dispersion formation. We will discuss the theoretical basis of mechanochemistry, key equipment, and stress modes, and connect these foundations to practical pharmaceutical applications such as neat, liquid-assisted, and polymer-assisted grinding, as well as emerging technologies, which include reactive extrusion and acoustic mixing. Particular focus is placed on mechanisms of disorder generation, stabilization of amorphous phases by polymers, critical process parameters, and reproducibility considerations. By integrating mechanistic insight with formulation and process perspectives, the chapter aims to clarify the capabilities, limitations, and future potential of mechanochemical routes as viable and sustainable alternatives for ASD manufacturing.

## 6.2 Mechanochemistry

Mechanochemistry constitutes a branch of chemistry concerned with physical and chemical changes of substances in any aggregate state that are induced by the direct absorption of mechanical energy [9–12]. While the term "mechanochemistry" was formally introduced by Wilhelm Ostwald in his Handbook of General Chemistry in 1919, classifying it as one of the four main sub-disciplines of chemistry alongside thermochemistry, photochemistry, and electrochemistry, the empirical utilization of mechanical force to induce chemical change predates recorded history [11]. Prehistoric humans utilized friction and percussion to generate fire, effectively performing the earliest mechanochemical reactions [11, 13]. The first documented scientific observation is attributed to Theophrastus of Eresus (ca. 315 B.C.E.), who described the reduction of cinnabar (mercury sulfide) to elemental mercury via grinding in a copper mortar [14]. However, systematic investigation only began in the late nineteenth century with M. Carey Lea, who demonstrated that mechanical action on silver halides produced chemical changes distinct from those induced by heat, establishing that mechanochemistry is not merely a thermal artifact [15–17].

Modern mechanochemistry is defined by the International Union of Pure and Applied Chemistry (IUPAC) as a chemical reaction induced by the direct absorption of mechanical energy [18]. This definition encompasses a vast array of phenomena, from the degradation of polymers under shear to the synthesis of complex pharmaceutical co-crystals [19, 20]. Unlike traditional thermal reactions, which rely on the statistical distribution of thermal energy to overcome activation barriers (Arrhenius kinetics), mechanochemical transformations are often non-equilibrium processes [21, 22]. Mechanical energy is channeled directionally into bond breaking, structural deformation, and the creation of metastable states with high defect densities [11]. High-energy milling, for example, generates extremely high defect densities and forced mixing that can destabilize thermodynamic equilibrium phases, resulting in materials that otherwise wouldn't be produced via conventional thermal or solution-based routes [13].

## *6.2.1 Theoretical Principles*

The theoretical background of mechanochemistry is complex because mechanical treatment involves physical processes occurring across multiple length and time scales: from atomic bond ruptures and dislocation motion to macroscopic heat transfer and bulk powder flow [11]. The distinction is one of energy distribution. Thermal chemistry depends on statistical fluctuations in temperature, whereas mechanochemistry uses directional mechanical force to lower activation barriers along specific reaction coordinates [13, 16].

### 6.2.1.1 Reactive Extrusion as an Emerging Green Chemistry Technology

In recent decades, mechanochemistry has experienced a renaissance, driven largely by the search for sustainable chemical processes. It aligns closely with the principles of Green Chemistry, satisfying requirements for waste prevention, safer solvents (or solvent-free conditions), and energy efficiency [13, 20]. Conventional chemical synthesis often relies on vast quantities of organic solvents for dissolving reactants and thermal energy for driving reactions. Mechanochemistry eliminates the need for bulk solvents, thereby reducing the environmental burden associated with solvent handling and disposal.

A critical technological advancement in scaling up these green principles from the laboratory to the industrial level is Reactive Extrusion, specifically Twin Screw Extrusion (TSE). While ball milling is the standard for laboratory-scale mechanochemistry, it is inherently a batch process characterized by limitations in scalability, such as idle times for loading/unloading and difficulties in heat management [11, 13]. TSE offers a continuous flow alternative. In a TSE system, solid reactants are fed into a barrel containing two intermeshing, counter-rotating (or co-rotating) screws. These screws convey the material through a series of modular segments designed for specific tasks: i) the conveying elements move the material forward, while ii) kneading elements apply intense shear and compression forces to induce mixing and reaction [20, 22].

TSE process allows for precise control over reaction parameters such as residence time, temperature profiles along the barrel, and the intensity of shear stress [11]. This method has been successfully applied to the continuous, solvent-free manufacturing of metal-organic frameworks (MOFs), deep eutectic solvents, and pharmaceutical co-crystals at rates of kilograms per hour [23]. For example, the synthesis of hydrazone-based APIs and porous organic cages has been demonstrated using TSE, bridging the gap between milligram-scale mechanosynthesis and industrial production [24]. The efficiency of TSE lies in its ability to maintain high reactant concentrations (due to the absence of solvent) and provide continuous mechanical energy input, leading to Space-Time Yields (STY) that often exceed those of solvent-based batch processes.

#### 6.2.1.2 Tribochemistry/Tribology

The term tribochemistry (or tribology) is often used interchangeably with mechanochemistry, particularly in the context of lubrication and wear, but strictly refers to chemical and physicochemical changes occurring at the interface of interacting surfaces in relative motion. It focuses on the chemical consequences of energy dissipated at sliding contacts, involving friction, lubrication, and wear [13, 25].

Historically, two main theoretical frameworks have been proposed to explain the high reactivity observed at tribological interfaces: the "Hot Spot" (or Magma--Plasma) theory and the Stress-Augmented Thermal Activation theory [20]. The Magma-Plasma Model, proposed by Thiessen in 1967, postulates that energetic impacts and friction create localized "hot spots" where temperatures can transiently exceed $10^3$–$10^4$ K [17, 20]. At these temperatures, material may locally melt or even reach a plasma state, leading to the emission of energetic species like electrons and ions, and facilitating bond rupture and reformation [17, 23]. While this model explains certain high-energy phenomena in inorganic materials (e.g., triboluminescence), it has been critiqued for organic systems. It is considered unlikely that such extreme temperatures are the primary drivers of reactivity in molecular organic systems, where extensive thermal decomposition is not typically observed [13, 20].

Conversely, the Stress-Augmented Thermal Activation theory, rooted in the work of Zhurkov, Bell, and Eyring, provides a more atomistic explanation relevant to organic mechanochemistry [16]. This model suggests that mechanical stress lowers the activation energy barrier for a chemical reaction. According to the Bell model (Eq. 6.1), the rate constant $k(f)$ of a bond-breaking reaction under an applied force, $F$, is given by the following equation:

$$k(F) = k_0 e^{\frac{F\Delta x^{\ddagger}}{k_B T}} \tag{6.1}$$

where $\Delta x^{\ddagger}$ is constant and represents the activation length (the distance the bond must be stretched to reach the transition state), $k_B$ is the Boltzmann constant, and $T$ is the temperature [16, 25]. This equation implies that the reaction rate increases exponentially with the applied force. This framework has been successfully applied to explain plastic deformation, friction, and the kinetics of mechanochemical reactions in polymers and biological molecules [25]. Recent studies using Atomic Force Microscopy (AFM) to slide single tips over surfaces have confirmed that tribochemical reaction rates follow this exponential dependence on stress, supporting the concept that mechanical force directly distorts the potential energy surface of the reactants [25, 26].

#### 6.2.1.3 Mode of Stress Application (Continuous vs. Pulsed) and Stress Types (Impact vs. Friction)

The outcome of a mechanochemical reaction is heavily dependent on the nature of the mechanical stress applied. Stress application can be classified based on its temporal nature (continuous vs. pulsed) and the vector of the applied force (impact vs. friction/shear) [11, 27].

In typical milling devices, the mechanical energy is delivered in pulses. A particle is subjected to a stress event (collision) characterized by a duration $t_1$, followed by a relaxation period $t_2$ before the next event. The ratio of these times is critical. If the relaxation time $t_2$ is long compared to the characteristic relaxation time of the excited states (e.g., heat dissipation, defect annealing), the material may return to equilibrium between impacts. However, if the frequency of impacts is high (short $t_2$ ), the material accumulates defects and excess energy, remaining in a metastable, "activated" state [28]. This accumulation of energy is essential for driving reactions that require crossing high activation barriers. In contrast, continuous stress application, such as in hydrostatic compression (e.g., in a diamond anvil cell), maintains the material under constant stress, which can lead to reversible polymorphic transitions or amorphization that differ from those induced by pulsed milling.

Mechanical forces can be categorized into impact (normal/perpendicular force) and shear/friction (tangential force). Different mechanoreactors provide different ratios of these forces. For instance, shaker mills and jet mills are dominated by impact forces, whereas mortars, extruders, and certain rolling mills are dominated by shear forces. Impact involves high-velocity collisions that generate fractures and deep structural defects. Shear involves the sliding of surfaces across one another, generating surface defects and causing plastic deformation.

This distinction is chemically significant because the symmetry of the mechanical stress field can influence the reaction pathway. An interesting example is the mechanochemical co-crystallization of piroxicam and succinic acid, where impact forces promote the formation of the co-crystal, whereas shear forces induce the decomposition of the co-crystal back into its starting components [27]. Similarly, different polymorphs of drugs like chlorpropamide or paracetamol may be accessed selectively, depending on whether the stress is hydrostatic, shear, or impact based. For instance, grinding (shear + impact) transforms paracetamol form II to form I, while hydrostatic compression transforms form I to the elusive form II [29, 30].

### *6.2.2 Equipment*

The equipment used in mechanochemistry varies widely, determining the type and intensity of mechanical energy transferred to the sample [22].

(i) *Mortar and Pestle*: The most ancient and simplest tool, primarily utilizing friction and shear forces. While useful for preliminary screening, the energy input is low, operator-dependent, and difficult to reproduce for kinetic studies.
(ii) *Ball Mills*: These are the most used equipment of laboratory mechanochemistry. They consist of a milling jar containing grinding media (balls). They are further divided into:
   - *Vibratory/Mixer Mills* (e.g., *SPEX*)*:* The jar is shaken back and forth at high frequencies (e.g., 10–60 Hz). The balls collide with the jar walls and against each other, generating high-energy impact. These mills generally provide higher power densities than planetary mills.
   - *Planetary Ball Mills:* The jar rotates around its own axis while revolving around a central axis in the opposite direction (like a planet). This complex motion generates high centrifugal forces, resulting in a combination of strong shear friction (balls rolling on the wall) and impact.

(iii) *Extruders (Twin Screw)*: As discussed in Sect. 6.2.1.1, these devices use rotating screws to apply continuous shear and compressive stress. They are essential for scaling up mechanochemical processes to industrial levels, offering continuous processing capabilities, not available in batch ball mills.
(iv) *Resonant Acoustic Mixers (RAM)*: A newer technology that uses low-frequency acoustic energy to induce intensive mixing [31]. Unlike ball mills, RAM can operate without grinding media, relying on the resonance of the entire vessel-spring system to generate acceleration forces up to 100 g. This creates chaotic bulk flow and micro-mixing zones, efficient for mixing and solid-state reactions (see Sect. 6.3.2).

The choice of the milling jar and ball material (stainless steel, zirconia, agate, PMMA, Teflon) affects the reaction outcome due to different contamination (abrasion), density (impact energy), and thermal properties [32, 33]. For example, replacing a stainless steel jar for a polymer jar can significantly alter reaction kinetics and even the polymorphic outcome because polymers insulate heat (leading to higher bulk temperatures) and are softer, absorbing more impact energy [34].

### 6.2.3 *Influence of Reactor Environment and Milling Media*

While the thermodynamic potential of a chemical reaction is determined by the molecular structure of the reactants, the kinetic realization of mechanochemical transformations is intrinsically governed by the physical environment within the reactor. The "stochastic" nature of ball milling implies that the reaction rate and product selectivity are functions of the probability and intensity of collisions. Consequently, parameters that are often considered trivial in solution chemistry, such as the material of the vessel, the surface cleanliness of the grinding media, and the geometric configuration of the reactor, become governing variables in mechanochemistry.

#### 6.2.3.1 Milling Media: Size, Mass, and Material Composition

The efficiency of energy transfer from the milling device to the sample is mediated entirely by the grinding media (balls). The kinetic energy ($E_k$) delivered during a single impact event is a function of the mass of the ball ($m$) and its velocity ($v$), described by $E_k = ½mv^2$. Therefore, the pycnometric density and diameter of the balls are critical variables that dictate the "impact severity."

There exists a fundamental trade-off between the individual impact-energy and the collision event-frequency. Increasing the diameter of the balls (and thus their mass) enhances the kinetic energy of each impact, which is often required to induce deep structural changes, fracture hard inorganic crystals, or overcome high activation energy barriers [35–39]. However, for a fixed volume of the milling jar, using larger balls necessitates reducing their number, which decreases the total number of contact points and the specific surface area available for reaction.

Recent studies on the mechanochemical behavior of caffeine and other organic systems have demonstrated that for certain transformations, the specific surface area of the milling media is more critical than the impact energy. Using a larger number of smaller balls increases the likelihood of ball-sample-ball or ball-sample-wall collision events [35, 37, 39]. This suggests that in softer organic materials, where high-energy impact might not be necessary to break bonds, the reaction rate is limited by the probability of the powder being trapped in a collision zone.

The material of the milling beads defines their density ($\rho$) and hardness. Common materials range from high-density tungsten carbide ($\rho = 15.6$ g/cm$^3$) and stainless steel ($\rho = 7.8$ g/cm$^3$) to lower-density zirconia ($\rho = 5.7$ g/cm$^3$), agate, and polymers. Denser media deliver significantly higher impact forces for the same energy input to the mill. For example, replacing steel balls with ceramic or polymer balls of the same size drastically reduces the energy input, often resulting in slower kinetics or a complete failure to react [34].

The milling media are rarely inert. "Abrasive contamination" is a well-documented phenomenon where wear from the balls and jar introduces foreign material (Fe, Cr, Ni, Zr) into the sample [40, 41]. For instance, the choice of jar/ball material (steel, $ZrO_2$, WC, polymers) changes both performance and contamination, which has been highlighted as a "major challenge" in mechanochemical chitin amorphization [42]. While usually undesirable, this effect has been exploited for "direct mechanocatalysis," where copper or palladium milling beads serve as the reservoir for catalytic species in coupling reactions [43–45].

#### 6.2.3.2 The "Polymer Effect" and 3D Printed Jars for In Situ Monitoring

The rapid advancement of in situ analytical techniques, particularly X-ray powder diffraction (XRPD) using synchrotron radiation, has necessitated a redesign of mechanochemical reactors. Standard stainless-steel jars are opaque to X-rays, preventing real-time observation of phase transformations. To overcome this obstacle, researchers have increasingly turned to milling jars made of organic polymers, such as polymethylmethacrylate (PMMA, Perspex®) or polycarbonate, which are transparent to high-energy X-rays [46, 47]. A pivotal advancement in the field was the adaptation of commercial vibratory ball mills for use at synchrotron facilities. As shown in Fig. 6.1, Halasz and co-workers successfully modified a standard Retsch MM200 shaker mill to accommodate in situ X-ray powder diffraction (XRPD) experiments [46]. This setup allowed the X-ray beam to pass through the oscillating jar, probing the sample continuously, while it was being milled. This technological leap enabled the first direct observation of rapid phase transformations, the detection of short-lived intermediates (such as cocrystals or unstable polymorphs), and the quantitative monitoring of reaction kinetics under realistic milling conditions.

The advent of fused deposition modeling (FDM) 3D printing has further democratized this approach, allowing researchers to fabricate custom jars from materials like polylactic acid (PLA), polyethylene terephthalate glycol (PETG), and acrylonitrile butadiene styrene (ABS) [48]. 3D printing allows for internal geometries that optimize powder flow and enhance the efficiency of comminution of the end

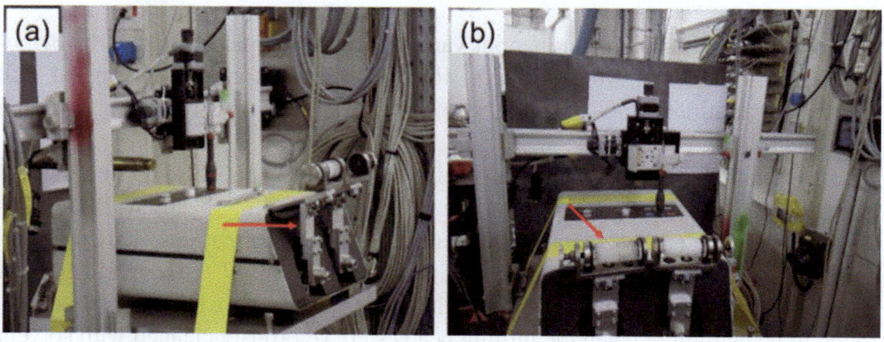


**Fig. 6.1** (**a**) Positioning of a modified Retsch MM200 mill on a motorized, movable x-,y-,z-stage at the ESRF beamline ID15B with modified arms (red arrow). (**b**) Modified arms position the polymer vessels (red arrow) in the X-ray beam. (Reprinted with permission from [46])

product and reduce the milling time or restrict the ball motion to specific "probing zones" for analysis [49].

It has been rigorously demonstrated that substituting steel for polymer is not a benign change; the material of the jar significantly influences reaction kinetics and outcomes [34]. A primary concern is mechanical damping, since polymers possess significantly lower elastic moduli than steel, and tend to deform elastically upon impact. This deformation absorbs a substantial portion of the grinding media's kinetic energy, effectively "cushioning" the strike and reducing the total mechanical dose delivered to the powder [49, 50].

Compounding this mechanical loss is the thermal divergence between the materials. While stainless steel's high thermal conductivity allows for rapid heat dissipation, polymers act as insulators. This causes heat generated by friction and impact to remain trapped within the reactor, leading to elevated bulk temperatures that can shift the thermodynamic landscape, potentially favoring high-temperature polymorphs or triggering undesirable thermal degradation [51].

Finally, the chemical compatibility of polymer vessels, particularly those produced via 3D printing, remains a bottleneck for liquid assisted grinding (LAG). Common solvents can compromise the polymer matrix integrity, such as acetone's effect on ABS or chlorinated solvents on PMMA, risking both jar structural failure and sample contamination. While materials like PLA exhibit broader solvent resilience, they remain mechanically inferior to steel, underscoring the inherent trade-offs in choosing non-metallic reaction vessels [48, 52, 53].

An interesting example of the "polymer effect" is the polymorphic transformation of $\beta$-glycine to $\alpha$-glycine [34]. This transformation proceeds rapidly in stainless steel jars but is significantly delayed or completely inhibited in PMMA or ABS jars under identical milling parameters, highlighting the risk of directly extrapolating in situ kinetic data collected in polymer jars to standard laboratory syntheses in stainless steel equipment.

#### 6.2.3.3 Surface Conditions and the "Fingerprint Effect"

One of the most subtle yet influential variables in mechanochemistry is the tribological state of the milling surfaces. Recent investigations have revealed that "minor" variations in cleaning protocols can drastically alter reaction reproducibility.

Experimental evidence has identified that the presence of trace amounts of grease, specifically, sebaceous lipids left by handling milling beads with bare fingers, can significantly accelerate mechanochemical transformations. This phenomenon has been termed "fingerprint-assisted grinding" [54]. In the mechanical treatment of carbamazepine form III, the polymorphic transformation rate was found to be negligible when the milling beads were rigorously cleaned with solvent and handled with forceps. However, when the same balls were briefly rolled between bare fingers, the transformation proceeded rapidly, producing the rather unexpected outcome of the trigonal form II.

The mechanism governing this effect is as much chemical as it is physical. While efficient mechanochemical activation requires the powder to be trapped within the collision zone to ensure maximum energy transfer, the role of trace grease (e.g., such as a fingerprint) extends beyond simple adhesion. The saturated fatty acids present in sebum modify the surface energy of the milling media, but they also act as a non-polar micro-environment. This promotes the transition to the trigonal Form II, likely through the transient formation of a non-stoichiometric solvate that subsequently desolvates [55]. Thus, the fingerprint does not merely facilitate a "direct hit" by the milling bead; it actively lowers the polymorphic transformation barrier by providing the non-polar conditions necessary for the nucleation of energetically less favorable crystalline phases.

This finding has profound implications for the standardization of mechanochemical protocols. It suggests that inconsistencies in reported yields or kinetics between different laboratories (or even different researchers in the same laboratory) may stem from variations in equipment cleaning and handling procedures. It underscores the necessity for strict monitoring and reporting of "pre-milling" conditions, including the cleaning solvents used and the method of loading the reactor [34, 54].

## 6.3 Mechanochemical Treatment of Pharmaceuticals

In pharmaceutical science, mechanochemistry is utilized not merely for particle size reduction (comminution) but as a potent tool for "crystal engineering" to modify the solid-state properties of Active Pharmaceutical Ingredients (APIs) [56, 57]. The poor aqueous solubility of many modern drug candidates (BCS Class II and IV) presents a major challenge for bioavailability. Mechanochemical treatment can overcome high lattice energies to create disordered states, amorphous phases, salts, and co-crystals with enhanced solubility and dissolution rates [11, 56].

### *6.3.1 Ball Milling*

Ball milling is the most prevalent technique for the mechanical activation of pharmaceuticals [12]. The process involves the collision of grinding media with the powder charge, leading to a sequence of events: particle fracture (comminution), creation of fresh surfaces, plastic deformation, and the generation of crystal defects (dislocations, vacancies).

#### 6.3.1.1 Neat Grinding (NG)

Neat grinding (NG), or dry grinding, refers to the milling of reactants in the absence of any added solvent. It is the simplest form of mechanochemistry and is widely used for amorphization and the synthesis of some co-crystals.

Milling crystalline drugs below their $T_g$ can lead to a complete loss of long-range order, resulting in an amorphous state [58]. This state has a higher free energy and consequently higher apparent solubility than the crystalline form. However, NG is often limited by slow kinetics and incomplete conversion. As the reaction progresses, a product layer can coat the reactant particles, creating a diffusion barrier that halts the reaction [12]. Furthermore, NG tends to generate products with broad diffraction peaks indicating low crystallinity or high disorder, which can be detrimental to physical stability (tendency to recrystallize).

In highly exothermic inorganic systems, NG can trigger self-propagating high-temperature synthesis (SHS), but this is rare in pharmaceuticals due to the generally lower exothermicity of organic reactions [59, 60].

#### 6.3.1.2 Liquid-Assisted Grinding (LAG)

Liquid-assisted grinding (LAG), also known as solvent-drop grinding, is a technique where a small, catalytic amount of liquid is added to the milling jar. The quantity of liquid is defined by the parameter $\eta$, expressed as the ratio of liquid volume (μL) to the total weight of the solids (mg) [61, 62].

LAG is typically defined by the range $0 < \eta < 1$ μL/mg. If $\eta = 0$, it is neat grinding; if $\eta > 1$, the system is considered a slurry. The addition of this small amount of liquid has profound effects. It acts as a lubricant, facilitating mass transfer and diffusion between solid particles, which significantly accelerates reaction kinetics compared to NG. It can enable reactions that do not proceed at all under neat conditions. A critical advantage of LAG is the ability to control polymorphism. By varying the polarity of the liquid additive, one can selectively direct the formation of specific polymorphs [62]. For instance, in the co-crystallization of caffeine and glutaric acid, non-polar liquids yield Form I, while polar liquids yield Form II. This level of control is often unattainable in solution crystallization or neat grinding.

Unlike NG, which promotes amorphization, LAG often promotes the formation of highly crystalline products [63, 64]. The liquid facilitates the continuous recrystallization of the intermediate amorphous phase, yielding products with sharp diffraction peaks and better physical stability.

#### 6.3.1.3 Polymer-Assisted Grinding (PoLAG)

Polymer-assisted grinding (POLAG) involves milling APIs in the presence of solid polymers (e.g., polyethylene glycol (PEG), polyvinylpyrrolidone (PVP), hydroxypropyl methylcellulose (HPMC)) [13, 63, 65]. The polymer amount is sometimes described by a parameter $\delta$, analogous to $\eta$ in LAG.

One of the primary uses of POLAG is to produce and stabilize amorphous solid dispersions. The polymer interacts with the drug (often via hydrogen bonding) and raises the $T_g$ of the mixture, thereby restricting molecular mobility and preventing recrystallization of the high-energy amorphous drug during storage. Polymers can also act as heteronuclei or templates. In co-crystal synthesis, specific polymers can direct the crystallization of specific polymorphs that might be difficult to isolate otherwise [13, 63]. For example, milling with PEG can catalyze the formation of specific co-crystal forms by acting as a transient solvent or surface template. Polymers can absorb water released during the milling of hydrates (scavenging), preventing the powder from becoming a sticky paste (caking) and ensuring efficient energy transfer during milling.

### *6.3.2 Acoustic Mixing*

Resonant Acoustic Mixing (RAM) represents a paradigm shift in mechanochemical processing. Unlike ball mills that rely on the stochastic collision of media RAM utilizes low-frequency acoustic energy (typically around 60 Hz) to induce mixing [31].

The RAM device consists of a vessel mounted on a resonant spring system. An oscillator drives the system at its resonant frequency, generating acceleration forces up to 100 g (where g is gravitational acceleration, 9.81 m/s$^2$) [31]. The physics of the system is described by the equation of motion (Eq. 6.2):

$$F_0 \sin(\omega t) = m\frac{d^2}{dt^2}x(t) + c\frac{d}{dt}x(t) + kx(t) \tag{6.2}$$

At mechanical resonance, the inertial forces ($m$) and spring forces ($k$) cancel each other out, allowing the input force ($F_0$) to be applied directly to the mixing of the material ($c$, the damping coefficient).

Inside the sealed vessel, the intense vertical acceleration creates a complex, chaotic flow regime that facilitates "contactless" mixing. Figure 6.2 provides a cross-sectional schematic representation of the fluid dynamics occurring within the material charge during operation. The bottom of the vessel acts as an oscillating boundary, transmitting energy into the powder or slurry bed. The high-frequency vibration generates longitudinal acoustic waves that propagate through the medium. Because the material is mechanically decoupled from the driver, these waveforms

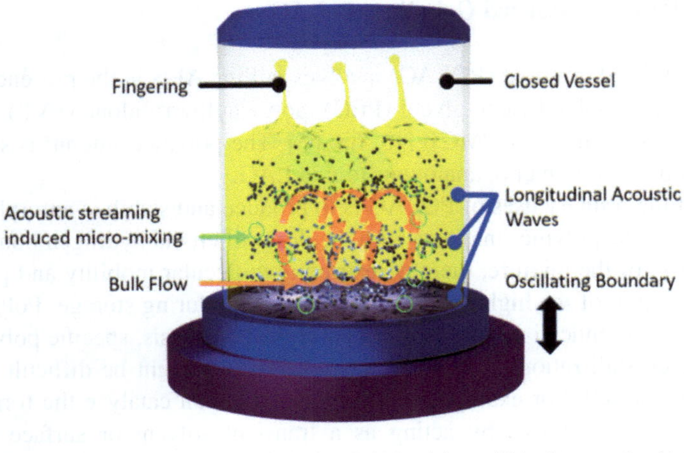


**Fig. 6.2** The processes which take place during RAM mixing within the sample. (Adapted with permission from [66])

break and scatter within the sample. The attenuation of these acoustic waves generates a steady momentum flux known as acoustic streaming. This induces intense micro-mixing zones (indicated by the small circular vectors in Fig. 6.2) distributed throughout the bulk of the material [66]. These zones are responsible for the intimate mixing of components at the particle level, facilitating mass transfer for co-crystallization or chemical reaction without the need for grinding media. At the free surface of the material, the intense acceleration often results in Faraday instabilities, visualized as "fingering" (Fig. 6.2), where material is ejected upward and tossed back into the bulk. This "free-fall" mechanism contributes further to the randomization and homogenization of the mixture [31, 66].

This combination of bulk chaotic flow and localized acoustic streaming allows RAM to mix materials with vastly different rheological properties (e.g., mixing solids into viscous pastes) and drive mechanochemical reactions uniformly, avoiding the localized "hot spots" typical of ball milling impacts.

This unique environment has been successfully applied to produce ASDs through coprecipitation, demonstrating good phase mixing and similar $T_g$s compared to other methods like spray drying (SD). RAM prepared ASDs showed superior mechanical properties such as tabletability and flow, indicating its suitability for downstream pharmaceutical processing [67]. While RAM is well-established for co-crystals, its use in amorphous material synthesis is emerging, with evidence supporting its scalability and high potential to tune material properties by controlling mixing parameters [67, 68]. The technology's gentle, blade-free mixing reduces mechanical stress, which may help maintain the amorphous state without inducing crystallization or degradation [69, 70]. However, detailed studies specifically focusing on RAM for purely amorphous materials beyond ASDs are limited, suggesting that while promising, more research is needed to fully establish its effectiveness for broader amorphous material synthesis. Overall, RAM shows potential as a scalable and efficient method for preparing amorphous pharmaceutical formulations with improved mechanical characteristics [67, 68].

#### 6.3.2.1 Advantages

A key feature of RAM is that it can operate effectively without grinding media (balls), for which it has been termed "ball-free mechanochemistry" [70]. The elimination of balls reduces the risk of localized hot spots and frictional sparks. This makes RAM particularly safer for processing energetic materials (explosives, propellants) where traditional milling poses significant detonation risks. Moreover, it eliminates contamination arising from the wear and tear of milling media (e.g., iron or zirconia contamination), a common issue in ball milling [31, 71]. RAM technology is readily scalable from laboratory vials to industrial drums (hundreds of kilograms) because the mixing physics remains consistent across scales, addressing one of the main limitations of traditional mechanochemistry [72, 73].

#### 6.3.2.2 Applications and Kinetic Control

RAM has been successfully applied to the synthesis of pharmaceutical co-crystals (e.g., carbamazepine-nicotinamide), MOFs, and energetic material co-crystals (e.g., CL-20:HMX) with 100% yield [74]. However, the use of RAM specifically for amorphous materials is less documented.

In co-crystallization, the mechanism can switch based on mixing intensity. At low accelerations (e.g., 50 g), reaction products may coat the reactant particles (core-shell effect), halting the reaction. However, at high accelerations (100 g), the intense shear and particle-particle collisions facilitate the removal of this coating layer, exposing fresh reactant surfaces and driving the reaction toward the final co-crystal product. This phenomenon of "kinetic switching" was explicitly observed in the liquid-assisted mechanochemical synthesis of the pharmaceutical co-crystal carbamazepine-nicotinamide (CBZ-NIC) using Resonant Acoustic Mixing (RAM) [72]. In situ monitoring using time-resolved X-ray diffraction revealed that the reaction pathway is heavily dependent on the acceleration applied (G-force), acting as a switch between kinetic trapping and thermodynamic completion.

When the synthesis is conducted at lower accelerations (50 g) in the presence of water (liquid-assisted grinding), the reaction does not proceed directly to the co-crystal. Instead, carbamazepine (CBZ) reacts rapidly with the added water to form carbamazepine dihydrate (CBZ·DH). This dihydrate phase forms a passive coating over the surface of the unreacted CBZ particles. Because the mixing intensity is insufficient to abrade or strip away this layer, the system becomes kinetically trapped. The CBZ·DH layer acts as a diffusion barrier (core-shell effect), preventing nicotinamide from contacting the underlying CBZ core, thereby halting the co-crystallization process at a low conversion yield (approx. 10%).

When the acceleration is increased to 100 g, the mechanical dynamics within the jar change significantly. Although the CBZ·DH phase still forms initially, it behaves as a transient reactive intermediate rather than a kinetic trap. Higher acceleration generates intense shear forces and more energetic particle-particle collisions. These forces are sufficient to continuously strip away the dihydrate product layer as it forms, a process known as surface regeneration [72]. By removing the coating, fresh surfaces of the reactants are continuously exposed, maintaining mass transfer. Consequently, the intermediate CBZ·DH is consumed, and the reaction proceeds to full conversion, yielding the thermodynamically stable CBZ-NIC co-crystal.

This mechanism highlights a distinct advantage of RAM over traditional ball milling. In RAM, the "impact severity" can be fine-tuned via the G-force parameter to control particle interaction regimes (coating vs. stripping) without introducing the contamination or high-energy damage associated with grinding media (balls). At high accelerations, RAM promotes the generation of reaction zones through comminution and the renewal of particle interfaces, preventing the agglomeration that typically inhibits reactions at lower energy inputs.

## 6.4 Mechanisms of Amorphization by Mechanical Activation

Mechanical activation via milling introduces energy into the solid system, primarily through impact and shear forces. This energy input leads to the disruption of the long-range order characteristic of the crystal lattice [2, 5]. While melting and solvent-based methods achieve amorphization by transitioning through a thermodynamically stable disordered intermediate (liquid or solution), mechanical activation drives a direct transition from the crystalline to the disordered amorphous solid state [75]. This process is not necessarily thermal in origin; although milling generates heat, the amorphization often occurs far below the melting point of the crystal, suggesting that lattice disruption is driven by the application of mechanical stress rather than transient melting [5].

The fundamental driver for solid-state amorphization is the accumulation of crystal defects. As mechanical stress is applied, defects such as vacancies, dislocations, and grain boundaries accumulate within the lattice [5]. When the density of these defects reaches a critical threshold, the free energy of the defective crystal exceeds that of the amorphous state, driving a spontaneous collapse of the lattice structure into a disordered arrangement [3, 5]. This mechanism explains why amorphization is often gradual, requiring specific milling durations to build sufficient defect density to trigger the phase transition.

Milling concurrently reduces particle size and increases specific surface area. As particle size decreases down to the micron and sub-micron range, the ratio of surface area to volume increases significantly. Molecules at the surface of a particle have higher mobility and higher energy states compared to the bulk [76–78]. This high surface energy contributes to the overall destabilization of the crystalline state. However, the creation of a large amount of free surface can also facilitate surface-enhanced crystallization, where nucleation and crystal growth occur significantly faster at the surface than in the bulk, potentially countering the amorphization process [77, 78].

Mechanical activation can induce transformations beyond simple amorphization, including polymorphic transitions mediated by transient amorphous phases. In some cases, such as these of fananserin or indomethacin, milling induces a transformation from a stable to a metastable polymorph via a detectable intermediate amorphous phase [79, 80]. The mechanical energy drives the system to explore the energy landscape, eventually pushing the material to an "effective temperature" where barriers between different solid-state forms can be overcome [81, 82].

## 6.4.1 Drug–Polymer Systems in Mechanochemical ASDs

In the context of ASDs, co-milling a drug with a polymer facilitates intimate mixing at the molecular level. Mechanical energy promotes the development of intermolecular interactions, such as hydrogen bonding and ionic interactions, between the drug and the carrier [83, 84]. For example, milling ketoconazole with cross-linked poly(acrylic acid) allows for the formation of a stable amorphous system where the polymer restricts drug mobility [85]. Acid-base reactions can also be induced thermomechanically; milling naproxen with Eudragit E PO results in the formation of ionic polyelectrolyte complexes that stabilize the amorphous drug [86].

The final state of a milled material is determined by a kinetic competition between mechanically induced disordering (ballistic jumps) and thermally activated restoration (recrystallization) [5]. If milling is performed at temperatures significantly below the $T_g$ of the system, the relaxation and recrystallization kinetics are slow, favoring the accumulation of disorder and the formation of a stable amorphous product [5, 80]. Conversely, if the milling temperature approaches or exceeds the $T_g$ (which can occur due to heat generation during processing), molecular mobility increases, potentially leading to rapid recrystallization of the generated amorphous material into stable or metastable crystalline forms [79, 80].

The selection of a suitable polymer is critical for stabilizing the high-energy amorphous drug. Key criteria include the polymer's ability to inhibit crystallization, its miscibility with the drug, and its glass transition temperature [3]. Polymers must be chosen to interact with the specific functional groups of the API (e.g., hydrogen bond donors/acceptors) to reduce the chemical potential of the drug and hinder nucleation [83, 84]. Additionally, the polymer should have a sufficiently high $T_g$ to raise the $T_g$ of the resulting composite ASD, thereby reducing molecular mobility at storage temperatures [87, 88].

### 6.4.2 Polymer Properties Influencing Milling Outcomes

The selection of a polymer for mechanochemical amorphization is dictated by a complex interplay of thermal, molecular, and mechanical properties. A primary consideration is the $T_g$; polymers with high $T_g$ values (e.g., PVP or HPMC) are generally preferred for their robust anti-plasticizing effect. By increasing the viscosity of the drug-polymer matrix, these carriers slow down the molecular diffusion processes that lead to recrystallization. Predictively, the $T_g$ of the resulting binary mixture typically follows the Gordon-Taylor equation, and deviations from this model often serve as a diagnostic indicator of specific molecular interactions, such as hydrogen bonding, between the components [89, 90].

This stabilization is further modulated by the molecular weight of the polymer. While higher molecular weight grades enhance viscosity and provide superior steric hindrance against crystal growth, they can present significant processing challenges due to decreased flowability [91]. Furthermore, the hygroscopicity of the selected matrix is a critical environmental variable. Because water is a potent plasticizer, even minor moisture absorption during milling or storage can drastically lower the $T_g$ and increase molecular mobility. Consequently, managing the effect of moisture sorption (polymers are highly hygroscopic) is essential to prevent moisture-induced recrystallization [92–94].

Beyond thermal stability, the intrinsic mechanical behavior of the polymer, specifically its brittleness versus ductility, directly impacts milling efficiency. Polymers that exhibit brittle fracture are more conducive to mechanochemistry as they facilitate rapid particle size reduction and intimate mixing. In contrast, ductile or rubbery polymers may undergo plastic deformation, effectively absorbing the impact energy of the grinding media without achieving the necessary fracturing or homogeneous phase distribution [5].

### 6.4.3 Process Parameters and Their Influence

The transition from a crystalline lattice to a disordered amorphous state is not merely a function of total energy, but a competition between ballistic disordering and thermal restoration. Milling time and intensity (frequency/speed) are the primary drivers of this kinetics, while longer milling durations generally increase defect accumulation and amorphous content, they eventually reach a steady state. For fast-crystallizing compounds like acetaminophen, a dynamic equilibrium is established where the amorphization rate is perfectly balanced by the recrystallization rate, preventing 100% conversion [6, 8].

This balance is heavily influenced by the milling intensity and the effective temperature of the system. High-intensity milling, driven by high speeds and optimized milling media (ball size, number, and density), imparts the kinetic energy necessary for "ballistic jumps" that force molecules into a higher-energy, disordered state [5].

However, this same energy generates significant heat via friction and impact. Temperature control is therefore paramount; if the local temperature surpasses the $T_g$, the material enters a rubbery state where molecular mobility is high enough to trigger recrystallization [95]. This necessitates the use of cryomilling for thermally sensitive compounds to suppress these restoration pathways [96].

Beyond mechanical settings, the formulation environment dictates the stability of the resulting Amorphous Solid Dispersion (ASD):

1. *Drug-to-Polymer Ratio*: The polymer acts as both a physical barrier (dilution effect) and a thermal stabilizer (raising the composite $T_g$) [97, 98]. There is typically a threshold concentration required to stabilize the amorphous drug; for instance, acetaminophen requires at least 60% PVP to prevent surface-mediated crystallization during the milling process [8].
2. *Atmospheric Control*: The presence of moisture can act as a potent plasticizer, drastically lowering the $T_g$. Consequently, milling is frequently conducted under dry, inert atmospheres to mitigate moisture-induced instability [93].
3. *Grinding Aids*: The use of LAG or POLAG can fundamentally alter the reaction coordinate. While neat grinding relies on dry impact, LAG utilizes catalytic amounts of solvent to accelerate structural transitions, often facilitating the formation of glassy solutions that are otherwise difficult to reach through purely mechanical means [6].

Ultimately, these parameters must be carefully calibrated during scale-up. Moving from laboratory-scale ball mills to industrial scales (e.g., vibrational mills, attritors) presents challenges in maintaining consistent energy input and temperature control. The "effective temperature" concept suggests that maintaining the same balance between ballistic intensity and thermal relaxation is key to reproducing the solid form at larger scales [5]. Additionally, downstream processing of milled ASDs, such as compression, must be carefully managed, as mechanical stress during tableting can induce phase separation or crystallization.

## 6.5 Advantages and Limitations of Mechanochemical Methods

### *6.5.1 A Solvent-Free Method: Safety and Environmental Impact*

The primary advantage of mechanochemical processing is the elimination of organic solvents, which addresses a critical safety and stability bottleneck in pharmaceutical manufacturing [5]. By operating in the solid state, researchers circumvent the risk of residual solvents acting as plasticizers, a common issue in solvent-based methods that lowers the $T_g$ and triggers premature recrystallization [3]. This "neat" processing is particularly beneficial for hydrolysis-sensitive compounds and removes the need for energy-intensive secondary drying steps, which often risk the thermal degradation of delicate APIs.

From a regulatory and "Green Chemistry" perspective, these solvent-free routes significantly reduce hazardous waste and simplify compliance with VOC and residual solvent limits (e.g., ICH Q3C) [99, 100]. Furthermore, the absence of flammable solvents eliminates the inherent explosion risks associated with technologies like spray drying, thereby reducing the capital expenditure required for specialized, explosion-proof facilities.

### *6.5.2 Energy Dynamics and Efficiency*

While mechanochemistry is environmentally "greener," its energy efficiency is highly process-dependent. Methods like HME or KinetiSol® require substantial energy to generate shear and frictional heat, yet they can be more efficient than spray drying when compared to the massive latent heat of vaporization required to remove solvents [101]. KinetiSol®, for instance, can generate the necessary heat in under 30 s, drastically minimizing the thermal history of the drug compared to traditional extrusion [102, 103]. Conversely, high-energy ball milling may require prolonged processing times to achieve full amorphization, making it more energy-intensive on a per-gram basis.

### *6.5.3 Operational Challenges: Reproducibility, Contamination, and Scale-Up*

The transition from laboratory to industrial scale introduces significant hurdles, primarily regarding process reproducibility. The final solid state is a result of a delicate "ballistic vs. thermal" competition. Small fluctuations in milling intensity or barrel temperature can shift the "effective temperature" of the system, leading to inconsistent steady states such as partial amorphization or unintended polymorphic shifts [5, 6, 102, 103].

Moreover, the intense mechanical forces required for amorphization pose risks of equipment wear and product contamination. Impact and friction in ball mills can introduce microscopic metallic or ceramic "shrapnel" from the milling media (e.g., zirconia or steel) into the product [40, 41]. Additionally, the cohesive and "sticky" nature of amorphous solids can lead to significant adhesion on equipment surfaces, reducing yields and complicating the cleaning validation required for GMP-compliant, multi-product facilities [4].

Finally, scale-up remains underexplored for many mechanochemical techniques. While HME is a continuous process and relatively scalable, batch processes like high-energy ball milling struggle to meet commercial throughput requirements. The adoption of these novel techniques is currently limited not by their chemical efficacy, but by a lack of large-scale, GMP-compliant equipment capable of maintaining consistent shear histories and heat transfer across varying volumes [101].

## 6.6 Conclusions

Mechanochemical processing has emerged as a powerful and versatile platform for the preparation of amorphous solid dispersions, offering a fundamentally different pathway to disorder compared with thermal and solvent-based techniques. By directly coupling mechanical energy into solid-state systems, milling, extrusion, and acoustic mixing enable crystal lattice disruption, defect accumulation, and molecular-level mixing without requiring bulk solvents or prolonged thermal exposure. This capability is particularly valuable for poorly water-soluble and thermolabile APIs, where conventional spray drying or melt processing may be unsuitable or environmentally burdensome.

A central theme across mechanochemical ASD preparation is the balance between mechanically induced disorder and thermally driven relaxation. The final solid-state outcome is governed not by a single parameter but by a network of interdependent variables, including stress mode, impact severity, temperature evolution, polymer properties, and reactor environment. Concepts such as effective temperature, defect accumulation, and stress-augmented activation provide a useful framework for understanding why small variations in milling intensity, milling media or material can shift outcomes between amorphization, polymorphic transformation, co-crystallization, or recrystallization. For drug–polymer systems, successful ASD formation depends strongly on polymer selection, drug-polymer miscibility, intermolecular interactions, and $T_g$ elevation, all of which jointly determine the physical stability of the amorphous phase.

Polymer-assisted grinding and related hybrid approaches demonstrate that mechanochemistry is not limited to particle size reduction but can be used as a true solid-state formulation tool. Mechanical activation promotes intimate mixing and interaction between API and carrier, enabling the direct formation of glass solutions and ionic or hydrogen-bonded complexes. Complementary strategies such as liquid-assisted and polymer-assisted grinding expand kinetic control and phase selectivity, while cryogenic and humidity-controlled milling provide routes to stabilize otherwise labile amorphous products. At the same time, recent findings on reactor and surface effects, including media composition, polymer jar damping, and surface contamination, highlight the need for stricter protocol standardization and reporting to ensure reproducibility.

Emerging technologies broaden the mechanochemical toolbox beyond traditional ball mills. Reactive extrusion offers a continuous, scalable route consistent with green manufacturing principles, while resonant acoustic mixing provides media-free, low-contamination processing with tunable mixing intensity and promising scale-up characteristics. These newer platforms address several long-standing limitations of batch milling, particularly with respect to throughput, contamination risk, and process control, and they are likely to play an increasing role in future ASD manufacturing strategies.

Despite clear advantages, challenges remain for widespread industrial adoption. Energy efficiency, temperature management, equipment wear, contamination, and scale translation must be carefully managed, and predictive models linking process parameters to ASD quality attributes are still developing. Greater integration of in situ and operando analytical techniques, together with data-driven process modeling, will be essential to move mechanochemical ASD production from empirically optimized practice toward rational, design-based control.

Overall, mechanochemical approaches provide a solvent-minimized, structurally direct, and increasingly scalable route to amorphous solid dispersions. As understanding of stress-driven phase behavior, reactor physics, and drug–polymer interactions continues to deepen, these methods are positioned to become a core component of the advanced formulation toolbox for next-generation poorly soluble drugs.

## References

1. Karagianni A, Kachrimanis K, Nikolakakis I. Co-amorphous solid dispersions for solubility and absorption improvement of drugs: composition, preparation, characterization and formulations for oral delivery. Pharmaceutics. 2018;10(3):98.
2. Vasconcelos T, Marques S, das Neves J, Sarmento B. Amorphous solid dispersions: rational selection of a manufacturing process. Adv Drug Deliv Rev. 2016;100:85–101.
3. Vo CL, Park C, Lee BJ. Current trends and future perspectives of solid dispersions containing poorly water-soluble drugs. Eur J Pharm Biopharm. 2013;85(3 Pt B):799–813.
4. Bhujbal SV, Mitra B, Jain U, Gong Y, Agrawal A, Karki S, et al. Pharmaceutical amorphous solid dispersion: a review of manufacturing strategies. Acta Pharm Sin B. 2021;11(8):2505–36.
5. Descamps M, Willart JF. Perspectives on the amorphisation/milling relationship in pharmaceutical materials. Adv Drug Deliv Rev. 2016;100:51–66.
6. Willart JF, Descamps M. Solid state amorphization of pharmaceuticals. Mol Pharm. 2008;5(6):905–20.
7. Mesallati H, Tajber L. Polymer/amorphous salt solid dispersions of ciprofloxacin. Pharm Res. 2017;34(11):2425–39.
8. Capece M, Davé R. Enhanced physical stability of amorphous drug formulations via dry polymer coating. J Pharm Sci. 2015;104(6):2076–84.
9. Baláž P. Mechanochemistry and nanoscience. In: Baláž P, editor. Mechanochemistry in nanoscience and minerals engineering. Berlin/Heidelberg, Springer Berlin Heidelberg; 2008. p. 1–102.
10. Boldyrev VV. Mechanochemistry and mechanical activation. Mater Sci Forum. 1996;225–227:511–20.
11. Boldyreva E. Fundamentals of mechanochemistry. In: Török B, editor. Encyclopedia of Green Chemistry. 1st ed. Oxford: Elsevier, 2025; 2:419–436.

12. Kaupp G. Mechanochemistry: the varied applications of mechanical bond-breaking. CrystEngComm. 2009;11(3):388–403.
13. Michalchuk AAL, Boldyreva EV, Belenguer AM, Emmerling F, Boldyrev VV. Tribochemistry, mechanical alloying, mechanochemistry: what is in a name? Front Chem. 2021;9:685789.
14. Marchini M, Montanari G, Casali L, Martelli M, Raggetti L, Baláž M, et al. "What makes every work perfect is cooking and grinding": the ancient roots of mechanochemistry. RSC Mechanochem. 2024;1(1):123–9.
15. Chadwick K, Davey R, Cross W. How does grinding produce co-crystals? Insights from the case of benzophenone and diphenylamine. CrystEngComm. 2007;9(9):732–4.
16. Makarov DE. Perspective: mechanochemistry of biological and synthetic molecules. J Chem Phys. 2016;144(3):030901.
17. Takacs L. The historical development of mechanochemistry. Chem Soc Rev. 2013;42(18):7649–59.
18. Chalk S, McEwen L. The IUPAC gold book. Chem Int. 2017;39:25–30.
19. Auvray T, Friščić T. Shaking things from the ground-up: a systematic overview of the mechanochemistry of hard and high-melting inorganic materials. Molecules [Internet]. 2023; 28(2):[897 p.].
20. James SL, Adams CJ, Bolm C, Braga D, Collier P, Friščić T, et al. Mechanochemistry: opportunities for new and cleaner synthesis. Chem Soc Rev. 2012;41(1):413–47.
21. Alrbaihat M, Khalil Al-Zeidaneen F, Abu-Afifeh Q. Reviews of the kinetics of mechanochemistry: theoretical and Modeling aspects. Mater Today: Proc. 2022;65:3651–6.
22. Lapshin OV, Boldyreva EV, Boldyrev VV. Role of mixing and milling in mechanochemical synthesis (review). Russ J Inorg Chem. 2021;66(3):433–53.
23. Leonardi M, Villacampa M, Menéndez JC. Multicomponent mechanochemical synthesis. Chem Sci. 2018;9(8):2042–64.
24. Floyd K, Gonnet L, Friščić T, Batteas J. The role of the milling environment on the copper-catalysed mechanochemical synthesis of tolbutamide. RSC Mechanochem. 2024;1(3):289–95.
25. Spikes H. Stress-augmented thermal activation: tribology feels the force. Friction. 2018;6(1):1–31.
26. Gosvami NN, Bares JA, Mangolini F, Konicek AR, Yablon DG, Carpick RW. Tribology. Mechanisms of antiwear tribofilm growth revealed in situ by single-asperity sliding contacts. Science. 2015;348(6230):102–6.
27. Tumanov IA, Achkasov AF, Myz SA, Boldyreva EV, Boldyrev VV. Different effect of impact and shear mechanical treatment on mechanochemical cocrystallization of piroxicam and succinic acid. Dokl Chem. 2014;457(2):154–9.
28. Boldyreva E. Spiers memorial lecture: mechanochemistry, tribochemistry, mechanical alloying—retrospect, achievements and challenges. Faraday Discuss 2023;241(0):9–62.
29. Boldyreva EV, Shakhtshneider TP, Ahsbahs H, Sowa H, Uchtmann H. Effect of high pressure on the polymorphs of paracetamol. J Therm Anal Calorim. 2002;68(2):437–52.
30. Boldyreva EV, Shakhtshneider TP, Vasilchenko MA, Ahsbahs H, Uchtmann H. Anisotropic crystal structure distortion of the monoclinic polymorph of acetaminophen at high hydrostatic pressures. Acta Crystallogr B. 2000;56(Pt 2):299–309.
31. Michalchuk AAL, Boldyreva EV. Resonant acoustic mixing: applications to organic and materials synthesis. In: Török B, editor. Encyclopedia of Green Chemistry. 1st ed. Oxford: Elsevier; 2025. p. 532–41.
32. Germann LS, Arhangelskis M, Etter M, Dinnebier RE, Friščić T. Challenging the Ostwald rule of stages in mechanochemical cocrystallisation. Chem Sci. 2020;11(37):10092–100.
33. Kulla H, Becker C, Michalchuk AAL, Linberg K, Paulus B, Emmerling F. Tuning the apparent stability of polymorphic cocrystals through mechanochemistry. Cryst Growth Des. 2019;19(12):7271–9.
34. Losev E, Arkhipov S, Kolybalov D, Mineev A, Ogienko A, Boldyreva E, et al. Substituting steel for a polymer in a jar for ball milling does matter. CrystEngComm. 2022;24(9):1700–3.

35. Fischer F, Fendel N, Greiser S, Rademann K, Emmerling F. Impact is important—systematic investigation of the influence of milling balls in mechanochemical reactions. Org Process Res Dev. 2017;21(4):655–9.
36. Kulla H, Fischer F, Benemann S, Rademann K, Emmerling F. The effect of the ball to reactant ratio on mechanochemical reaction times studied by in situ PXRD. CrystEngComm. 2017;19(28):3902–7.
37. Michalchuk AAL, Tumanov IA, Boldyreva EV. The effect of ball mass on the mechanochemical transformation of a single-component organic system: anhydrous caffeine. J Mater Sci. 2018;53(19):13380–9.
38. Michalchuk AAL, Tumanov IA, Boldyreva EV. Ball size or ball mass—what matters in organic mechanochemical synthesis? CrystEngComm. 2019;21(13):2174–9.
39. Yuan Y, Wang L, Li D, Deng Z, Zhang H. How many parameters can affect the solid form of cocrystallization products in mechanochemical reactions? A case study. Cryst Growth Des. 2018;18(12):7244–7.
40. Gammage RB, Glasson DR. Wear of mill components during the ball milling of calcium carbonate. J Colloid Interface Sci. 1979;71(3):522–5.
41. Juhnke M, Märtin D, John E. Generation of wear during the production of drug nanosuspensions by wet media milling. Eur J Pharm Biopharm. 2012;81(1):214–22.
42. Di Nardo T, Moores A. Mechanochemical amorphization of chitin: impact of apparatus material on performance and contamination. Beilstein J Org Chem. 2019;15:1217–25.
43. Hwang S, Grätz S, Borchardt L. A guide to direct mechanocatalysis. Chem Commun. 2022;58(11):1661–71.
44. Kubota K, Ito H. Mechanochemical cross-coupling reactions. Trends Chem. 2020;2(12):1066–81.
45. Pickhardt W, Grätz S, Borchardt L. Direct mechanocatalysis: using milling balls as catalysts. Chemistry. 2020;26(57):12903–11.
46. Halasz I, Puškarić A, Kimber SAJ, Beldon PJ, Belenguer AM, Adams F, et al. Real-time in situ powder X-ray diffraction monitoring of mechanochemical synthesis of pharmaceutical cocrystals. Angew Chem Int Ed. 2013;52(44):11538–41.
47. Weidenthaler C. In situ analytical methods for the characterization of mechanochemical reactions. Crystals [Internet]. 2022; 12(3):[345 p.].
48. Tumanov N, Ban V, Poulain A, Filinchuk Y. 3D-printed jars for ball-milling experiments monitored in situ by X-ray powder diffraction. J Appl Crystallogr. 2017;50(4):994–9.
49. Broseghini M, D'Incau M, Gelisio L, Pugno NM, Scardi P. Numerical and experimental investigations on new jar designs for high efficiency planetary ball milling. Adv Powder Technol. 2020;31(7):2641–9.
50. Broseghini M, D'Incau M, Gelisio L, Pugno NM, Scardi P. Effect of jar shape on high-energy planetary ball milling efficiency: simulations and experiments. Mater Des. 2016;110:365–74.
51. Linberg K, Emmerling F, Michalchuk AAL. Unintended rate enhancement in mechanochemical kinetics by using poly(methyl methacrylate) jars. Cryst Growth Des. 2023;23(1):19–23.
52. Erokhin KS, Gordeev EG, Ananikov VP. Revealing interactions of layered polymeric materials at solid-liquid interface for building solvent compatibility charts for 3D printing applications. Sci Rep. 2019;9(1):20177.
53. Heikkinen ITS, Kauppinen C, Liu Z, Asikainen SM, Spoljaric S, Seppälä JV, et al. Chemical compatibility of fused filament fabrication-based 3-D printed components with solutions commonly used in semiconductor wet processing. Addit Manuf. 2018;23:99–107.
54. Zheltikova D, Losev E, Boldyreva E. To touch or not to touch? Fingerprint-assisted grinding of carbamazepine form III. CrystEngComm. 2023;25(34):4879–88.
55. Kioseoglou J, Boldyreva E, Kachrimanis K. Molecular dynamics simulations of the mechanically induced crystallization of carbamazepine's trigonal polymorph, CBZ(II). Struct Chem. 2026;
56. Delori A, Friščić T, Jones W. The role of mechanochemistry and supramolecular design in the development of pharmaceutical materials. CrystEngComm. 2012;14(7):2350–62.

57. Solares-Briones M, Coyote-Dotor G, Páez-Franco JC, Zermeño-Ortega MR, de la O Contreras CM, Canseco-González D, et al. Mechanochemistry: a green approach in the preparation of pharmaceutical cocrystals. Pharmaceutics [Internet]. 2021; 13(6):[790 p.].
58. Iyer J, Brunsteiner M, Modhave D, Paudel A. Role of crystal disorder and mechanoactivation in solid-state stability of pharmaceuticals. J Pharm Sci. 2023;112(6):1539–65.
59. Mossino P. Some aspects in self-propagating high-temperature synthesis. Ceram Int. 2004;30(3):311–32.
60. Munir ZA, Anselmi-Tamburini U. Self-propagating exothermic reactions: the synthesis of high-temperature materials by combustion. Mater Sci Rep. 1989;3(7):277–365.
61. Do J-L, Friščić T. Mechanochemistry: a force of synthesis. ACS Cent Sci. 2017;3(1):13–9.
62. Hasa D, Miniussi E, Jones W. Mechanochemical synthesis of multicomponent crystals: one liquid for one polymorph? A myth to dispel. Cryst Growth Des. 2016;16(8):4582–8.
63. Hasa D, Schneider Rauber G, Voinovich D, Jones W. Cocrystal formation through mechanochemistry: from neat and liquid-assisted grinding to polymer-assisted grinding. Angew Chem Int Ed. 2015;54(25):7371–5.
64. Mukherjee A, Rogers RD, Myerson AS. Cocrystal formation by ionic liquid-assisted grinding: case study with cocrystals of caffeine. CrystEngComm. 2018;20(27):3817–21.
65. Germann LS, Emmerling ST, Wilke M, Dinnebier RE, Moneghini M, Hasa D. Monitoring polymer-assisted mechanochemical cocrystallisation through in situ X-ray powder diffraction. Chem Commun. 2020;56(62):8743–6.
66. Yalcin D, Rajesh S, White J, Howard SC, Pigram PJ, Tran N, et al. Resonant acoustic mixing method to produce lipid-based liquid-crystal nanoparticles. J Phys Chem C. 2021;125(19):10653–64.
67. Hou HH, Rajesh A, Pandya KM, Lubach JW, Muliadi A, Yost E, et al. Impact of method of preparation of amorphous solid dispersions on mechanical properties: comparison of coprecipitation and spray drying. J Pharm Sci. 2019;108(2):870–9.
68. Fowler HR, O'Shea R, Sefton J, Howard SC, Muir BW, Stockman RA, et al. Rapid, highly sustainable ring-opening polymerization via resonant acoustic mixing. ACS Sustain Chem Eng. 2025;13(5):1916–26.
69. Kefei J, Zihan Z, Kesheng L. Research progress of acoustic resonance hybrid technology and its application. Acad J Eng Technol Sci. 2025;8(2):90–5.
70. Michalchuk AAL, Hope KS, Kennedy SR, Blanco MV, Boldyreva EV, Pulham CR. Ball-free mechanochemistry: in situ real-time monitoring of pharmaceutical co-crystal formation by resonant acoustic mixing. Chem Commun. 2018;54(32):4033–6.
71. Lennox CB, Borchers TH, Gonnet L, Barrett CJ, Koenig SG, Nagapudi K, et al. Direct mechanocatalysis by resonant acoustic mixing (RAM). Chem Sci. 2023;14(27):7475–81.
72. am Ende DJ, Anderson SR, Salan JS. Development and scale-up of cocrystals using resonant acoustic mixing. Org Process Res Dev. 2014;18(2):331–41.
73. Nagapudi K, Umanzor EY, Masui C. High-throughput screening and scale-up of cocrystals using resonant acoustic mixing. Int J Pharm. 2017;521(1):337–45.
74. Anderson SR, am Ende DJ, Salan JS, Samuels P. Preparation of an energetic-energetic cocrystal using resonant acoustic mixing. Propellants Explos Pyrotech. 2014;39(5):637–40.
75. Patterson JE, James MB, Forster AH, Lancaster RW, Butler JM, Rades T. The influence of thermal and mechanical preparative techniques on the amorphous state of four poorly soluble compounds. J Pharm Sci. 2005;94(9):1998–2012.
76. Cai T, Zhu L, Yu L. Crystallization of organic glasses: effects of polymer additives on bulk and surface crystal growth in amorphous nifedipine. Pharm Res. 2011;28(10):2458–66.
77. Kestur US, Ivanesivic I, Alonzo DE, Taylor LS. Influence of particle size on the crystallization kinetics of amorphous felodipine powders. Powder Technol. 2013;236:197–204.
78. Wu T, Yu L. Surface crystallization of indomethacin below Tg. Pharm Res. 2006;23(10):2350–5.

79. De Gusseme A, Neves C, Willart JF, Rameau A, Descamps M. Ordering and disordering of molecular solids upon mechanical milling: the case of fananserine. J Pharm Sci. 2008;97(11):5000–12.
80. Descamps M, Willart JF, Dudognon E, Lefort R, Desprez S, Caron V. Phase transformations induced by grinding: what is revealed by molecular materials. MRS Online Proc Library. 2011;979(1):606.
81. Baláž P, Achimovičová M, Baláž M, Billik P, Cherkezova-Zheleva Z, Criado JM, et al. Hallmarks of mechanochemistry: from nanoparticles to technology. Chem Soc Rev. 2013;42(18):7571–637.
82. Enrique RA, Bellon P. Nonequilibrium fluctuations, effective temperature, and effective interactions driven by irradiation of alloys. Phys Rev B. 2004;70(22):224106.
83. Mesallati H, Umerska A, Paluch KJ, Tajber L. Amorphous polymeric drug salts as ionic solid dispersion forms of ciprofloxacin. Mol Pharm. 2017;14(7):2209–23.
84. Paluch KJ, McCabe T, Müller-Bunz H, Corrigan OI, Healy AM, Tajber L. Formation and physicochemical properties of crystalline and amorphous salts with different stoichiometries formed between ciprofloxacin and succinic acid. Mol Pharm. 2013;10(10):3640–54.
85. Mistry P, Suryanarayanan R. Strength of drug–polymer interactions: implications for crystallization in dispersions. Cryst Growth Des. 2016;16(9):5141–9.
86. Kindermann C, Matthée K, Strohmeyer J, Sievert F, Breitkreutz J. Tailor-made release triggering from hot-melt extruded complexes of basic polyelectrolyte and poorly water-soluble drugs. Eur J Pharm Biopharm. 2011;79(2):372–81.
87. Vasanthavada M, Tong W-Q, Joshi Y, Kislalioglu MS. Phase behavior of amorphous molecular dispersions II: role of hydrogen bonding in solid solubility and phase separation kinetics. Pharm Res. 2005;22(3):440–8.
88. Yoshioka M, Hancock BC, Zografi G. Crystallization of indomethacin from the amorphous state below and above its glass transition temperature. J Pharm Sci. 1994;83(12):1700–5.
89. Marsac PJ, Shamblin SL, Taylor LS. Theoretical and practical approaches for prediction of drug-polymer miscibility and solubility. Pharm Res. 2006;23(10):2417–26.
90. Taylor LS, Zografi G. Spectroscopic characterization of interactions between PVP and indomethacin in amorphous molecular dispersions. Pharm Res. 1997;14(12):1691–8.
91. Sahoo A, Kumar NSK, Suryanarayanan R. Crosslinking: an avenue to develop stable amorphous solid dispersion with high drug loading and tailored physical stability. J Control Release. 2019;311–312:212–24.
92. Ahlneck C, Zografi G. The molecular basis of moisture effects on the physical and chemical stability of drugs in the solid state. Int J Pharm. 1990;62(2):87–95.
93. Tong P, Zografi G. Effects of water vapor absorption on the physical and chemical stability of amorphous sodium indomethacin. AAPS PharmSciTech. 2004;5(2):e26.
94. Yoshioka S, Aso Y. Correlations between molecular mobility and chemical stability during storage of amorphous pharmaceuticals. J Pharm Sci. 2007;96(5):960–81.
95. Shakhtshneider TP. Phase transformations and stabilization of metastable states of molecular crystals under mechanical activation. Solid State Ionics. 1997;101–103:851–6.
96. Crowley KJ, Zografi G. Cryogenic grinding of indomethacin polymorphs and solvates: assessment of amorphous phase formation and amorphous phase physical stability. J Pharm Sci. 2002;91(2):492–507.
97. Huang J, Wigent RJ, Schwartz JB. Drug–polymer interaction and its significance on the physical stability of nifedipine amorphous dispersion in microparticles of an ammonio methacrylate copolymer and ethylcellulose binary blend. J Pharm Sci. 2008;97(1):251–62.
98. Konno H, Taylor LS. Influence of different polymers on the crystallization tendency of molecularly dispersed amorphous felodipine. J Pharm Sci. 2006;95(12):2692–705.
99. ICf H. International Council for Harmonization. Q3C impurities: residual solvents.6.
100. Yue H, Nicholson SJ, Young JD, Hsieh D, Ketner RJ, Hall RG, et al. Development of a control strategy for benzene impurity in HPMCAS-stabilized spray-dried dispersion drug products using a science-based and risk-based approach. Pharm Res. 2015;32(8):2636–48.

101. Mendonsa N, Almutairy B, Kallakunta VR, Sarabu S, Thipsay P, Bandari S, et al. Manufacturing strategies to develop amorphous solid dispersions: an overview. J Drug Deliv Sci Technol. 2020;55:101459.
102. Keen JM, McGinity JW, Williams Iii RO. Enhancing bioavailability through thermal processing. Int J Pharm. 2013;450(1):185–96.
103. LaFountaine JS, Jermain SV, Prasad LK, Brough C, Miller DA, Lubda D, et al. Enabling thermal processing of ritonavir–polyvinyl alcohol amorphous solid dispersions by KinetiSol® dispersing. Eur J Pharm Biopharm. 2016;101:72–81.

# Part III
# Emerging and Alternative ASD Platforms

# Chapter 7
# Evolving Perspectives on Co-amorphous Drug Delivery Systems: From Fundamentals to Recent Advancements (2021–2025)

**Ioannis Partheniadis, Inês C. B. Martins, Ioannis Nikolakakis , and Thomas Rades**

**Abstract** Co-amorphous drug delivery systems (CAMS) have emerged as a robust alternative to traditional polymer-based amorphous solid dispersions (ASDs) for addressing the poor aqueous solubility and limited bioavailability of Biopharmaceutics Classification System (BCS) Class II and IV drug candidates. While polymer-based systems frequently require high amounts of excipients and have limited drug capacity, CAMS provide a distinct alternative. CAMS are characterized as single-phase amorphous systems containing low-molecular-weight components in either drug-to-drug or drug-to-excipient combinations. They achieve stability through specific intermolecular interactions including hydrogen bonding and ionic forces. This chapter provides a comprehensive review of the CAMS landscape, beginning with an overview of fundamental preparation methods and characterization techniques. It highlights the transition from conventional analytical tools like X-ray powder diffraction (XRPD) and differential scanning calorimetry (DSC) to sophisticated modern approaches, including solid-state NMR (ssNMR) and computational molecular dynamics (MD) simulations, which offer deeper insights into molecular packing and stability. The discussion centers on the critical quality attributes (CQAs) of CAMS: coformability, physical stability, and dissolution performance. Finally, the chapter analyzes major developments in the field from 2021 to 2025, identifying

I. Partheniadis · I. C. B. Martins · T. Rades (✉)
Department of Pharmacy, Faculty of Health and Medical Sciences, University of Copenhagen, Copenhagen, Denmark
e-mail: ioannis.partheniadis@sund.ku.dk; ines.martins@sund.ku.dk; thomas.rades@sund.ku.dk

I. Nikolakakis
Laboratory of Pharmaceutical Technology, School of Pharmacy, Faculty of Health Sciences, Aristotle University of Thessaloniki, Thessaloniki, Greece
e-mail: yannikos@pharm.auth.gr

 
I. Nikolakakis, I. Partheniadis (eds.), *Amorphous Solid Dispersions*, AAPS Introductions in the Pharmaceutical Sciences 5,
https://doi.org/10.1007/978-3-032-24883-1_7

research gaps and future directions to support the clinical translation of these innovative delivery systems into viable pharmaceutical products.

**Keywords** Physical stability · Molecular interactions · Coformer selection · In vivo bioavailability · Downstream processing

## 7.1 Introduction

The persistent challenge of poor aqueous solubility among many new drug candidates, particularly those classified under the Biopharmaceutics Classification System (BCS) Classes II and IV, continues to limit their oral bioavailability and therapeutic efficacy [1–3]. Amorphous solid dispersions (ASDs) have long been employed to address these limitations, with polymer-based ASDs now being an established approach to enhance drug solubility and increase amorphous drug stability ( [4]). However, polymer-based ASDs face some drawbacks such as a usually high excipient content, limited drug-loading capacity, and concerns regarding long-term stability [1, 5, 6].

Co-amorphous drug delivery systems (CAMS) have emerged as a promising alternative within the ASD landscape. CAMS are single-phase amorphous blends composed of two or more low-molecular-weight components, either drug–drug or drug–excipient combinations, stabilized through intermolecular interactions such as ionic forces, hydrogen bonding, or π-π stacking [1, 2, 6]. This strategy enables higher drug loading, reduced excipient use, and improved physical stability, while maintaining or enhancing dissolution and bioavailability [1, 5, 7].

A comprehensive understanding of CAMS requires insight into both their preparation and their characterization techniques. The former include ball milling, solvent evaporation, melt quenching, and hot melt extrusion, each influencing molecular arrangement and amorphous stability. The latter include conventional analytical tools such as X-ray powder diffraction (XRPD), differential scanning calorimetry (DSC), and Fourier transform infrared (FTIR) spectroscopy, which remain essential for confirming amorphous formation, determination of the formation of a single-phase amorphous system, and probing molecular interactions [1, 5, 6]. However, the field is rapidly advancing with the integration of more sophisticated analytical and computational methods. Solid-state nuclear magnetic resonance (ssNMR) provides atomic level insights into molecular packing, hydrogen bonding, and phase composition, while ssNMR relaxometry and dielectric spectroscopy allow the assessment of molecular mobility, offering direct correlations with physical stability and crystallization tendencies [1, 6, 8]. In parallel, molecular dynamics (MD) simulations, density functional theory (DFT) calculations, and other computational approaches are increasingly used to predict miscibility and drug–coformer interactions, and also to interpret experimental spectra, thereby enabling a deeper mechanistic understanding of CAMS at the atomic scale [9–11].

This chapter begins with an overview of the methods used for the preparation and characterization of CAMS, with an emphasis on emerging analytical and computational tools. This is followed by an examination of the critical quality attributes (CQAs) of CAMS, as defined by Liu et al. (2021a): coformability, physical stability, and dissolution performance. The final part of the chapter provides a critical account of advances made in the field up to 2021 followed by an analysis of developments reported between 2021 and 2025. By identifying prevailing trends, research gaps, and future directions, this chapter aims to guide ongoing innovation and support the translation of CAMS into clinically relevant drug products.

## 7.2 Molecular Basis and Formation of Co-amorphous Systems

The formation of CAMS is fundamentally governed by molecular-level interactions and the thermodynamic miscibility between the selected components. Understanding these aspects is crucial for designing physically stable and pharmaceutically relevant amorphous formulations.

### *7.2.1 Prediction of Miscibility Between Components*

Prediction of miscibility in pharmaceutical CAMS is a critical step in the rational design of stable and effective drug formulations. Traditionally, miscibility between drugs and coformers or excipients has been estimated using solubility parameters, such as the Hansen or Hildebrand solubility parameters, and thermodynamic models like the Flory–Huggins interaction parameter ($\chi$) [5, 12, 13]. These approaches provide an initial estimation, where a small difference in solubility parameters ($\Delta\delta_t < 7.00$ MPa$^{1/2}$) or a negative Flory–Huggins parameter favors miscibility [13].

Recent advances have expanded predictive capabilities by integrating computational modeling, statistical learning, and experimental validation. Machine learning (ML) models, including partial least squares discriminant analysis and stacking methods that use molecular descriptors, have shown high accuracy in predicting drug–excipient compatibility. These models sometimes outperform traditional thermodynamic approaches and enable efficient virtual screening on a large scale [14–16]. Quantum mechanical calculations and molecular dynamics (MD) simulations further help to clarify the strength and nature of drug–coformer interactions, including hydrogen bonding and other noncovalent forces, which are essential for the stability of CAMS [17, 18]. For example, MD simulations have been employed to calculate solubility parameters and predict the miscibility of drugs such as indomethacin with different carriers. Experimental validation by DSC confirmed the computational predictions [15].

Nevertheless, experimental verification remains indispensable to confirm miscibility predictions. Techniques such as DSC, XRPD, FTIR spectroscopy, and increasingly ssNMR are used to identify single-phase systems, characterize the glass transition ($T_g$) temperatures, and probe intermolecular interactions, providing evidence for or against computational predictions [13, 17]. In addition, activity coefficient models and more advanced thermodynamic approaches, including the Margules, Non-Random Two-Liquid (NRTL), and Redlich Kister models, have been applied to describe composition-dependent $T_g$ behavior and phase equilibria in CAMS more accurately [19, 20].

Overall, the prediction of miscibility in pharmaceutical CAMS now relies on a synergistic strategy that combines computational screening, data driven modeling, and experimental validation. This integrated approach accelerates formulation development, reduces time wasting on trial and error experiments, and enhances the probability of producing physically stable and pharmaceutically relevant co-amorphous drug products [16, 17, 21].

### 7.2.2 Physicochemical Attributes of Coformers

The physicochemical properties of coformers are central to the formation, stability, and performance of CAMS. Among these, one of the most critical attributes is the ability of the coformer to participate in intermolecular interactions, particularly hydrogen bonding or ionic interactions with the drug [1, 5, 6]. These interactions, often referred to as supramolecular synthons, can be quantified and visualized through spectroscopic and computational techniques. For example, studies on CAMS formed between lurasidone hydrochloride and different coformers have shown that the strength and nature of hydrogen bonding directly correlated with physical stability [18]. Systems characterized by charge-assisted hydrogen bonds and high binding energies display the most stable amorphous phases [22].

Configurational entropy, which reflects the number of possible molecular arrangements, also plays an important thermodynamic role. A higher degree of configurational entropy, supported by strong intermolecular interactions, enhances resistance to crystallization and improves overall stability, even when molecular mobility is not significantly reduced [23].

The molecular shape and flexibility of coformers further influence CAMS formation. Coformers with irregular or bulky molecular shapes, such as sodium taurocholate, can hinder efficient molecular packing, thus reducing the tendency for recrystallization and phase separation. This structural effect, combined with the capacity to form diverse hydrogen bonding networks, contributes to the long-term stability of CAMS across a wide range of drugs [24, 25]. Another practical indicator of stability is the $T_g$ temperature of the CAMS. Systems with high $T_g$ temperatures can be achieved often by incorporating coformers with intrinsically high $T_g$ values such as lactose. These are generally more resistant to crystallization during storage [26, 27].

The acid base characteristics of the coformer, especially its pKa value, significantly influence both the solubility and the stabilization mechanisms in CAMS. Coformers with higher pKa values improve drug solubility, and when the pKa difference between the drug and coformer is sufficiently large, salt formation may occur, further stabilizing the amorphous phase [28, 29]. In addition, molecular descriptors such as molecular weight, van der Waals volume, and the number of aromatic rings have been identified as useful predictive factors for successful CAMS formation and stability. Multivariate statistical approaches can assist in rationally selecting coformers based on these descriptors [28].

In summary, optimal selection of coformer for CAMS requires careful consideration of hydrogen bonding potential, molecular geometry, $T_g$, and acid base properties. Together, these and other molecular descriptors determine the miscibility, physical stability, and dissolution behavior, forming the scientific foundation for the rational design of co-amorphous drug delivery systems.

### 7.2.3 Mechanistic Insights into Co-amorphization

CAMS represent a promising strategy for enhancing the solubility and stability of poorly water-soluble drugs. Mechanistically, the stabilization of these systems arises from the combined influence of thermodynamic and kinetic factors.

From a thermodynamic perspective, the formation of a single-phase amorphous blend, often achieved by combining a drug with a suitable low-molecular-weight conformer, results in increased configurational entropy and an elevated $T_g$ temperature compared to the individual amorphous components. This rise in the $T_g$ temperature is a hallmark of enhanced stability, as it reflects reduced molecular mobility and greater resistance to recrystallization during storage [30, 31]. For example, studies on a co-amorphous tolbutamide system reported a dramatic increase in $T_g$ from about 4 °C for the neat drug to over 83 °C for its CAMS, translating into excellent physical stability over extended periods of time [31]. However, an elevated $T_g$ temperature is not a strict prerequisite, and CAMS have been formed without increasing the $T_g$ of the drug [32–35].

Kinetic stabilization in CAMS is primarily achieved through the inhibition of crystallization, which is often facilitated by strong intermolecular interactions such as hydrogen bonding, salt formation, or π–π stacking between the drug and coformer. These interactions can be identified using spectroscopic methods and computational modeling, and they serve to spatially separate like molecules, thereby hindering nucleation and crystal growth [1, 5, 18, 23]. Notably, even in the absence of strong specific interactions, intimate molecular mixing can still enhance stability by disrupting the molecular regularity required for crystallization [5, 6].

The choice of coformer and the molar ratio between components play critical roles, as they affect both thermodynamic miscibility and kinetic suppression of crystallization. Optimal component ratios tend to yield the highest stability, while deviations from these ratios may promote phase separation or accelerate recrystallization, particularly under humid conditions [36]. Furthermore, the physical stability of CAMS cannot be explained by a single parameter such as relaxation time or fragility index. A complete understanding requires simultaneous consideration of thermodynamic parameters, including configurational entropy, and kinetic factors such as molecular mobility [18, 23, 37, 38].

In summary, the stabilization of CAMS depends on the interplay among thermodynamic compatibility, intermolecular forces, and dynamic inhibition of crystallization. A thorough understanding of these mechanisms provides the foundation for the rational design of co-amorphous drug formulations with superior physical stability and enhanced biopharmaceutical performance.

## 7.3 Preparation Methods of Co-amorphous Systems

The preparation method used is a critical step that determines their structural homogeneity, physical stability, and performance in dissolution of CAMS. Various preparation techniques have been explored to achieve intimate molecular mixing between the drug and coformer, thereby promoting the formation of a single amorphous phase. The choice of method depends on the physicochemical properties of the components, their thermal stability, and the desired scale of production. Broadly, preparation methods can be divided into the following three categories. Each of these having their advantages and limitations: (a) mechanical, (b) thermal, and (c) solvent-based methods.

### *7.3.1 Mechanical Methods*

Mechanical activation is one of the most commonly used routes for the preparation of CAMS, typically achieved through ball milling. This technique relies on high-energy collisions between the milling balls and solids, inducing local temperature and pressure changes that disrupt the crystal lattice and promote amorphization [1, 6, 39].

Ball milling offers several advantages, including milling in the absence of solvents and the ability to process thermolabile compounds. However, prolonged milling can lead to degradation or incomplete amorphization if the components are not fully miscible. The efficiency of the process is influenced by parameters such as milling time, rotational speed, ball size, and the drug-to-conformer ratio [39]. Importantly, maintaining the milling temperature below $T_g$ of the resulting amorphous phase is essential. Exceeding $T_g$ increases molecular mobility and allows recrystallization, preventing stable CAMS formation [40].

Vibrational milling and cryogenic milling have also been employed to control the thermal input and reduce the risk of degradation. These approaches are particularly useful for compounds with low melting points or low thermal stability.

### 7.3.2 *Thermal Methods*

Thermal approaches such as melt quenching and hot melt extrusion are widely applied for preparing CAMS when the drug and coformer possess adequate thermal stability. In melt quenching, the powder physical mixture is heated until a homogeneous molten phase is obtained and then rapidly cooled to prevent crystallization [2, 6]. Due to the rapid cooling, molecules are trapped in a disordered state, forming an amorphous matrix. The cooling rate, mixing efficiency, and miscibility of components all play key roles in determining the success of thermal methods.

Hot melt extrusion (HME) offers a scalable and continuous process suitable for industrial applications [2]. In this method, the components are mixed and melted under controlled temperature and shear conditions before being extruded and solidified. HME enables precise control over processing parameters, ensures uniform molecular dispersion, and allows integration with downstream processing such as pelletization or film casting. However, HME is not directly applicable to thermolabile compounds (e.g., amino acids) and modification/optimization is required to avoid degradation [41]. To achieve this aim, solvent-assisted HME feed pretreatment approach can facilitate the successful processing of drugs with amino acids for the preparation of CAMS [35, 42, 43].

### 7.3.3 *Solvent-Based Methods*

Solvent evaporation and spray drying are solvent-mediated approaches that allow co-amorphization under relatively mild thermal conditions [6]. In solvent evaporation methods, both components are dissolved in a common volatile solvent, followed by evaporation of the solvent under reduced pressure. Rapid solvent removal prevents crystallization, favoring formation of a uniform amorphous matrix. Spray drying represents a technologically relevant realization of solvent evaporation, where the solution is atomized into fine droplets that undergo instantaneous solvent evaporation in a hot gas stream, resulting in amorphous particles [44, 45].

Freeze drying and supercritical fluid processing have also been investigated for their ability to produce CAMS with controlled particle size and morphology [46, 47]. These methods are particularly useful for moisture-sensitive or heat-sensitive compounds.

## 7.4 Characterization Methods of Co-amorphous Systems

CAMS are single-phase amorphous drug/coformer mixtures developed to enhance solubility, stability, and bioavailability of the drug. Comprehensive characterization is essential to confirm amorphous formation, elucidate molecular interactions, and predict long-term stability. A combination of solid-state analytical (diffractometric, thermal, and spectroscopic) and computational methods provides a multidimensional understanding of these complex systems.

### 7.4.1 Solid-State Characterization

X-ray powder diffraction (XRPD) remains the most widely used method for detecting amorphous content [48]. Recent studies have applied XRPD to distinguish subtle solid-state form differences and monitor stability during storage [49]. For example, XRPD has been used to verify the amorphous nature and long-term stability of CAMS, including both binary drug and coformer pairs as well as ternary drug combinations, under a variety of storage conditions [50]. When combined with complementary techniques, diffraction techniques enable real-time in situ monitoring of phase transitions and recrystallization pathways [51, 52].

Differential scanning calorimetry (DSC) is frequently employed together with dynamic mechanical analysis (DMA) to confirm the presence of amorphous content and probe both $\alpha$ and $\beta$ relaxations, providing deeper insight into molecular mobility and phase behavior [53]. It is essential to distinguish between the types of molecular mobility captured by these techniques. While DSC primarily detects the $\alpha$-relaxation (the glass transition, $T_g$), DMA is often more sensitive to secondary ($\beta$-) relaxations occurring at sub-$T_g$ temperatures. In the context of CAMS, the $\alpha$-relaxation governs the global stability of the system, whereas $\beta$-relaxations are increasingly recognized as potential predictors of long-term physical stability and recrystallization events. Analysis of both transitions has been shown to help optimize drug to coformer ratios for maximum physical stability [54]. Theoretical models such as the Gordon-Taylor equation are now routinely applied to predict and interpret experimental $T_g$ values in complex multicomponent systems.

### 7.4.2 Advanced Analytical and Spectroscopic Techniques

Fourier transform infrared (FTIR) spectroscopy and Raman spectroscopy are standard tools for identifying molecular interactions such as hydrogen bonding and confirming amorphous formation. Solid-state nuclear magnetic resonance (ssNMR) provides atomic level insight into molecular packing and phase composition [55]. Near-infrared and terahertz spectroscopy further support the identification of amorphous content and intermolecular interactions. The application of chemometrics and multivariate analysis has improved interpretation of complex spectra, allowing more precise mapping of interactions and phase behavior.

Recent advances include the use of multivariate analysis and quantum theory of atoms in molecules to obtain detailed information on hydrogen bonding and other noncovalent interactions in CAMS [26, 56]. Stimulated Raman scattering microscopy now allows label-free, spatially resolved mapping of solid-state forms at submicron-level resolution, enabling detection of trace crystallinity and local heterogeneity that may not be visible with conventional techniques [57, 58].

ssNMR, including $^{19}$F and $^{15}$N methods, continues to offer powerful atomistic insight into molecular packing, hydrogen bonding, and phase composition [59, 60]. Improvements in sensitivity and resolution have enabled detection of subtle interactions and quantification of amorphous content even at low drug loadings. In addition, the integration of chemometric analysis with ssNMR data provides a robust framework for interpreting complex mixtures.

Broadband dielectric spectroscopy and dynamic mechanical analysis are increasingly used to study molecular mobility and relaxation dynamics, both of which are directly related to crystallization kinetics and physical aging [61–63]. These methods help to clarify the link among molecular mobility, configurational entropy, and physical stability.

### 7.4.3 Computational Approaches

Computational approaches have become essential to the design and understanding of CAMS [18, 56, 64]. Density functional theory, molecular dynamics simulations, and virtual screening are widely used to predict miscibility, assess stability, and interpret experimental observations. These methods allow for the systematic evaluation of coformers and reduce reliance on empirical trial and error.

Artificial intelligence and machine learning are now being incorporated into computational pipelines to accelerate coformer screening and predict the formation of stable CAMS [65]. By combining data-driven algorithms with established thermodynamic and molecular modeling frameworks, these tools are transforming formulation development and enabling more efficient discovery of robust co-amorphous drug products.

## 7.5 Critical Quality Attributes of Co-amorphous Systems

The performance of CAMS depends on a set of fundamental physicochemical parameters known as critical quality attributes (CQAs). These attributes determine whether a formulation will remain physically stable, achieve sufficient dissolution, and deliver reproducible therapeutic outcomes. For CAMS, three CQAs are generally recognized: coformability, physical stability, and dissolution performance [6]. Understanding and optimizing these parameters are central to the rational design and successful translation of CAMS formulations into medicines.

## 7.5.1 Coformability

Coformability refers to the ability of a drug and coformer to combine and form a homogeneous, single-phase amorphous system. This property is fundamental to the success of CAMS, as it directly impacts their physical stability, dissolution performance, and overall pharmaceutical utility.

The coformability of CAMS is influenced by several factors, including the molecular properties of both drug and coformer, the molar ratio between them, and the preparation method used. Traditionally, a 1:1 molar ratio has been assumed optimal for achieving a stable and homogeneous co-amorphous phase, especially when strong molecular interactions resulting in salt formation are expected. However, recent research has shown that 1:1 molar ratio can deviate from this assumption [66, 67]. For example, studies with carvedilol and amino acid coformers like *L*-aspartic acid and *L*-glutamic acid found that the highest $T_g$ values and best physical stability were achieved at molar ratios closer to 1:1.5, rather than exactly 1:1 [67]. This highlights the importance of systematically optimizing the drug-to-coformer ratio for each specific system, rather than relying on standard ratios.

The choice of coformer is also critical. Amino acids, carboxylic acids, sulphonic acids, and dipeptides have all been explored, with varying success [68, 69]. Sulphonic acids and dipeptides, for instance, often show superior coformability and stability due to formation of strong intermolecular hydrogen bonding or ionic bonding with drugs. In contrast, some amino acids, particularly acidic ones, may be less effective as coformers in certain systems [70].

Predictive models and screening methods are increasingly being used to identify suitable coformers [14, 71]. These approaches consider variables such as molecular weight, hydrogen bonding capacity, and mixing enthalpy to forecast the likelihood of successful CAMS formation. Such models have demonstrated high predictive accuracy, streamlining the development process and reducing the need for extensive experimental screening.

The choice of preparation method, such as ball milling, solvent evaporation, and quench cooling, also affects coformability [6, 68]. Ball milling, in particular, has been shown to be effective for a wide range of drug-to-coformer combinations, but the success of amorphization depends on the specific properties of the components and the conditions used.

## 7.5.2 Physical Stability

Physical stability is the resistance of CAMS to recrystallization during storage and processing. It is enhanced by strong intermolecular interactions (e.g., hydrogen bonding, salt formation) and a high $T_g$. Stable systems can maintain their amorphous form for extended periods, sometimes up to 1 or 2 years [1, 6].

Physical stability of CAMS is achieved by reducing the molecular mobility of the components, often reflected in a higher $T_g$. Strong and specific interactions between drug and coformer are essential, and the presence of excess of either component can lead to phase separation and subsequent recrystallization [54, 72, 73]. Surface diffusion studies have shown that even small amounts of coformer can significantly slow down molecular movement, correlating with improved stability [72]. Incorporating small amounts of polymers or additional coformers can further enhance physical stability by strengthening intermolecular networks and reducing molecular mobility, as seen in ternary CAMS [74–76], Addition of plasticizer (Compritol/Kolliphor 75/25) during HME of griseofulvin/*L*-leucin CAMS slightly reduced the concentration of dissolved drug at plateau and altered the dissolution profile from spring-and-parachute to rise-to-maximum [42].

An optimal drug-to-coformer molar ratio is crucial for maximizing stability. For example, in carvedilol/*L*-aspartic acid systems, the 1:1 ratio provided the highest physical stability under humid conditions, with samples remaining amorphous for at least 47 weeks, while excess drug or coformer led to fast recrystallization of the excess component [36]. It should be noted, however, that the optimal ratio can be shifted depending on the storage conditions. Especially moisture can disrupt molecular interactions and promote salt disproportionation or phase separation [36, 54].

Physical stability is highly sensitive to humidity and temperature. Most CAMS are more stable under dry conditions, with moisture reducing the $T_g$ and increasing molecular mobility, which accelerates recrystallization. Elevated temperatures further stress the system, and stability testing at both ambient and accelerated conditions is common practice [6, 36].

### 7.5.3 Dissolution Performance

The dissolution performance of CAMS is the third CQA driving their use in improving the oral bioavailability of poorly water-soluble drugs. CAMS typically combine two or more low-molecular-weight components into a single-phase amorphous matrix, resulting in enhanced solubility, faster dissolution rates, and the ability to achieve and maintain drug supersaturation in solution.

CAMS have consistently demonstrated superior dissolution rates compared to both crystalline and single-component amorphous forms. For example, indomethacin CAMS with paracetamol or nicotinamide showed significantly improved solubility, supersaturation, and intrinsic dissolution rates in biorelevant media, with the co-amorphous formulations maintaining their amorphicity and dissolution advantage over several months of storage [77]. Similarly, drug–drug CAMS such as toltrazuril/ponazuril and naproxen/cimetidine have exhibited enhanced dissolution rates and a pronounced spring and parachute effect, where rapid initial supersaturation is followed by sustained elevated drug concentrations, attributed to strong intermolecular interactions that inhibit recrystallization during dissolution [10, 78, 79].

The choice of coformer, drug-to-coformer ratio, and preparation method all influence dissolution performance. Binary drug-amino acid CAMS have been shown to provide long-lasting supersaturation and improved dissolution in both buffer and biorelevant media, although the magnitude of improvement may be less pronounced in simulated intestinal fluids compared to simple aqueous buffers [80]. Incorporating small amounts of polymers, such as hydroxypropyl methylcellulose (HPMC), can further optimize dissolution by moderating the initial release rate and prolonging supersaturation without compromising physical stability [75].

Despite these advances, it is important to note that in vitro dissolution improvements do not always directly translate to enhanced bioavailability in vivo, due to the complexity of physiological conditions. Only a minority of studies have systematically linked in vitro dissolution data with pharmacokinetic outcomes, highlighting a need for more comprehensive in vivo investigations [6].

## 7.6 Recent Advancements

In this section, recent advancements in the field are cited, focusing on developments from the past 5 years. To this end, we conducted a systematic literature review of published articles between January 2021 and October 2025, using the Scopus® (Elsevier) and PubMed® (NLM: United States National Library of Medicine) bibliographic databases. The search algorithm for the different databases is presented below:

- Scopus: TITLE-ABS-KEY (co-amorphous) OR TITLE-ABS-KEY (coamorphous) AND PUBYEAR >2020 AND (EXCLUDE (DOCTYPE, "re")) AND (LIMIT-TO (LANGUAGE, "English")) AND (LIMIT-TO (SUBJAREA, "PHAR"))
- PubMed: (("co amorphous"[All Fields] OR "co-amorphous"[All Fields] OR "coamorphous"[All Fields] AND ("system"[All Fields] OR "systems"[All Fields])) AND (2021:3000/12/12[pdat]) NOT (Review[Publication Type])

### *7.6.1 Research Interest (2021–2025)*

The term "co-amorphous" was first introduced in 2009 [81], although binary amorphous systems, the precursors to CAMS, had been reported before 2000, with most early studies focusing primarily on thermal analysis [82–84]. Since then, research interest in CAMS has grown rapidly, with nearly 200 studies published by 2020 [1, 6]. However, most of these investigations remained case based, with limited systematic exploration of drug–coformer combinations [6].

The search algorithm identified a total of 368 studies, which were screened by title and abstract, resulting in exclusion of 20 studies. After removing 129

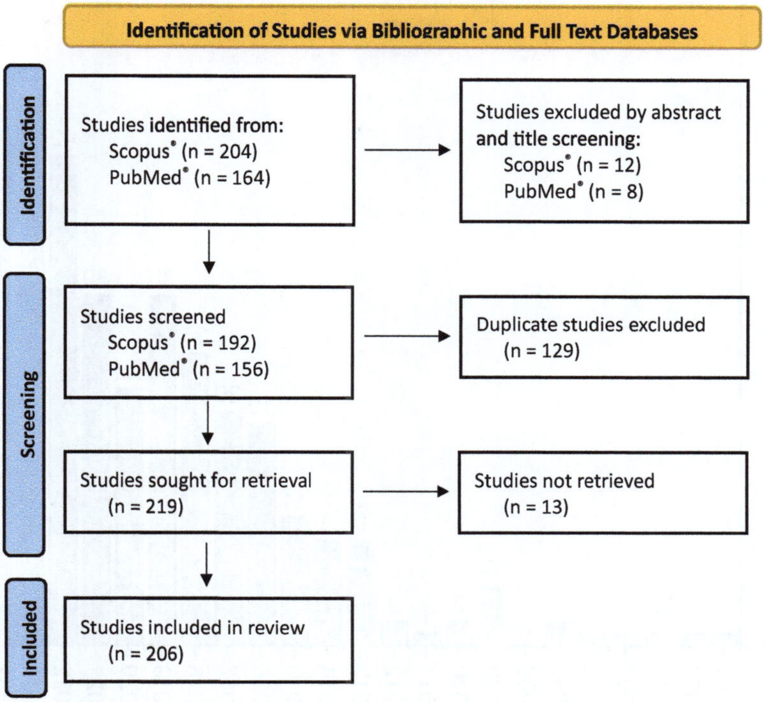


**Fig. 7.1** PRISMA flowchart of the study selection process

duplicates, 219 studies remained for screening. Of these, 13 could not be retrieved. Ultimately, 206 studies met the inclusion criteria and were included in the review. The PRISMA flowchart illustrating the study selection process is presented in Figure 7.1, and the distribution of studies by publication year is shown in Figure 7.2.

As shown in Figure 7.2, research activity in the field of CAMS has increased notably over the last 5 years (2021 to October 2025), resulting in significantly more publications ($n = 206$) than the total reported up to 2020 ($n = 162$). To have an idea of how systematically CAMS have been investigated during the last 5 years, the following subsections provide a structured analysis of the identified studies.

### *7.6.2 Classes of Investigated CAMS*

As reported by Liu et al., amino acids were the most frequently used coformers in CAMS up to 2020, accounting for approximately 46% of all reported systems [6]. However, as shown in Figure 7.3, research over the past 5 years has increasingly focused on drug–drug CAMS, which now represent 44.3% of all studies reporting binary CAMS. Drug–amino acid CAMS rank second, comprising 21.7% of the total. This shift may be attributed to the fact that drug–drug CAMS not only enhance

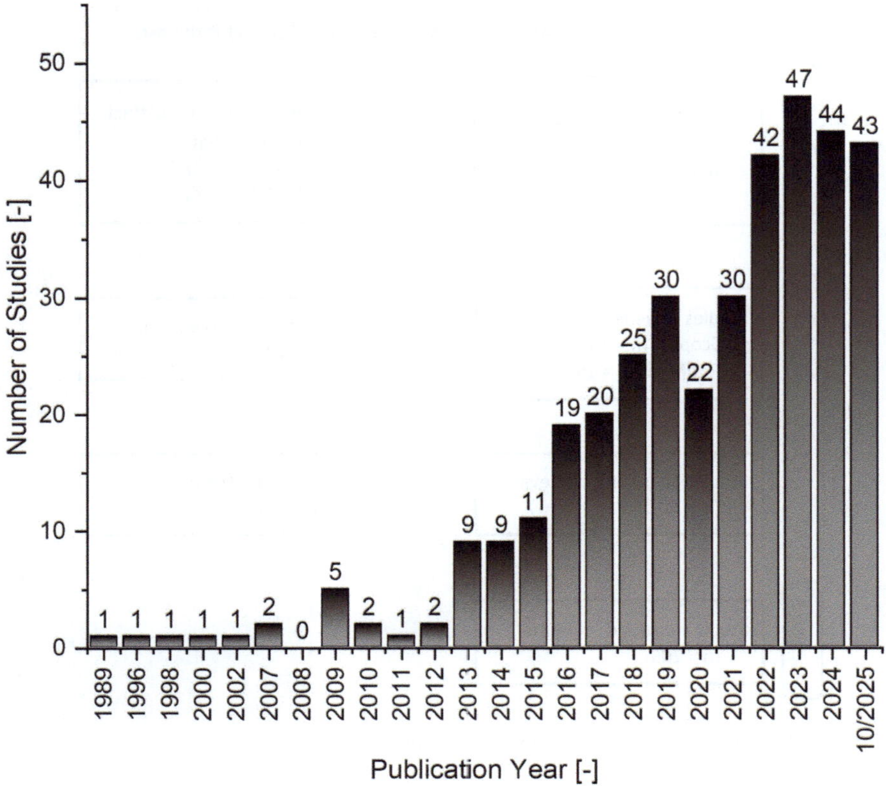


**Fig. 7.2** Column diagram showing the number of studies by year from 1989 to October 2025

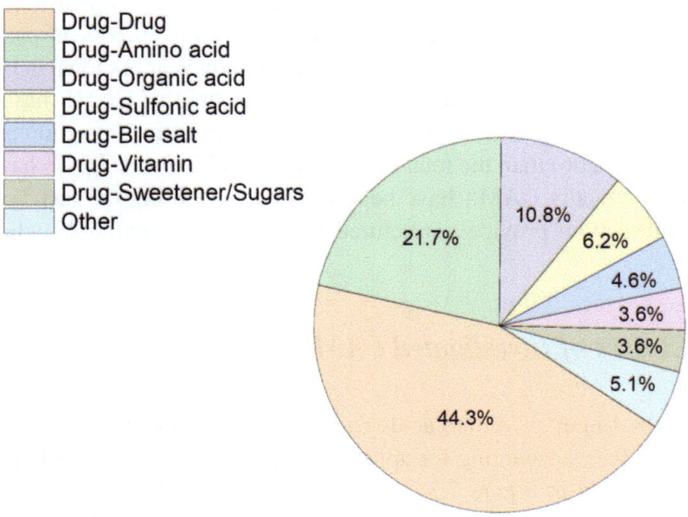


**Fig. 7.3** Percentage of the different classes of co-amorphous drug delivery systems (CAMS) reported in the identified literature

dissolution and stability but also offer the possibility of developing combination therapies.

### *7.6.3 Drugs and Coformers Investigated for the Formation of CAMS*

Regarding the drugs investigated for CAMS formation, a total of 187 different compounds were identified. More than half of these (55.6%) were reported only once (Fig. 7.4a), and only 5.9% appeared in more than five studies. This pattern underscores the ongoing lack of systematic exploration and the predominance of

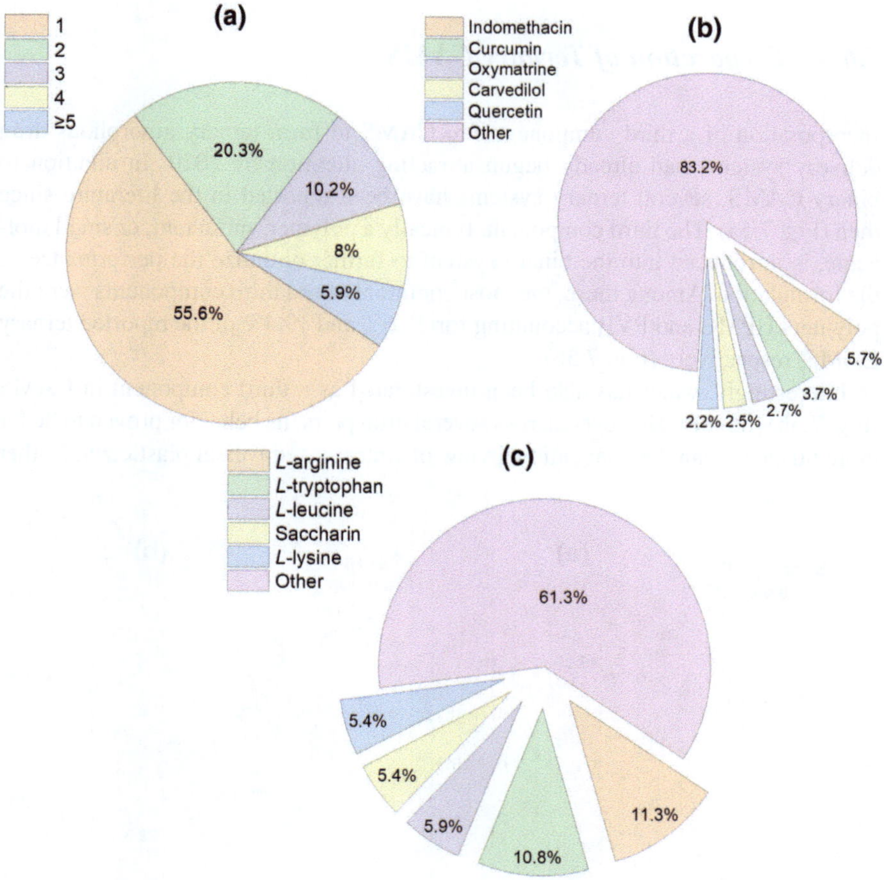


**Fig. 7.4** Percentage of (**a**) how often a specific drug has been reported in different studies; (**b**) how often one of the top five used drugs was reported in studies compared to other drugs; (**c**) how often one of the top five used coformers was reported in studies compared to other coformers

case-by-case investigations, consistent with the trends noted by Liu et al. for studies published up to 2020 [6].

The five most frequently examined drugs were indomethacin, curcumin, oxymatrine, carvedilol, and quercetin, which together accounted for 16.8% of all screened drug/coformer combinations (a coformer may also be a drug here) (Fig. 7.4b). All five belong to BCS Class II, indicating a focus on poorly soluble compounds rather than on broader, property-driven screening strategies. Notably, two of these drugs, curcumin and quercetin, are of natural origin, suggesting a growing interest in incorporating bioactive natural molecules as candidates for CAMS formulation.

The five most frequently examined coformers were *L*-arginine, *L*-tryptophan, *L*-leucine, saccharin, and *L*-lysine, which together accounted for 38.7% of all screened coformers (a coformer cannot be a drug here) (Fig. 7.4c). Four out of the top five coformers are amino acids, while the fifth is a sultam (saccharin).

### 7.6.4 Preparation of Ternary CAMS

Incorporation of a third component into CAMS to form ternary amorphous drug delivery systems had already begun attracting attention by 2020. In addition to binary CAMS, several ternary systems have been reported in the literature since then (Fig. 7.5a). The third component, typically a polymer, surfactant, or small molecule, is introduced into the binary system to further optimize the performance of the formulation. Among these, the most commonly used third components were the polymers HPMC and PVP, accounting for 22.6% and 19.4% of the reported ternary CAMS, respectively (Fig. 7.5b).

Interestingly, water has also been investigated as a third component in CAMS (Fig. 7.5b) [85–88]. However, across several drug pairs, its behavior proved to be far more nuanced than the conventional view of water as a universal plasticizer. Rather

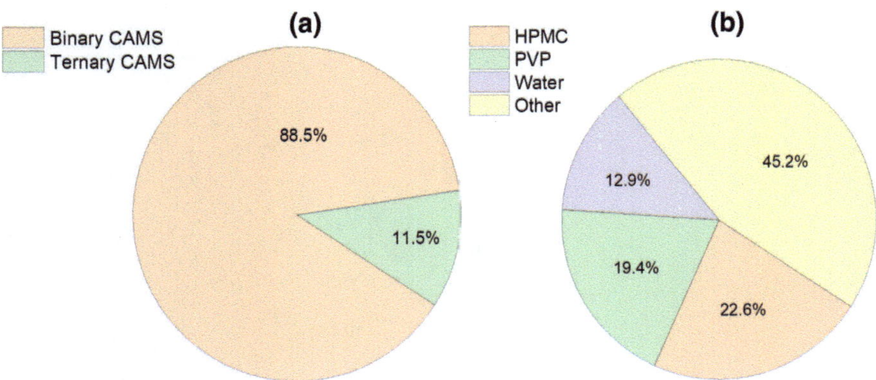


**Fig. 7.5** Percentage of (**a**) binary and ternary CAMS reported in the identified literature and (**b**) of the three most commonly used third components in ternary CAMS

than uniformly lowering $T_g$ and accelerating molecular mobility, its influence depended strongly on the specific interaction environment created by the coformers.

In mixtures containing prilocaine (PRL) and lidocaine (LID), hydration resulted in modest $T_g$ increase, suggesting that water became part of a stabilizing interaction network rather than causing disruption [86]. Since amorphous LID is difficult to characterize independently, this behavior was inferred from mixed systems and supported by deviations from the standard Gordon–Taylor predictions. Under these conditions, water behaved as an antiplasticizing agent. Spectroscopic analyses of hydrated PRL–LID mixtures further supported this interpretation, showing that water bound preferentially to carbonyl groups in both drugs and was distributed uniformly throughout the co-amorphous matrix [88]. These interactions were associated with enhanced physical stability and a reduced tendency toward crystallization.

A similar composition-dependent effect was observed when nicotinamide (NIC) was used as a coformer with PRL [87]. Systems with higher NIC content showed the expected plasticizing response to hydration, with lower $T_g$ values. However, at lower NIC proportions, water reinforced the amorphous structure, leading to increased $T_g$ values and reduced molecular mobility, indicating the formation of a more rigid network.

A comparable stabilizing influence was also reported for flurbiprofen–lidocaine (FLB–LID) systems [85]. Upon hydration, the amorphous matrix became less prone to crystallization, and both dynamic and thermodynamic measurements showed that water increased the energetic barriers associated with molecular rearrangement.

Together, these findings demonstrate that water can act as a purposeful structural component in ternary co-amorphous systems. When it engages in specific, stabilizing interactions with the coformers, occurring at low water contents, it can increase matrix rigidity, reduce molecular mobility, and improve resistance to crystallization.

### 7.6.5 Coformer Selection for the Formation of CAMS

The selection of an appropriate coformer is crucial for the preparation of a desired CAMS, as it determines their first critical quality attribute (CQA), coformability. The criteria used for coformer selection in the reported CAMS are summarized in Figure 7.6. The largest proportion of coformers (45.2%) was chosen based on considerations of combination therapy, which is to be expected given that drug–drug CAMS represent the most extensively studied class (Fig. 7.3). Factors such as bioactivity, structural compatibility between the drug and coformer, and physicochemical properties accounted for 22.6% of selections. In contrast, a systematic coformer screening approach was applied in only 17.5% of the studies (Fig. 7.6). Thus, systematic, predictive, and computational strategies for coformer selection remain largely underexplored.

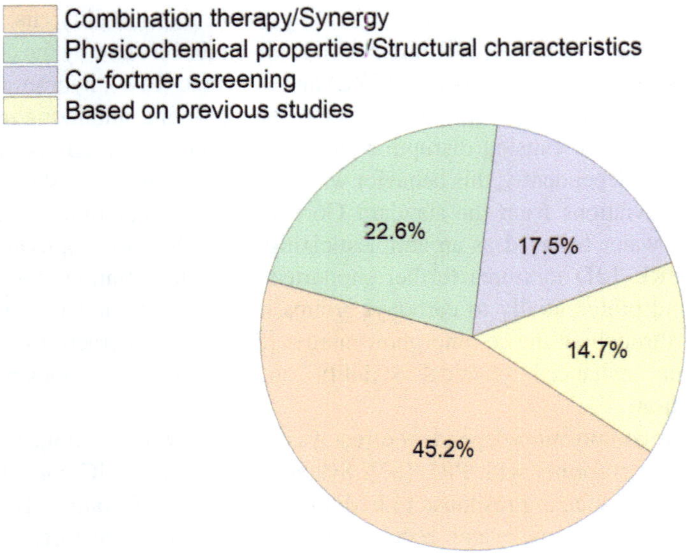


**Fig. 7.6** Different reasons (expressed as percentage) on which coformer selection was based on in CAMS reported in the identified literature

Interestingly, specific molecular interactions between the drug and the coformer have been shown to be decisive contributors to the physical stability of CAMS [89]. Yarlagadda et al. prepared drug-drug CAMS using various combinations of indomethacin, ketoprofen, flurbiprofen, flufenamic acid, aripiprazole, bifonazole, and clotrimazole and showed that CAMS incorporating at least one drug with a carboxylic acid group consistently displayed higher than expected $T_g$, thus showing how strong, directional hydrogen bonds can reinforce the amorphous matrix and slow molecular mobility. In formulations where this effect was amplified by secondary interactions, such as π–π stacking, halogen contacts, or even ionic pairing enabled by contrasting *pK*a values, the resulting CAMS showed pronounced stabilization beyond ideal mixing behavior. These interactions not only elevated the $T_g$ but also narrowed the transition width. By contrast, weaker interaction types in the absence of a -COOH group contributed little to stability.

Moreover, a recent work on the prediction of CAMS formation has illustrated how data-driven methods can streamline the identification of suitable drugs/coformer combinations to successfully prepare CAMS [90]. Using a gradient boosting framework trained on several categories of various binary mixtures, including drug–drug combinations and systems with amino or organic acids, Fink et al. were able to forecast whether a given pair would form a co-amorphous phase. The model showed good predictive power when tested against external data and proved particularly useful to identify promising drug–drug CAMS for inhalation therapies, where multidrug combinations are frequently employed. Experimental validation through milling of three predicted pairs confirmed the computational results, demonstrating how such modeling approaches can reduce empirical trial-and-error and support more efficient early-stage screening of co-amorphous candidates.

### 7.6.6 *Molar Ratio Optimization of Investigated CAMS*

Following coformer selection, stoichiometry plays a key role in the successful formation of CAMS, as it directly influences the three critical quality attributes (CQAs). As shown in Figure 7.7, the 1:1 molar ratio is the most commonly used, with approximately 51.0% of reported CAMS prepared at this equimolar ratio. Only 8.2% of CAMS were based on a molar ratio optimization process, typically with respect to the second and/or third CQA, while 35.0% of studies employed various molar ratios (e.g., 1:2, 1:1, or 2:1) without providing explicit justification.

A strategic starting point for optimization is the identification of eutectic mixtures. These provide a thermodynamic framework by identifying compositions that inhibit individual crystallization, resulting in a single melting point lower than that of the pure components. This "eutectic point" often marks the stoichiometry where drug–coformer hetero-interactions are maximized, effectively spatially separating similar molecules and disrupting the formation of a crystal lattice [91, 92]. While the 1:1 ratio is a common eutectic point, studies show that the optimal stoichiometry

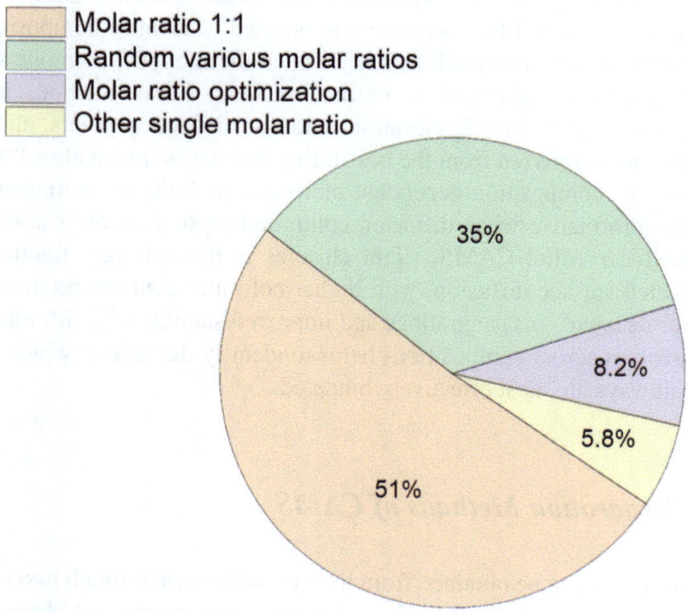


**Fig. 7.7** Percentage of studies using certain molar ratios, various molar ratios, and molar ratio optimization in the CAMS reported in the identified literature

for solubility and stability is case specific and does not always coincide with a simple equimolar mixture [93, 94].

Optimization of molar ratio can be based on how the selected composition influences intermolecular interactions, thermal behavior, molecular mobility, and ultimately physical stability. One study has shown that the optimal molar ratio for highest CAMS physical stability differs between samples stored under different environmental conditions, dry or humid [36]. Recent studies described below have introduced several complementary strategies that have been used to identify the most favorable ratios.

One approach has focused on identifying the composition that maximizes favorable interactions between drug and coformer [95]. By combining FTIR with multivariate principal component analysis (PCA), shifts in hydrogen-bond-related signals can be tracked as the ratio changes. In the carvedilol–tryptophan system (CAR-TRP), this type of spectral deviation peaked near 30% (mol/mol) drug, which coincided with the most stable CAMS, illustrating that the combination of FTIR with PCA is an effective approach to guide optimal molar ratio selection. Moreover, predictive multivariate models have been trained on larger co-amorphous screening datasets [14]. Discriminant-analysis (DA) approaches applied across various drug/coformer combinations have been successfully used to find optimal molar ratios or even predict which pairs are likely to form CAMS in the first place. Although these predictions do not always correlate directly with the long-term stability, they provide useful guidance during early screening and help to reduce the experimental workload.

Thermal modeling has also been used. The $T_g$ often rises nonlinearly in strongly interacting mixtures, and this behavior can be used to identify compositions with enhanced resistance to recrystallization. Refined thermodynamic models, such as activity-coefficient-based extensions of the Couchman–Karasz method, can be used to identify these synergistic $T_g$ elevations and their inflection points, allowing the optimal ratio to be inferred from the best-fitting interaction parameters [20].

Additionally, composition-dependent molecular mobility measurements can be particularly informative when diffusion-controlled crystallization is concerned. In aspartic acid–carvedilol CAMS, slight changes in the coformer fraction significantly affected surface diffusion, with higher coformer contents producing markedly slower decay of surface gratings and improved stability [72]. Monitoring how mobility evolves across compositions helps to identify the ratio at which recrystallization pathways are most effectively hindered.

### 7.6.7 Preparation Methods of CAMS

Amorphous drugs can be obtained from its crystalline form through two fundamentally different pathways: the kinetic and the thermodynamic [6]. Approximately 31.5% of the reported CAMS were prepared via the kinetic pathway (Fig. 7.8a), of

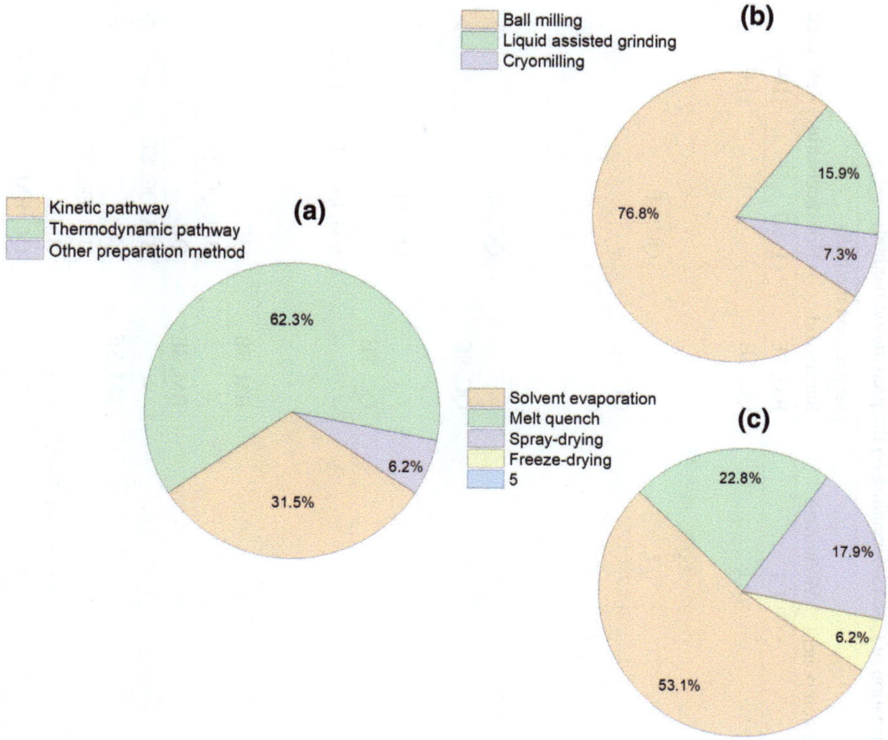


**Fig. 7.8** Percentage of the different methods used to obtain CAMS reported in the identified literature using (**a**) the thermodynamic or kinetic pathway or other preparation methods, (**b**) different preparation methods of the kinetic pathways, and (**c**) different preparation methods of the thermodynamic pathways

which 76.8% were obtained by ball milling, 15.9% by liquid-assisted grinding, and 7.3% by cryo-milling (Fig. 7.8b). More CAMS were prepared via the thermodynamic pathway, making up 62.3% of the cases, of which the majority (53.1%) were produced by solvent evaporation. Other commonly employed thermodynamic methods included melt quenching (22.8%), spray-drying (17.9%), and freeze-drying (6.2%) (Fig. 7.8c). Additional preparation techniques, such as hot-melt extrusion (mainly thermodynamic pathway) [35, 42, 43, 96–99], in situ co-amorphization [20], nanoprecipitation (thermodynamic pathway) [100], antisolvent precipitation (thermodynamic pathway) [101], high-shear mixing (mainly kinetic pathway) [102], and supercritical fluid technology (thermodynamic pathway) [103], have also been reported. Some of these methods are difficult to be assigned definitively to a single pathway.

Successful preparation of CAMS depends on the applied preparation method [6]. In Table 7.1 an overview of different successful or unsuccessful attempts of CAMS preparation, depending on the preparation method is presented.

**Table 7.1** Overview of the studies reporting successful (or unsuccessful) formation of CAMS depending on the preparation method

| CAMS class | Drug | Coformer | Third component | Molar ratio | Preparation methods | | Reference |
|---|---|---|---|---|---|---|---|
| | | | | | Successful | Not successful | |
| Drug-Drug | Valsartan | Nicotinamide | n.a. | 1:1 | BM, SE | LAG | [104] |
| Drug-Bislactam | Aspirin | bis-N-vinyl(caprolactam) | n.a. | 1:1 | QC, SE | – | [14] |
| | Chloramphenicol | | | | | | |
| | Chloropropamide | | | | | | |
| | Famotidine | | | | – | QC, SE | |
| | Flurbiprofen | | | | QC, SE | – | |
| | Furosemide | | | | | | |
| | Indomethacin | | | | | | |
| | Ketoprofen | | | | | | |
| | Mebendazole | | | | – | QC, SE | |
| | Paracetamol | | | | QC, SE | – | |
| | Phenobarbital | | | | | | |
| | Piroxicam | | | | – | QC, SE | |
| | Simvastatin | | | | QC, SE | – | |
| Drug-Organic acid | Olanzapine | Citric acid | n.a. | 1:1 | – | BM, QC, SE | [105] |
| | | Oxalic acid | | | | | |
| | | Tartaric acid | | | | | |
| Drug-Sulfonic acid | | Acesulfame | | | BM, SE | QC | |
| | | Potassium acesulfame | | | – | BM, QC, SE | |
| | | Cyclamic acid | | | BM, SE | QC | |
| | | Sodium cyclamate | | | – | BM, QC, SE | |
| | | Saccharin | | | BM, SE | QC | |
| | | Sodium saccharin | | | – | BM, QC, SE | |
| Drug-Amino acid | Indomethacin | *L*-arginine | PVP | 40:18.3:41.7 wt% | BM | HME | [96] |
| Drug-Drug | Raloxifene HCl | Quercetin | n.a. | 1:1 | SE | QC, BM | [106] |

| | | | | | | | |
|---|---|---|---|---|---|---|---|
| Drug-Amino acid | Sinapic acid | *L*-tryptophan | n.a. | 1:1, 1:2 | BM | SE, FD | [107] |
| | | *L*-arginine | n.a. | 1:1, 1:2 | SE, FD | BM | |
| | | *L*-lysine | n.a. | 1:1, 1:2 | BM, SE, FD | – | |
| Drug-Organic acid | Dipyridamole | p-hydroxybenzoic acid | n.a. | 1:2 | QC, SE | LAG | [108] |
| Drug-Amino acid | Indomethacin | Chiral tryptophan | n.a. | 1:1 | BM, SD | – | [109] |
| Drug-Drug | Pioglitazone HCl | Rosuvastatin Ca | n.a. | 1:1, 2:1 | BM | LAG | [110] |
| Drug-Drug | Pioglitazone HCl | Fluvastatin Na | n.a. | 1:1 | BM | LAG | [111] |
| Drug-Organic acid | Dasatinib | Gallic acid | n.a. | 1:1 | BM, LAG | – | [112] |
| Drug–Phospholipid | Celecoxib | Soybean PC | n.a. | 1:1 | QC, SE | BM | [113] |
| | | Hydrogenated PC | | | SE | BM, QC | |
| | | Mono-acyl PC | | | QC, SE | BM | |
| | Furosemide | Soybean PC | n.a. | 1:1 | SE | BM, QC | |
| | | Hydrogenated PC | | | SE | BM, QC | |
| | | Mono-acyl PC | | | SE | BM, QC | |
| | Nilotinib | Soybean PC | n.a. | 1:1 | – | BM, QC, SE | |
| | | Hydrogenated PC | | | – | BM, QC, SE | |
| | | Mono-acyl PC | | | – | BM, QC, SE | |
| Drug–Phospholipid | Ritonavir | Soybean PC | n.a. | 1:1 | QC, SE | BM | [113] |
| | | Hydrogenated PC | | | QC, SE | BM | |
| | | Mono-acyl PC | | | QC, SE | BM | |

(continued)

**Table 7.1** (continued)

| CAMS class | Drug | Coformer | Third component | Molar ratio | Preparation methods | | Reference |
|---|---|---|---|---|---|---|---|
| | | | | | Successful | Not successful | |
| Drug–Phospholipid | Indomethacin | Soybean PC | n.a. | 1:1 | QC, SE | BM | [114] |
| | | Hydrogenated PC | | | SE | BM, QC | |
| | | Mono-acyl PC | | | QC, SE | BM | |
| Drug-Organic acid | Azithromycin | 2-aminobenzoic acid | n.a. | 1:1 | SE | LAG | [115] |
| | | 3-aminobenzoic acid | n.a. | 1:1 | | | |
| | | 4-aminobenzoic acid | | | | | |
| Drug-Drug | | Salicylic acid | | | | | |
| Drug-Fatty acid | | Caprylic acid | | | | | |
| Drug-Drug | Lumefantrine | Salicylic acid | n.a. | 1:1 | SE | LAG | [116] |
| | | Acetylsalicylic acid | | | | | |
| Drug-Drug | Rifampicin | Ethambutol | n.a. | 1:1 | BM, SD | – | [117] |
| Drug-Organic acid | Lumefantrine | α-ketoglutaric acid | n.a. | 1:1 | SE, QC, LAG | – | [49] |
| Drug-Amino acid | Griseofulvin | *L*-methionine | n.a. | 1:1 | SE, HME | QC, BM, LAG | [98] |
| | | *L*-lysine | | | | | |
| | | *L*-valine | | | | | |
| Drug-Amino acid | Cilostazol | *L*-tryptophan | n.a. | 1:1 | BM, SD | – | [118] |
| | | | | 1:1.5, 1:2 | BM | SD | |

*BM* ball milling, *QC* quench cooling, *SE* solvent evaporation, *LAG* liquid assisted grinding, *SD* spray drying, *FD* freeze drying, *HME* hot melt extrusion, *PC* phosphatidylcholine

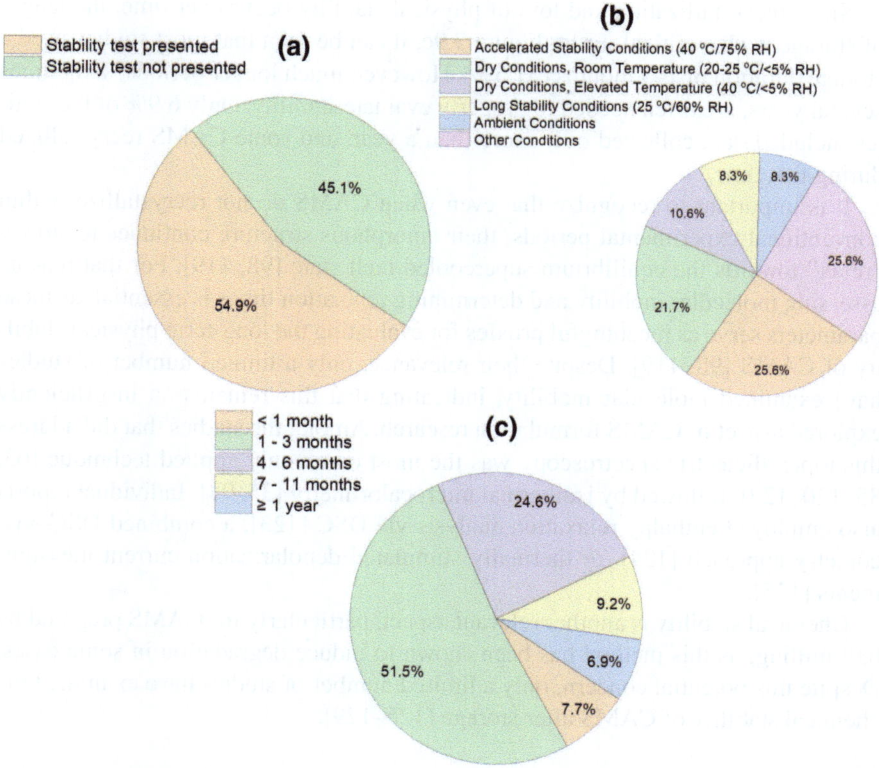


**Fig. 7.9** Percentage of (**a**) studies that presented physical stability tests; (**b**) studies that presented physical stability tests with respect to environmental conditions; (**c**) studies that presented physical stability tests with respect to the timescale of the storage period

## 7.6.8 *Physical Stability of CAMS*

Physical stability is the second CQA of CAMS, since the purpose of their development is to confer physical stability of the amorphous drug form, besides dissolution improvement. From Figure 7.9a, it appears that slightly more than half of the studies (54.9%) included physical stability testing. This indicates that, despite the central importance of stability for CAMS, nearly half of the published work still does not evaluate this essential CQA, highlighting a persistent gap in the systematic characterization of these systems.

Most of the reported stability studies (25.6%) were performed under accelerated conditions (45 °C/75% RH). A considerable number (32.3%) were carried out under dry conditions (21.7% at 25 °C and dry RH conditions, and 10.6% at 40 °C and dry RH conditions) (Fig. 7.9b). Additionally, 8.3% of studies used long-term storage conditions (25 °C/60% RH), and another 8.3% were performed at ambient temperature and humidity (Fig. 7.9b).

Since recrystallization and loss of physical stability occur over time, the length of storage is also critical. From Figure 7.9c, it can be seen that most studies used a storage duration of 1–3 months (51.5%). However, much longer periods, sometimes several years, are often needed to properly evaluate stability. Only 6.9% of the studies included data collected over more than a year, and some CAMS recrystallized during that time.

It is important to recognize that even when CAMS do not recrystallize within conventional experimental periods, their amorphous structure continues to slowly "relax" towards the equilibrium supercooled melt state [98, 119]. For that reason, assessing molecular mobility and determining relaxation times is essential, as these parameters serve as meaningful proxies for evaluating the long-term physical stability of CAMS [98, 119]. Despite their relevance, only a limited number of studies have examined molecular mobility, indicating that this remains an insufficiently explored aspect of CAMS formulation research. Among the studies that did address this topic, dielectric spectroscopy was the most commonly applied technique [63, 85, 120–122], followed by isothermal microcalorimetry [35, 98]. Individual reports also employed enthalpy relaxation analysis via DSC [123], a combined DSC–viscometry approach [124], or thermally stimulated depolarization current measurements [125].

Chemical stability is another relevant aspect, particularly for CAMS prepared by ball milling, as this method has been shown to induce degradation in some cases. Despite this potential concern, only a limited number of studies have examined the chemical stability of CAMS after storage [126–129].

### *7.6.9 In Vitro and in vivo Performance of CAMS*

Dissolution performance is the third CQA of CAMS. While these systems are developed to enhance the physical stability of an amorphous drug, their broader goal is to increase apparent solubility, improve dissolution, and ultimately enhance bioavailability. As a result, it is necessary to evaluate their in vitro and in vivo performance. Dissolution can be characterized and assessed by examination of the rate at which the drug dissolves, the solubility and extent of supersaturation achieved, and from the length of time that supersaturation is sustained. Still, it is widely recognized that in vitro results do not always align with in vivo outcomes [6, 130]. CAMS may show a strong improvement in dissolution under controlled testing, yet this does not guarantee better bioavailability, as the test conditions may not reflect the complexity of the biological environment [131]. Therefore, formulations that perform well in vitro must be validated through pharmacokinetic studies in vivo.

Figure 7.10 displays the percentage of studies with or without in vitro dissolution/solubility and in vivo data. From the total number of reviewed studies 62.9% evaluated CAMS solely through in vitro dissolution performance, while 21.5% did not consider dissolution, focusing only on coformability and physical stability.

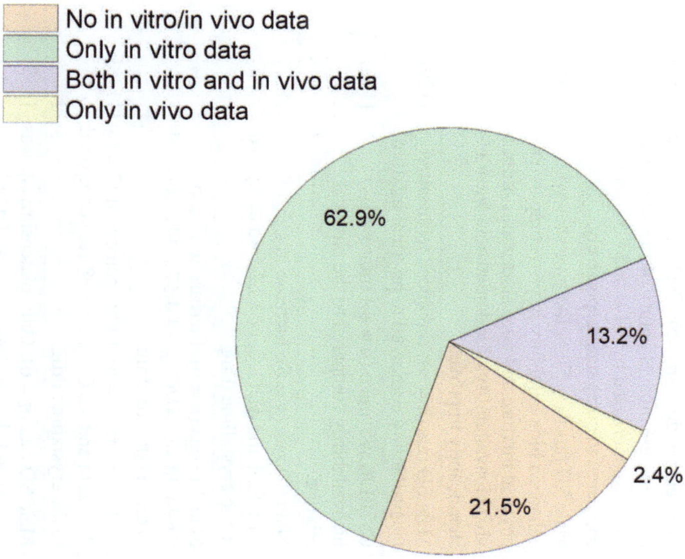


**Fig. 7.10** Percentage of studies with or without in vitro dissolution/solubility data and with or without in vivo data

Notably, 13.2% of the studies included both in vitro dissolution and in vivo pharmacokinetic profiles, linking the two-performance metrics. This proportion is almost 2.5 times higher than that reported in studies that were published before 2020, where only 4.9% included in vivo evaluation [6].

An updated account of the studies that have examined both the in vitro and in vivo performance of CAMS is shown in Table 7.2. All in vivo investigations were conducted in rats. Notably, of the five most frequently explored drugs (indomethacin, curcumin, oxymatrine, carvedilol, and quercetin, Fig. 7.4b) only CAMS of two drugs, indomethacin and curcumin have been evaluated both in vitro and in vivo, despite an abundance of data (Table 7.2). Research on the other three drugs has largely focused on other CQAs or on in vitro dissolution only.

Overall, from Table 7.2, it appears that there is more information needed on how CAMS behave in living systems and especially humans. Unfortunately, clear regulatory guidelines for CAMS are not available, and translation into clinical trials remains limited, hindering progress into potential products. Also, more research is needed on drug delivery routes beyond the oral, such as dermal and pulmonary. Furthermore, from Table 7.2, it appears that curcumin has been evaluated in two different cases for its CAMS in vivo performance [101, 132, 133]. As a BCS Class IV compound, curcumin presents multiple challenges, including low solubility, limited dissolution, and poor permeability. Thus, permeability studies for BCS Class IV drugs are of interest.

Studies presenting in vitro or ex vivo permeability studies are presented in Table 7.3.

**Table 7.2** Overview of the studies reporting both in vitro and in vivo performance

| | Drug | Coformer | Third component | Molar ratio | In vitro study | In vivo study | Reference |
|---|---|---|---|---|---|---|---|
| Drug–drug | Sacubitril | Valsartan | n.a. | 1:1 | Regardless of buffer, enhanced intrinsic dissolution rate | Male beagle dogs ($n = 6$): Compared with the cocrystal, the bioavailability of sacubitril and valsartan increased by 3.14- and 1.31-fold compared to cocrystal, respectively. | [134] |
| Drug–drug | Felodipine | Indo-methacin | n.a. | 1:1 | Enhanced drug release of felodipine. For indomethacin only in nonionized media | Male SD rats ($n = 5$): Felodipine showed enhanced bioavailability in vivo compared to physical mixture and pure amorphous drug. In vivo bioavailability of indomethacin wasn't significantly improved. | [135] |
| Drug–drug | Telmisartan | Hydrochloro-thiazide | n.a. | 1:3 | Enhanced telmisartan drug release | Male SD rats ($n = 6$): Enhanced telmisartan bioavailability compared to physical mixture. | [129] |
| Drug–drug | Simvastatin | Ezetimibe | n.a. | 1:1 | Enhanced drug release | C57BL/6 WT mice ($n = 8$): Improved bioavailability compared to the physical mixture. | [136] |
| Drug–drug | Epigallo-catechin-3-gallate | Simvastatin | n.a. | 1:3 | Enhanced solubility and drug release | Male SD rats ($n = 6$): Increase in $C_{max}$ of 1.81-fold and $AUC_{0\text{-}24h}$ of 1.62-fold compared to the crystalline drug. | [137] |
| | | Nifedipine | | | | Male SD rats ($n = 6$): Increase in $C_{max}$ of 5.69-fold and $AUC_{0\text{-}24h}$ of 4.57-fold compared to the crystalline drug. | |
| Drug–amino acid | Tadalafil | *L*-alanine | Mannitol | 1:1:4 | Enhanced solubility and drug release | Male SD rats ($n = 6$): Enhancement of $C_{max}$ by 2.48-fold and $AUC_{0\text{-}24h}$ by 1.98-fold compared to the crystalline drug. | [138] |
| Drug–drug | Platensi-mycin | Berberine chloride | n.a. | 1:1 | Decreased equilibrium solubility and slower drug release due to agglomeration | Male SD rats ($n = 6$): Only platensimycin was studied and showed lower $C_{max}$ and $AUC$ values compared to physical mixture or crystalline drug. | [139] |

| | | | | | | | |
|---|---|---|---|---|---|---|---|
| Drug–drug | Telmisartan | Amlodipine | n.a. | 1:1 | Enhanced telmisartan solubility and decreased amlodipine solubility | Male SD rats ($n = 4$–5): Oral bioavailability of telmisartan increased by up to 4.1 times, whereas that of amlodipine remained unchanged. | [140] |
| Drug–drug | Paclitaxel | Curcumin | n.a. | 1:1 | Enhanced drug release and solubility | Male SD rats ($n = 5$): Increase in $C_{max}$ by 2.58-and 4.07-fold and $AUC_{0\text{-}24h}$ by 2.16- and 4.57-fold, respectively. | [132] |
| Drug–sugar | Tacrolimus | Sucrose acetate isobutyrate | n.a. | 1:1 | Enhanced drug release | Male and female beagle dogs ($n = 2 + 2$): Increase in $C_{max}$ of 1.7-fold and *AUC* of 1.5-fold compared to the crystalline drug. | [141] |
| Drug–organic acid | Posacon-azole | Citric acid | n.a. | 1:1 | Enhanced drug release and solubility | Male SD rats ($n = 5$): Increase in $C_{max}$ of 1.51-fold and *AUC* of 2.81-fold compared to the crystalline drug. | [142] |
| Drug–drug | Ceritinib | Rutin | n.a. | 2:1, 1:1, 1:2 | Enhanced drug release and solubility | Male Wistar rats ($n = 3$): Increase in $C_{max}$ by 2.05-, 1.41-, and 1.41-fold and $AUC_{0\text{-}t}$ by 2.59-, 1.99-, and 1.41- fold for 2:1, 1:1, and 1:2 molar ratios, respectively. | [143] |
| Drug–drug | Curcumin | Piperine | n.a. | 1:1 | Enhanced drug release | Male SD rats ($n = 6$): Increase in $C_{max}$ by 6.65- and 2.76-fold and $AUC_{0\text{-}24h}$ by 5.60- and 2.38-fold, respectively. | [101] |
| Drug–sulfonic acid | Nintedanib | Camphor-sulfonic acid | n.a. | 1:1.07 | Enhanced drug release of the SNEDDS of CAMS | Male Wistar rats ($n$ = no info): SNEDDS of CAMS with 2% HPMCP-50 significantly improved drug's absorption and reduced the interindividual variation in oral absorption behavior compared to Ofev®. | [144] |

(continued)

**Table 7.2** (continued)

| | Drug | Coformer | Third component | Molar ratio | In vitro study | In vivo study | Reference |
|---|---|---|---|---|---|---|---|
| Drug–drug | Olaparib | Diclofenac | n.a. | 1:1 | Enhanced drug release and solubility | Male SD rats ($n$ = 6): Increase in $AUC_{0\text{-}t}$ by 6.14-fold compared to the crystalline drug. | [145] |
| Drug–drug | Atenolol | Irbesartan | n.a. | 2:1, 1:1, 1:2 | Enhanced drug release | Male SD rats ($n$ = 5): The relative oral bioavailability of irbesartan was 2.78-fold higher than crystalline irbesartan. | [146] |
| Drug–drug | Ketocon-azole | Disodium glycyrrhizin | n.a. | 1:1, 1:2.5, 1:5, 1:10 | Enhanced drug release | Male SD rats ($n$ = 6): Increase in $AUC_{0\text{-}t}$ by 2.13-, 2.30-, 2.16-, and 1.86-fold for the 1:1, 1:2.5, 1:5, and 1:10, respectively, compared to the crystalline drug. | [147] |
| Drug–drug | Ceritinib | Naringin | n.a. | 2:1 | Enhanced drug release | Male Wistar rats ($n$ = 3): Increase in $C_{max}$ of 2.4-fold and $AUC_{0\text{-}t}$ of 2.1-fold compared to the crystalline drug. | [148] |
| Drug–sulfonic acid | Albend-azole | Acesulfame | n.a. | 1:2 | Enhanced drug release and solubility | Male Balb/c mice ($n$ = 5): Increase in $C_{max}$ of 11.0-fold and $AUC_{0\text{-}t}$ of 7.6-fold compared to the crystalline drug. | [149] |
| Drug–drug | Wogonin | Aloperine | n.a. | 1:1 | Enhanced drug release and solubility | Male SD rats ($n$ = 6): Increase in $C_{max}$ of 17.82-fold and $AUC_{0\text{-}t}$ of 2.71-fold compared to the crystalline wogonin. | [150] |
| Drug–drug | Enalapril | Lacidipine | $SiO_2$ | 2:1 | Enhanced drug release | Male SD rats ($n \geq 5$): Compared to CAMS powder, remarkable enhancement of bioavailability of both drugs was observed by the CAMS-loaded $SiO_2$. | [151] |
| Drug–drug | Paclitaxel | Resveratrol | n.a. | 1:1 | Enhanced drug release | Male SD rats ($n$ = 5): Increase in $C_{max}$ of 1.99-fold and $AUC_{0\text{-}t}$ of 1.74-fold compared to the crystalline paclitaxel. | [152] |

| | | | | | | | |
|---|---|---|---|---|---|---|---|
| Drug–organic acid | Lume-fantrine | α-ketoglutaric acid | n.a. | 1:1 | Enhanced drug release and solubility | Male SD rats ($n = 6$): Increase in $C_{max}$ of 10.86-, 9.24-, and 8.46-folds of CAMS prepared by LAG, SE, and QC, respectively, compared to crystalline drug. | [49] |
| Drug–drug | Raloxifene HCl | Naringin | n.a. | 2:1, 1:1, 1:2 | Enhanced drug release and permeability | Male Wistar rats ($n = 6$): Increase in $AUC_{0\text{-}t}$ by 2.80-, 1.43-, and 1.794-fold for the 2:1, 1:1, and 1:2, respectively, compared to the crystalline raloxifene. | [106] |
| Drug–amino acid | Bezafibrate | *L*-arginine | n.a. | 2:1 | Enhanced drug release and solubility | Male Wistar rats ($n = 3$): Increase in $C_{max}$ of 2.41-fold and $AUC_{0\text{–}36}$ of 2.04-fold compared to the crystalline drug. | [153] |
| Drug–drug | Pazopanib | Naringin | n.a. | 1:1 | Enhanced drug release and solubility | Male SD rats ($n = 6$): A 1.26-fold bioavailability increase of pazopanib compared to the crystalline drug. | [154] |
| Drug–drug | Curcumin | Myricetin | n.a. | 1:1 | Enhanced drug release | Male Wistar rats ($n = 6$): Increase in $C_{max}$ of 16.69-fold and *AUC* of 15.86-fold compared to the crystalline curcumin. | [133] |

*SD rat* Sprague–Dawley rat, $C_{max}$ maximum concentration; *AUC* area under the curve, $SiO_2$ mesoporous silica, *LAG* liquid-assisted grinding, *SE* solvent evaporation, *QC* quench cooling

**Table 7.3** Overview of the studies reporting in vitro or ex vivo permeability data

| CAMS class | Drug | Coformer | Third component | Molar ratio | Permeability study | Outcome | Reference |
|---|---|---|---|---|---|---|---|
| Drug–drug | Nifedipine | Ketoconazole | n.a. | 1:1 | Caco-2 cells | Permeated amounts of nifedipine and ketoconazole were 2.3- and 3.6- fold higher than those of the crystalline drugs, respectively. | [155] |
| Drug–amino acid | Furosemide | *L*-arginine | n.a. | 1:1 | PAMPA membranes | Only *L*-arginine coformer enhanced permeation of the drug. Piperine, quercetin, and verapamil showed similar or lower permeation. | [156] |
| Drug–drug | Darunavir | Ritonavir | n.a. | 2:1, 1:1, 1:2 | Everted gut sack | Permeability of darunavir was greatly reduced compared to the amorphous drug alone. Similarly, the permeability of ritonavir was also reduced. | [157] |
| | | Indomethacin | | | | Permeability of darunavir was slightly reduced compared to the amorphous drug alone. The permeability of indomethacin was also reduced. | |
| Drug–drug | Pyrazinamide | Naringin | n.a. | 1:1 | Noneverted gut sack | Apparent permeability was found to be 3.41x higher than that of the crystalline pyrazinamide. | [158] |
| Drug–drug | Fexofenadine | Naringin | n.a. | 2:1, 1:1, 1:2 | Everted gut sack | Permeability was found improved by up to 5-fold and 3.5-fold for fexofenadine and naringin, respectively. | [159] |
| Drug–amino acid | Genisein | *L*-arginine | n.a. | 1:1 | PAMPA membrane | Increased apparent solubility resulted in decreased permeability, which was more pronounced for the CAMS with *L*-lysine. | [160] |
| | | *L*-lysine | | | | | |
| Drug–drug | Telmisartan | Amlodipine | n.a. | 1:1 | PAMPA membrane | Decreased $P_{app}$ in the presence of a thicker unstirred water layer without stirring but increased in the presence of a thinner layer with stirring. | [140] |
| Drug–drug | Naringenin | Hesperitin | n.a. | 1:1 | Strat-M® membrane | Microemulsions of CAMS increased skin permeation of both drugs by more than four times. | [161] |
| Drug–drug | Dasatinib | Gallic acid | n.a. | 1:1 | Noneverted gut sack | Apparent permeability was found to be 1.9x higher than that of the crystalline drug. | [112] |

| | | | | | | | |
|---|---|---|---|---|---|---|---|
| Drug–drug | Ketoconazole | Disodium glycyrrhizin | n.a. | 1:1, 1:2.5, 1:5, 1:10 | PAMPA membrane | Permeation rate of ketoconazole under the concentration of disodium glycyrrhizin in solution above the CMC showed a significant downward trend compared to that below CMC. | [147] |
| Drug–drug | Ceritinib | Naringin | n.a. | 2:1 | Everted gut sack | Permeability improved by up to 5-fold for ceritinib compared to crystalline drug. | [148] |
| Drug–amino acid | Curcurmin | *L*-tryptophan | P(VP-co-VAc) | 10:5:85 %wt | PAMPA membrane | Permeability improved by more than 3-fold compared to crystalline drug. | [103] |
| Drug–drug | Palbociclib | Naringin | n.a. | 1:2 | Noneverted gut sack | Permeability improved by 3.6-fold for palbociclib compared to crystalline drug. | [162] |
| Drug–drug | Raloxifene HCl | Naringin | n.a. | 2:1 | Everted gut sack | Permeability improved by 10-fold for raloxifene compared to crystalline drug. | [106] |
| Drug–drug | Curcumin | Naringin | n.a. | 1:1, 1:2 | Everted gut sack | Permeability improved by 19- and 3.8-fold for the 1:2 and 1:1 molar ratios, respectively. | [163] |
| Drug–amino acid | Griseofulvin | L-lysine<br>L-methionine<br>L-valine | n.a. | 1:1 | Lipophilic membrane/ hydrophilic filter | *L*-methionine and *L*-valine led to enhanced permeability compared to crystalline drug. L-lysine did not significantly improve permeability. | [98] |
| Drug–drug | Pazopanib | Naringin | n.a. | 1:1 | Noneverted gut sack | $P_{app}$ was found improved by 2.23-fold for pazopanib compared to the crystalline drug. | [154] |

*PAMPA* parallel artificial membrane permeability assay, $P_{app}$ apparent permeability, *CMC* critical micelle concentration, *P(VP-co-VAc)* poly(1-vinylpyrrolidone)-co-(vinyl acetate)

### *7.6.10 Downstream Processing of CAMS*

Downstream processing of CAMS presents notable challenges due to their physical properties and stability requirements. Mechanical stress during tableting or exposure to environmental factors during formulation can induce recrystallization, thus reducing the solubility and stability [164–166]. In addition, CAMS powders often show greater cohesiveness and reduced flowability than their crystalline counterparts, complicating their processing [164, 165]. For instance, co-amorphous olanzapine blends exhibited a twofold increase in cohesion index, resulting in poor flowability [167]. Introducing inert carriers such as lactose or microcrystalline cellulose during spray drying has been reported to improve powdered CAMS' flowability and compressibility, enabling direct tablet compression [168].

Only a limited number of studies have investigated downstream processing of CAMS, with most them focusing on tablet [166, 167, 169–171] or pellet formulations [102, 172–175]. Other studies reported approaches including granulation [173, 176], CAMS-based microemulsions [161], CAMS filling into capsules [117], incorporation into SNEDDS [144], CAMS in fast-dissolving oral films [177], oleogels [178], and nanosized [179] CAMS.

#### 7.6.10.1 Tableting of CAMS

The tableting feasibility of CAMS is increasingly recognized as a decisive factor in determining their practical utilization for oral solid dosage forms. CAMS generally deform plastically under pressure, allowing for the formation of mechanically stable tablets without significant recrystallization [166, 167]. However, higher co-amorphous content can increase disintegration time and tablet elasticity [166]. The choice of coformer and excipients is critical for maintaining stability and optimizing tablet properties [168].

Although traditionally approached from a molecular/stabilization perspective, current evidence demonstrates that the same interactions responsible for maintaining the amorphous state also govern downstream processability and mechanical performance. Strong intermolecular forces, such as ionic interactions formed between loratadine and various organic acids or the hydrogen-bonding networks observed in systems containing carvedilol or olanzapine, appear to directly suppress molecular mobility and reinforce solid-state stability when subjected to process-induced stresses [167, 169, 171].

Manufacturing technologies can leverage these interactions to generate or maintain the amorphous form. High-pressure compaction has been shown to induce partial amorphization in physical mixtures of olanzapine–saccharin, indicating that tableting itself can act as an amorphization step when the interacting molecules are sufficiently compatible [167]. Moreover, of interest is the successful single-step production of fully co-amorphous tablets through injection molding, demonstrating that continuous, heat-assisted processing can bypass the recrystallization often associated with multistage powder handling [171].

The proportion of co-amorphous material incorporated into a tablet formulation exerts a clear and sometimes nonlinear influence on mechanical properties and disintegration. Increased co-amorphous content typically reduces elastic recovery and produces mechanically stronger tablets, although tensile strength may plateau once a minimal binder proportion is met [166, 169]. Interestingly, disintegration shows a complex relationship with co-amorphous content: moderate levels tend to prolong disintegration, likely due to densification and reduced pore structure, whereas higher fractions can accelerate disintegration, possibly through increased hydrophilicity or altered matrix dynamics [166]. This behavior suggests that CAMS are not a passive solubility enhancer but an active structural component that alters tablet architecture.

Additionally, although improved dissolution is a consistent advantage of co-amorphous formulations, its interplay with tablet disintegration must be considered carefully. For example, despite enhanced solubility, co-amorphous olanzapine tablets show reduced release rates relative to crystalline tablets, reflecting how slower disintegration can offset intrinsic solubility benefits [167].

#### 7.6.10.2 Formulation of CAMS into Pellets

The incorporation of CAMS into pellets has attracted significant attention as a strategy to overcome solubility and process-related limitations of poorly water-soluble drugs. Pellets offer advantages in flowability, content uniformity, and flexible dosing, while CAMS provide enhanced dissolution and improved physical stability. When combined, these two can create a platform capable of producing high-performance oral formulations with robust manufacturability. Recent work demonstrates that pelletization can serve not only as a shaping step but also as an active contributor to amorphization, enabling solventless, energy-efficient, and scalable production of CAMS [102]. The studies reviewed here illustrate how different pelletization methods, including extrusion/spheronization, high-shear layering, and bead coating, can be leveraged to generate stable CAMS.

A key development is the recognition that pelletization itself can promote in situ co-amorphization. In systems such as olanzapine–saccharin, water incorporation during extrusion/spheronization or bead coating was sufficient to induce the co-amorphization [172], while similar behavior occurred during wet granulation [173]. High-shear granulation of indomethacin and arginine showed that even moderate jacket temperatures can drive the formation of a uniform co-amorphous layer on pellet surfaces [102]. These observations shift the perception of pelletization from a downstream method to an enabling amorphization process, which may reduce or eliminate the need for separate processing steps.

The spatial distribution of the co-amorphous phase within the pellet also emerges as an important factor. When CAMS are located near the pellet surface, dissolution is accelerated relative to pellets in which CAMS are embedded in the matrix. This was evident in the olanzapine-coated beads, which delivered faster release than pellets which had the CAMS located internally [172]. Such findings highlight the potential to tailor release characteristics through controlled layering and microstructural design.

## 7.7 Conclusions and Future Perspectives

In the last 5 years (2021–2025), research on CAMS has continued to grow, though not uniformly across all aspects of the field. The rising number of published studies demonstrates sustained interest, yet the vast majority of them consist of case studies. As a result, many of the challenges highlighted in prior reviews (Liu et al., 2020) such as limited systematic screening, mechanistic investigation, and uneven characterization of all CQAs still exist in current literature.

That said, several encouraging developments distinguish the last 5 years. Most notable is the increased attention to in vivo performance outcomes. This shift signals a move toward translational relevance and offers a clearer picture of how CAMS behave under physiological conditions. Permeability studies, especially for BCS Class IV drugs, likewise underscore the growing recognition that the benefits of CAMS may extend beyond solubility enhancement.

At the same time, coformer selection and molar ratio optimization continue to be dominated by empirical choices. Systematic screening is still uncommon, and equimolar ratios remain the default choice, although there is evidence that often they do not represent the optimal composition. Recent computational and data-driven approaches demonstrate clear potential for more predictive designs, but these tools are only beginning to be adopted. Progress in this direction will be crucial for moving the field beyond exploratory formulation work and toward rational development strategies.

The second CQA, physical stability, illustrates both advancement and persistent gaps. More studies now examine storage stability under defined conditions, yet long-term data remain scarce. Investigations of molecular mobility, which provide a more mechanistic basis for predicting recrystallization, are limited to a small number of research groups. Expanding these analyses would not only refine stability predictions but also support more confident comparisons between CAMS and other amorphous solid dispersion technologies. Moreover, the growing exploration of ternary systems represents another promising trend. Incorporating polymers, surfactants, lipids, or even small amounts of water has revealed new ways to modulate intermolecular interactions and improve stability or dissolution behavior.

Downstream processing is likewise emerging as a key area of advancement. Tablet and pellet formulations, high-shear processes, and even continuous manufacturing approaches are increasingly being studied not merely as shaping steps but as potential enablers of amorphization. Understanding how mechanical stress, environmental conditions, and excipient interactions affect CAMS during manufacturing will be essential for enabling real-world pharmaceutical development.

To fully realize the promise of CAMS, future efforts should prioritize (i) systematic coformer screening supported by computational prediction and high-throughput experimentation, (ii) deeper mechanistic studies linking intermolecular interactions to long-term physical stability, (iii) broader and more standardized in vitro–in vivo assessment across diverse drug classes, (iv) rational exploration of CAMS for non-oral delivery routes, and (v) integration of co-amorphization into scalable manufacturing technologies.

Taken together, these directions could transform CAMS from a predominantly experimental formulation concept into a robust and widely applicable platform for addressing solubility, stability, and permeability challenges across pharmaceutical development.

## References

1. Karagianni A, Kachrimanis K, Nikolakakis I. Co-amorphous solid dispersions for solubility and absorption improvement of drugs: composition, preparation, characterization and formulations for oral delivery. Pharmaceutics. 2018;10(3)
2. Narala S, Nyavanandi D, Srinivasan P, Mandati P, Bandari S, Repka MA. Pharmaceutical cocrystals, salts, and co-amorphous systems: a novel opportunity of hot-melt extrusion. J Drug Deliv Sci Technol. 2021;61:102209.
3. Nyamba I, Sombié CB, Yabré M, Zimé-Diawara H, Yaméogo J, Ouédraogo S, et al. Pharmaceutical approaches for enhancing solubility and oral bioavailability of poorly soluble drugs. Eur J Pharm Biopharm. 2024;204:114513.
4. Moseson DE, Tran TB, Karunakaran B, Ambardekar R, Hiew TN. Trends in amorphous solid dispersion drug products approved by the U.S. Food and Drug Administration between 2012 and 2023. Int J Pharm X. 2024;7:100259.
5. Dengale SJ, Grohganz H, Rades T, Löbmann K. Recent advances in co-amorphous drug formulations. Adv Drug Deliv Rev. 2016;100:116–25.
6. Liu J, Grohganz H, Löbmann K, Rades T, Hempel N-J. Co-amorphous drug formulations in numbers: recent advances in co-amorphous drug formulations with focus on co-formability, molar ratio, preparation methods, physical stability, in vitro and in vivo performance, and new formulation strategies. Pharmaceutics. 2021;13(3)
7. Han J, Wei Y, Lu Y, Wang R, Zhang J, Gao Y, et al. Co-amorphous systems for the delivery of poorly water-soluble drugs: recent advances and an update. Expert Opin Drug Deliv. 2020;17(10):1411–35.
8. Marchetti A, Yin J, Su Y, Kong X. Solid-state NMR in the field of drug delivery: state of the art and new perspectives. Magn Reson Lett. 2021;1(1):28–70.
9. Bhatta HP, Han H-K, Maharjan R, Jeong SH. Recent techniques to improve amorphous dispersion performance with quality design, physicochemical monitoring, molecular simulation, and machine learning. Pharmaceutics. 2025;17(10)
10. Li B, Wang Y, Feng Y, Yuan D, Xu R, Jiang C, et al. Design and molecular insights of drug-active metabolite based co-amorphous formulation: a case study of toltrazuril-ponazuril co-amorphous. Int J Pharm. 2022;615:121475.
11. Salahshoori I, Golriz M, Nobre MAL, Mahdavi S, Eshaghi Malekshah R, Javdani-Mallak A, et al. Simulation-based approaches for drug delivery systems: navigating advancements, opportunities, and challenges. J Mol Liq. 2024;395:123888.
12. Marsac PJ, Shamblin SL, Taylor LS. Theoretical and practical approaches for prediction of drug–polymer miscibility and solubility. Pharm Res. 2006;23(10):2417–26.

13. Meng F, Dave V, Chauhan H. Qualitative and quantitative methods to determine miscibility in amorphous drug–polymer systems. Eur J Pharm Sci. 2015;77:106–11.
14. Chambers LI, Musa OM, Steed JW. Prediction and preparation of coamorphous phases of a bislactam. Mol Pharm. 2022;19(7):2651–61.
15. Gupta J, Nunes C, Vyas S, Jonnalagadda S. Prediction of solubility parameters and miscibility of pharmaceutical compounds by molecular dynamics simulations. J Phys Chem B. 2011;115(9):2014–23.
16. Hang NT, Long NT, Duy ND, Chien NN, Van Phuong N. Towards safer and efficient formulations: machine learning approaches to predict drug-excipient compatibility. Int J Pharm. 2024;653:123884.
17. Han J, Sun W, Chen J, Yue Z, Fang W, Liu X, et al. Design of coamorphous systems for flavonoid components coformed with meglumine by integrating theory-model-experiment techniques. Mol Pharm. 2025;22(6):3045–60.
18. Song Y, Chen Y, Heng W, Hu E, Shi Y, Gao Y, et al. The potential of supramolecular synthon to develop coamorphous systems with tailored physical stability: mechanistic insights integrating kinetics and thermodynamics. Int J Pharm. 2024;667:124857.
19. Medarević D, Djuriš J, Barmpalexis P, Kachrimanis K, Ibrić S. Analytical and computational methods for the estimation of drug-polymer solubility and miscibility in solid dispersions development. Pharmaceutics. 2019;11(8)
20. Zhao X, Cheng S, Koh YP, Kelly BD, McKenna GB, Simon SL. Prediction of the synergistic glass transition temperature of Coamorphous molecular glasses using activity coefficient models. Mol Pharm. 2021;18(9):3439–51.
21. Antipas GSE, Reul R, Voges K, Kyeremateng SO, Ntallis NA, Karalis KT, et al. System-agnostic prediction of pharmaceutical excipient miscibility via computing-as-a-service and experimental validation. Sci Rep. 2024;14(1):15106.
22. Qian S, Heng W, Wei Y, Zhang J, Gao Y. Coamorphous lurasidone hydrochloride–saccharin with charge-assisted hydrogen bonding interaction shows improved physical stability and enhanced dissolution with ph-independent solubility behavior. Cryst Growth Des. 2015;15(6):2920–8.
23. Heng W, Song Y, Luo M, Hu E, Wei Y, Gao Y, et al. Mechanistic insights into the crystallization of coamorphous drug systems. J Control Release. 2023;354:489–502.
24. Gniado K, MacFhionnghaile P, McArdle P, Erxleben A. The natural bile acid surfactant sodium taurocholate (NaTC) as a coformer in coamorphous systems: enhanced physical stability and dissolution behavior of coamorphous drug-NaTc systems. Int J Pharm. 2018;535(1–2, 132):–9.
25. Patel D, Wairkar S. In silico and in vitro investigation of bile salts as coformers for edaravone coamorphous dispersion – part I. Chem Phys Lipids. 2023;253:105302.
26. Sharma A, Dey S, Naga Sidhartha N, Shah S, Sapkal R, Srivastava S, et al. Coamorphous systems of rebamipide: selection of amino acid coformers based on protein-ligand docking, in vitro assessment and study of interactions by computational and multivariate analysis. Int J Pharm. 2024;667:124955.
27. Ueda H, Hirakawa Y, Miyano T, Imono M, Tse JY, Uchiyama H, et al. Design of a stable coamorphous system using lactose as an antiplasticizing agent for diphenhydramine hydrochloride with a low glass transition temperature. Mol Pharm. 2022;19(4):1209–18.
28. Deng Y, Deng W, Huang W, Zheng Z, Zhang R, Liu S, et al. Norfloxacin co-amorphous salt systems: effects of molecular descriptors on the formation and physical stability of co-amorphous systems. Chem Eng Sci. 2022;253:117549.
29. Hu Y, Guo Y, Li B, Xu R, Fang X, Cao Y, et al. Influence of the pKa value of cinnamic acid and P-hydroxycinnamic acid on the solubility of a lurasidone hydrochloride-based coamorphous system. ACS Omega. 2021;6(4):3106–19.
30. Martínez LM, Videa M, Sosa NG, Ramírez JH, Castro S. Long-term stability of new co-amorphous drug binary systems: study of glass transitions as a function of composition and shelf time. Molecules [Internet]. 2016;21(12).

31. Sá MR, Sarraguça JMG, de Sousa FF, Sarraguça MSC, Lopes JA, Lima ADSG, et al. Structural, thermal, vibrational, solubility and DFT studies of a tolbutamide co-amorphous drug delivery system for treatment of diabetes. Int J Pharm. 2022;615:121500.
32. Kasten G, Löbmann K, Grohganz H, Rades T. Co-former selection for co-amorphous drug-amino acid formulations. Int J Pharm. 2019;557:366–73.
33. Lu W, Rades T, Rantanen J, Chan H-K, Yang M. Amino acids as stabilizers for spray-dried simvastatin powder for inhalation. Int J Pharm. 2019;572:118724.
34. Lu W, Rades T, Rantanen J, Yang M. Inhalable co-amorphous budesonide-arginine dry powders prepared by spray drying. Int J Pharm. 2019;565:1–8.
35. Partheniadis I, Tsouka M, Nikolakakis I. Exploring hot melt extrusion in the formation of exemestane/amino acid co-amorphous systems. RSC Pharmaceutics. 2025;
36. Liu J, Rades T, Grohganz H. The influence of moisture on the storage stability of co-amorphous systems. Int J Pharm. 2021;605:120802.
37. Chieng N, Teo X, Cheah MH, Choo ML, Chung J, Hew TK, et al. Molecular dynamics and physical stability of pharmaceutical co-amorphous systems: correlation between structural relaxation times measured by Kohlrausch-Williams-Watts with the width of the glass transition temperature ($\Delta T_g$) and the onset of crystallization. J Pharm Sci. 2019;108(12):3848–58.
38. Graeser KA, Patterson JE, Zeitler JA, Gordon KC, Rades T. Correlating thermodynamic and kinetic parameters with amorphous stability. Eur J Pharm Sci. 2009;37(3):492–8.
39. Martínez LM, Cruz-Angeles J, Vázquez-Dávila M, Martínez E, Cabada P, Navarrete-Bernal C, et al. Mechanical activation by ball milling as a strategy to prepare highly soluble pharmaceutical formulations in the form of co-amorphous, co-crystals, or polymorphs. Pharmaceutics. 2022;14(10)
40. Jensen KT, Larsen FH, Cornett C, Löbmann K, Grohganz H, Rades T. Formation mechanism of coamorphous drug–amino acid mixtures. Mol Pharm. 2015;12(7):2484–92.
41. Lenz E, Löbmann K, Rades T, Knop K, Kleinebudde P. Hot melt extrusion and spray drying of co-amorphous indomethacin-arginine with polymers. J Pharm Sci. 2017;106(1):302–12.
42. Partheniadis I, Nikolakakis I. Development and characterization of co-amorphous griseofulvin/L-leucin by modified solvent processing hot-melt extrusion. Int J Pharm. 2024;652:123824.
43. Partheniadis I, Tsouka M, Nikolakakis I. Exploring the preparation of griseofulvin CAMS with amino acids of different hydrophobicity as co-formers using a modified hot-melt extrusion process. Int J Pharm. 2024;666:124818.
44. Alhajj N, O'Reilly NJ, Cathcart H. Development and characterization of a spray-dried inhalable ciprofloxacin-quercetin co-amorphous system. Int J Pharm. 2022;618:121657.
45. Singh A, Van den Mooter G. Spray drying formulation of amorphous solid dispersions. Adv Drug Deliv Rev. 2016;100:27–50.
46. Jakubowska E, Lulek J. The application of freeze-drying as a production method of drug nanocrystals and solid dispersions — a review. J Drug Deliv Sci Technol. 2021;62:102357.
47. Wostry M, Plappert H, Grohganz H. Preparation of co-amorphous systems by freeze-drying. Pharmaceutics. 2020;12(10):941.
48. Shah B, Kakumanu VK, Bansal AK. Analytical techniques for quantification of amorphous/crystalline phases in pharmaceutical solids. J Pharm Sci. 2006;95(8):1641–65.
49. Khemchandani R, Pardhi E, Goud S, Tomar DS, Medtiya P, Khushwaha BS, et al. Multi-method coamorphous systems of lumefantrine with alpha ketoglutaric acid: comprehensive characterization, biological evaluation and stability analysis. Int J Pharm. 2025;669:125056.
50. D'Abbrunzo I, Venier E, Selmin F, Škorić I, Bernardo E, Procida G, et al. Stability of ternary drug–drug–drug coamorphous systems obtained through mechanochemistry. Pharmaceutics. 2025;17(1):92.
51. Barbosa EJ, Andrade MAB, Gubitoso MR, Bezzon VDN, Smith PA, Byrn SR, et al. Acoustic levitation and high-resolution synchrotron X-ray powder diffraction: a fast screening approach of niclosamide amorphous solid dispersions. Int J Pharm. 2021;602:120611.

52. Newman A, Engers D, Bates S, Ivanisevic I, Kelly RC, Zografi G. Characterization of amorphous API:Polymer mixtures using X-ray powder diffraction. J Pharm Sci. 2008;97(11):4840–56.
53. Baird JA, Taylor LS. Evaluation of amorphous solid dispersion properties using thermal analysis techniques. Adv Drug Deliv Rev. 2012;64(5):396–421.
54. Kissi EO, Kasten G, Löbmann K, Rades T, Grohganz H. The role of glass transition temperatures in coamorphous drug–amino acid formulations. Mol Pharm. 2018;15(9):4247–56.
55. Bugay DE. Characterization of the solid-state: spectroscopic techniques. Adv Drug Deliv Rev. 2001;48(1):43–65.
56. de Oliveira Neto JG, Bezerra RDS, Domingos FNB, Lima ADSG, Souto EB, Lage MR, et al. A new coamorphous ethionamide with enhanced solubility: preparation, characterization, in silico pharmacokinetics, and controlled release by encapsulation. Int J Pharm. 2025;670:125159.
57. Harju EA, Tomberg T, Wurr L, van Rijckeghem A, Arbiol Enguita AM, Dordovic V, et al. Characterization of solid-state complexities in pharmaceutical materials via stimulated Raman scattering microscopy. Anal Chem. 2025;97(18):9627–37.
58. Schmitt PD. Recent advances in nonlinear optical analyses of pharmaceutical materials in the solid state. Mol Pharm. 2017;14(3):555–65.
59. Du Y, Su Y. 19F solid-state NMR characterization of pharmaceutical solids. Solid State Nucl Magn Reson. 2022;120:101796.
60. Geppi M, Mollica G, Borsacchi S, Veracini CA. Solid-state NMR studies of pharmaceutical systems. Appl Spectrosc Rev. 2008;43(3):202–302.
61. Ruggiero MT, Krynski M, Kissi EO, Sibik J, Markl D, Tan NY, et al. The significance of the amorphous potential energy landscape for dictating glassy dynamics and driving solid-state crystallisation. Phys Chem Chem Phys. 2017;19(44):30039–47.
62. Woodward WH. Broadband dielectric spectroscopy – a practical guide. Broadband dielectric spectroscopy: a modern analytical technique. ACS Publications; 2021. p. 3–59.
63. Zhang J, Shi Q, Qu T, Zhou D, Cai T. Crystallization kinetics and molecular dynamics of binary coamorphous systems of nimesulide and profen analogs. Int J Pharm. 2021;610:121235.
64. Queiroz LHS, Barros RS, de Sousa FF, Lage MR, Sarraguça MC, Ribeiro PRS. Preparation and characterization of a rifampicin coamorphous material with tromethamine coformer: an experimental–theoretical study. Mol Pharm. 2024;21(3):1272–84.
65. Deng Y, Liu S, Jiang Y, Martins ICB, Rades T. Recent advances in co-former screening and formation prediction of multicomponent solid forms of low molecular weight drugs. Pharmaceutics. 2023;15(9)
66. Jensen KT, Larsen FH, Löbmann K, Rades T, Grohganz H. Influence of variation in molar ratio on co-amorphous drug-amino acid systems. Eur J Pharm Biopharm. 2016;107:32–9.
67. Liu J, Rades T, Grohganz H. Determination of the optimal molar ratio in amino acid-based coamorphous systems. Mol Pharm. 2020;17(4):1335–42.
68. Bastos AC, Santos IA, Pinto JF, Fernandes AI. Co-formability, solubility enhancement and stability of olanzapine co-amorphous systems produced with different co-formers. Ann Med. 2021;53(sup1):S144–S5.
69. Wu W, Grohganz H, Rades T, Löbmann K. Comparison of co-former performance in co-amorphous formulations: single amino acids, amino acid physical mixtures, amino acid salts and dipeptides as co-formers. Eur J Pharm Sci. 2021;156:105582.
70. Kasten G, Grohganz H, Rades T, Löbmann K. Development of a screening method for co-amorphous formulations of drugs and amino acids. Eur J Pharm Sci. 2016;95:28–35.
71. Chambers LI, Grohganz H, Palmelund H, Löbmann K, Rades T, Musa OM, et al. Predictive identification of co-formers in co-amorphous systems. Eur J Pharm Sci. 2021;157:105636.
72. Liu J, Hwu E, Bannow J, Grohganz H, Rades T. Impact of molecular surface diffusion on the physical stability of co-amorphous systems. Mol Pharm. 2022;19(4):1183–90.

73. Wang M, Liu S, Jia L, Zhang J, Du S, Gong J. Exploring the physical stability of three nimesulide–indomethacin co-amorphous systems from the perspective of molecular aggregates. Eur J Pharm Sci. 2020;147:105294.
74. Fang X, Hu Y, Yang G, Shi W, Lu S, Cao Y. Improving physicochemical properties and pharmacological activities of ternary co-amorphous systems. Eur J Pharm Biopharm. 2022;181:22–35.
75. Liu J, Grohganz H, Rades T. Influence of polymer addition on the amorphization, dissolution and physical stability of co-amorphous systems. Int J Pharm. 2020;588:119768.
76. Wang Y, Grohganz H, Rades T. Effects of polymer addition on the non-strongly interacting binary co-amorphous system carvedilol-tryptophan. Int J Pharm. 2022;617:121625.
77. Fael H, Demirel AL. Indomethacin co-amorphous drug-drug systems with improved solubility, supersaturation, dissolution rate and physical stability. Int J Pharm. 2021;600:120448.
78. Allesø M, Chieng N, Rehder S, Rantanen J, Rades T, Aaltonen J. Enhanced dissolution rate and synchronized release of drugs in binary systems through formulation: amorphous naproxen–cimetidine mixtures prepared by mechanical activation. J Control Release. 2009;136(1):45–53.
79. Zhang Y-F, Yao Q, Lin X-Y, Ma Y-H, Zhang H-F, Yu H, et al. Co-amorphization, dissolution, and stability of quench-cooled drug–drug coamorphous supersaturating delivery systems with rt-unstable amorphous components. Pharmaceutics. 2024;16(12)
80. Heikkinen AT, DeClerck L, Löbmann K, Grohganz H, Rades T, Laitinen R. Dissolution properties of co-amorphous drug-amino acid formulations in buffer and biorelevant media. Pharmazie. 2015;70(7):452–7.
81. Chieng N, Aaltonen J, Saville D, Rades T. Physical characterization and stability of amorphous indomethacin and ranitidine hydrochloride binary systems prepared by mechanical activation. Eur J Pharm Biopharm. 2009;71(1):47–54.
82. Fukoka E, Makita M, Yamamura S. Glassy state of pharmaceuticals. III.: thermal properties and stability of glassy pharmaceuticals and their binary glass systems. Chem Pharm Bull. 1989;37(4):1047–50.
83. Yamamura S, Gotoh H, Sakamoto Y, Momose Y. Physicochemical properties of amorphous precipitates of cimetidine–indomethacin binary system. Eur J Pharm Biopharm. 2000;49(3):259–65.
84. Yamamura S, Momose Y, Takahashi K, Nagatani S. Solid-state interaction between cimetidine and naproxen. Drug Stab. 1996;1:173–8.
85. Xu X, Grohganz H, Knapik-Kowalczuk J, Paluch M, Rades T. Mechanistic investigation into crystallization of hydrated co-amorphous systems of flurbiprofen and lidocaine. Pharmaceutics. 2025;17(2)
86. Xu X, Grohganz H, Rades T. Influence of water on amorphous lidocaine. Mol Pharm. 2022;19(9):3199–205.
87. Xu X, Rades T, Grohganz H. Thermal investigation on hydrated co-amorphous systems of nicotinamide and prilocaine. Eur J Pharm Biopharm. 2023;186:1–6.
88. Xu X, Rades T, Grohganz H. Molecular interactions of hydrated co-amorphous systems of prilocaine and lidocaine. Int J Pharm. 2024;651:123807.
89. Yarlagadda DL, Kawakami K, Samavedi S. Leveraging molecular interactions to develop a generalized design framework for coamorphous drug–drug mixtures exhibiting elevated glass transition temperatures. Mol Pharm. 2025;22(6):3084–96.
90. Fink E, Brunsteiner M, Mitsche S, Schröttner H, Paudel A, Zellnitz-Neugebauer S. Data-driven prediction of the formation of co-amorphous systems. Pharmaceutics. 2023;15(2)
91. Kissi EO, Khorami K, Rades T. Determination of stable co-amorphous drug–drug ratios from the eutectic behavior of crystalline physical mixtures. Pharmaceutics. 2019;11(12):628.
92. Valenti S, Cazorla C, Romanini M, Tamarit JL, Macovez R. Eutectic mixture formation and relaxation dynamics of coamorphous mixtures of two benzodiazepine drugs. Pharmaceutics. 2023;15(1):196.

93. Czyrski GS, Kirkensgaard JJK, Rønholt S, Rades T, Heinz A. Terpene-based eutectic mixtures for cutaneous delivery: eutectic point vs. molar ratio — which matters more? J Mol Liq. 2024;411:125726.
94. Wang C, Wang X, Yin C, Gao L, He J, Zhang G, et al. Dissolution improvement and solubility enhancement of two tumor-associated enzyme inhibitors via the formation of eutectic mixtures. J Pharm Sci. 2025;114(9)
95. Di R, Liu J, Grohganz H, Rades T. A multivariate approach for the determination of the optimal mixing ratio of the non-strong interacting co-amorphous system carvedilol-tryptophan. Molecules 2021;26(4).
96. Iemtsev A, Zemánková A, Hassouna F, Mathers A, Klajmon M, Slámová M, et al. Ball milling and hot-melt extrusion of indomethacin–l-arginine–vinylpyrrolidone-vinyl acetate copolymer: solid-state properties and dissolution performance. Int J Pharm. 2022;613:121424.
97. Mujtaba MA, Fule R, Amin P, Elhassan GO, Almoutairi MMM, Kaleem M, et al. Development of hot melt extruded co-formulated artesunate and amodiaquine- Soluplus(®) solid dispersion system in fixed-dose form: amorphous state characterization and pharmacokinetic evaluation. Curr Drug Metab. 2024;25(7):505–22.
98. Partheniadis I, Martins ICB, Müllertz A, Nikolakakis I, Rades T. The role of feed pretreatment and thermal stress in the development of griseofulvin/amino acid CAMS by hot-melt extrusion — structure relaxation and release/permeability evaluation. Int J Pharm. 2025;679:125765.
99. Shen P, Zhang C, Hu E, Gao Y, Wei Y, Zhang J, et al. Switch between cocrystal and coamorphous forms depending on thermal modulation of hot-melt extrusion. Mol Pharm. 2023;20(7):3412–26.
100. Lim LM, Park JW, Hadinoto K. Benchmarking the solubility enhancement and storage stability of amorphous drug-polyelectrolyte Nanoplex against co-amorphous formulation of the same drug. Pharmaceutics. 2022;14(5)
101. Han N, Liu Y, Liu X, Li P, Lu Y, Du S, et al. The controlled preparation of a carrier-free nanoparticulate formulation composed of curcumin and Piperine using high-gravity technology. Pharmaceutics. 2024;16(6)
102. Kondo K, Rades T. Solventless amorphization and pelletization using a high shear granulator. Part II; preparation of co-amorphous mixture-layered pellets using indomethacin and arginine. Eur J Pharm Biopharm. 2022;181:183–94.
103. Garbiec E, Rosiak N, Sip S, Zalewski P, Cielecka-Piontek J. Curcumin solubility and bioactivity enhancement through amorphization with tryptophan via supercritical fluid technology. Int J Mol Sci. 2025;26(2)
104. Turek M, Różycka-Sokołowska E, Koprowski M, Marciniak B, Bałczewski P. Role of hydrogen bonds in formation of co-amorphous valsartan/nicotinamide compositions of high solubility and durability with anti-hypertension and anti-COVID-19 potential. Mol Pharm. 2021;18(5):1970–84.
105. da Costa NF, Santos IA, Fernandes AI, Pinto JF. Sulfonic acid derivatives in the production of stable co-amorphous Systems for solubility enhancement. J Pharm Sci. 2022;111(12):3327–39.
106. Navya Sree KS, Dengale SJ, Mutalik S, Bhat K. Raloxifene HCl — quercetin co-amorphous system: preparation, characterization, and investigation of its behavior in phosphate buffer. Drug Dev Ind Pharm. 2022;48(6):227–38.
107. Garbiec E, Rosiak N, Tykarska E, Zalewski P, Cielecka-Piontek J. Sinapic acid co-amorphous systems with amino acids for improved solubility and antioxidant activity. Int J Mol Sci [Internet]. 2023;24(6).
108. Guo W, Dong X, Li Y, Li C, Tian Y, Gao H, et al. Co-amorphous formulation of dipyridamole with p-hydroxybenzoic acid: underlying molecular mechanisms, physical stability, dissolution behavior and pharmacokinetic study. Eur J Pharm Biopharm. 2023;184:139–49.

109. Holzapfel K, Rades T, Leopold CS. Co-amorphous systems consisting of indomethacin and the chiral co-former tryptophan: solid-state properties and molecular mobilities. Int J Pharm. 2023;636:122840.
110. Muñoz Tecocoatzi MF, Páez-Franco JC, Rubio-Carrasco K, Núñez-Pineda A, Dorazco-González A, Fuentes-Noriega I, et al. Ball-milling preparation of the drug–drug solid form of pioglitazone-rosuvastatin at different molar ratios: characterization and intrinsic dissolution rates evaluation. Pharmaceutics. 2023;15(2)
111. Villeda-Villegas M, Páez-Franco JC, Coyote-Dotor G, Núñez-Pineda A, Dorazco-González A, Fuentes-Noriega I, et al. Diversity of solid forms promoted by ball milling: characterization and intrinsic dissolution studies of pioglitazone hydrochloride and fluvastatin sodium drug–drug systems. Pharmaceuticals. 2023;16(6)
112. Kanp T, Dhuri A, Aalhate M, Mahajan S, Munagalasetty S, Kumar Sah S, et al. Manifesting the Dasatinib-gallic acid co-amorphous system to augment anticancer potential: physicochemical characterization, in silico molecular simulation, ex vivo permeability, and in vitro efficacy. Int J Pharm. 2024;665:124672.
113. Khorami K, Farahani SD, Müllertz A, Rades T. Drug–phospholipid co-amorphous formulations: the role of preparation methods and phospholipid selection. Pharmaceutics. 2024;16(12)
114. Khorami K, Müllertz A, Rades T. Influence of preparation method and choice of phospholipid on co-amorphization, physical stability, and dissolution behavior of equimolar indomethacin-phospholipid systems — a case study. J Drug Deliv Sci Technol. 2024;93:105433.
115. D'Abbrunzo I, Battaiotto L, Abruzzo A, Bondi G, Bigucci F, Pagano C, et al. Structural insights into novel coamorphous systems of azithromycin with faster dissolution profile. Eur J Pharm Biopharm. 2025;216:114873.
116. Dayo Owoyemi BC, Zeller M, Pereira da Silva B, Akinyemi AO, Ando RA, de Araujo GLB, et al. Drug property optimization: design, synthesis, and characterization of novel pharmaceutical salts and cocrystal-salt of lumefantrine. Mol Pharm. 2025;22(2):1042–60.
117. Fröhlich E, Bordoni A, Mohsenzada N, Mitsche S, Schröttner H, Zellnitz-Neugebauer S. Development of co-amorphous systems for inhalation therapy — part 1: from model prediction to clinical success. Pharmaceutics. 2025;17(7)
118. Takayama T, Kaneko S, Palanisamy V, Ono M, Titapiwatanakun V, Higashi K, et al. Physical properties for novel cilostazol co-amorphous; effect of preparation method and molar ratio on the co-amorphous. Chem Pharm Bull. 2025;73(4):318–26.
119. Martins ICB, Larsen AS, Madsen AØ, Frederiksen OA, Correia A, Jensen KMØ, et al. Unveiling polyamorphism and polyamorphic interconversions in pharmaceuticals: the peculiar case of hydrochlorothiazide. Chem Sci. 2023;14(41):11447–55.
120. Han J, Li L, Su M, Heng W, Wei Y, Gao Y, et al. Deaggregation and crystallization inhibition by small amount of polymer addition for a co-amorphous curcumin-magnolol system. Pharmaceutics. 2021;13(10)
121. Li J, Duggirala NK, Kumar NSK, Su Y, Suryanarayanan R. Design of ternary amorphous solid dispersions for enhanced dissolution of drug combinations. Mol Pharm. 2022;19(8):2950–61.
122. Strojewski D, Lalik S, Danède F, Górska N, Deptuch A, Marzec M, et al. Bosentan monohydrate and sildenafil base as two companions in enabling formulations. Int J Pharm. 2024;661:124312.
123. Chen X, Li D, Zhang H, Duan Y, Huang Y. Co-amorphous systems of sinomenine with platensimycin or sulfasalazine: physical stability and excipient-adjusted release behavior. Mol Pharm. 2022;19(11):4370–81.
124. Luo M, Chen A, Shan S, Guo M, Cai T. Molar ratio-dependent crystallization in coamorphous celecoxib–carbamazepine systems: the interplay of thermodynamics and kinetics. Mol Pharm. 2025;22(6):3401–13.
125. Ekawa B, Diogo HP, Castro RAE, Caires FJ, Eusébio MES. Coamorphous systems of valsartan: thermal analysis contribution to evaluate intermolecular interactions effects on the structural relaxation. Molecules. 2023;28(17)

126. Adhikari BR, Sinha S, Gordon KC, Das SC. Amino acids improve aerosolization and chemical stability of potential inhalable amorphous spray-dried ceftazidime for Pseudomonas aeruginosa lung infection. Int J Pharm. 2022;621:121799.
127. Nascimento ALCS, Martins ICB, Spósito L, Morais-Silva G, Duarte JL, Rades T, et al. Indomethacin-omeprazole as therapeutic hybrids? Salt and co-amorphous systems enhancing physicochemical and pharmacological properties. Int J Pharm. 2024;653:123857.
128. Pu F, Wang S, Yang J, Yang J, Hong Y, Guo Y, et al. Study on co-amorphous emerging solubilization behavior after gelation during dissolution: the importance of complexation and anti-crystallization. Int J Pharm. 2024;664:124592.
129. Shi X, Zhou X, Shen S, Chen Q, Song S, Gu C, et al. Improved in vitro and in vivo properties of telmisartan in the co-amorphous system with hydrochlorothiazide: A potential drug-drug interaction mechanism prediction. Eur J Pharm Sci. 2021;161:105773.
130. Gao Y, Liao J, Qi X, Zhang J. Coamorphous repaglinide–saccharin with enhanced dissolution. Int J Pharm. 2013;450(1–2):290–5.
131. Dengale SJ, Hussen SS, Krishna B, Musmade PB, Shenoy GG, Bhat K. Fabrication, solid state characterization and bioavailability assessment of stable binary amorphous phases of Ritonavir with Quercetin. Eur J Pharm Biopharm. 2015;89:329–38.
132. Huang P, Yan H, Wu H, Liu L, Niu W, Zhai B, et al. Application of curcumin as a co-former and an efflux inhibitor in paclitaxel co-amorphous mixture. J Drug Deliv Sci Technol. 2023;84:104513.
133. Zhang J, Xu H, Xu X, Yan M, Yu Z, Qian X, et al. Engineering a curcumin-myricetin co-amorphous system: hydrogen bond-driven solubility & bioavailability enhancement with 16.7-fold oral exposure boost. J Drug Deliv Sci Technol. 2025;114:107439.
134. Hu C, Zhang F, Fan H. Evaluation of drug dissolution rate in co-amorphous and co-crystal binary drug delivery systems by thermodynamic and kinetic methods. AAPS PharmSciTech. 2021;22(1):21.
135. Li Y-W, Zhang H-M, Cui B-J, Hao C-Y, Zhu H-Y, Guan J, et al. "Felodipine-indomethacin" co-amorphous supersaturating drug delivery systems: "spring-parachute" process, stability, in vivo bioavailability, and underlying molecular mechanisms. Eur J Pharm Biopharm. 2021;166:111–25.
136. Bahetibieke S, Moinuddin SM, Baiyisaiti A, Liu X, Zhang J, Liu G, et al. Co-amorphous formation of simvastatin-ezetimibe: enhanced physical stability, bioavailability and cholesterol-lowering effects in LDLr−/−mice. Pharmaceutics. 2022;14(6)
137. Chen J, Li H, Li X, Yuan D, Cheng H, Ke Y, et al. Co-amorphous systems using epigallocatechin-3-gallate as a co-former: stability, in vitro dissolution, in vivo bioavailability and underlying molecular mechanisms. Eur J Pharm Biopharm. 2022;178:82–93.
138. Elmaradny H, Saab M. Spray-dried co-amorphous tadalafil ternary mixtures: a promising platform towards the enhancement of solubility and bioavailability. Braz J Pharm Sci. 2023:58.
139. Wang Z, Chen X, Li D, Bai E, Zhang H, Duan Y, et al. Platensimycin-berberine chloride co-amorphous drug system: sustained release and prolonged half-life. Eur J Pharm Biopharm. 2022;179:126–36.
140. Hatanaka Y, Uchiyama H, Kaneko S, Ueda K, Higashi K, Moribe K, et al. Designing a novel coamorphous salt formulation of telmisartan with amlodipine to enhance permeability and oral absorption. Mol Pharm. 2023;20(8):4071–85.
141. Mohamed EM, Dharani S, Nutan MTH, Cook P, Arunagiri R, Khan MA, et al. Application of sucrose acetate isobutyrate in development of co-amorphous formulations of tacrolimus for bioavailability enhancement. Pharmaceutics. 2023;15(5)
142. Wu H, Ma J, Qian S, Jiang W, Liu Y, Li J, et al. Co-amorphization of posaconazole using citric acid as an acidifier and a co-former for solubility improvement. J Drug Deliv Sci Technol. 2023;80:104136.
143. Yarlagadda DL, Anand VSK, Nair AR, Dengale SJ, Pandiyan S, Mehta CH, et al. A computational-based approach to fabricate Ceritinib co-amorphous system using a novel co-former Rutin for bioavailability enhancement. Eur J Pharm Biopharm. 2023;190:220–30.

144. Inoue T, Maehara S, Maruyama M, Higaki K. Combination of co-amorphization with SNEDDS outperforms Ofev® in the oral absorption of nintedanib. Int J Pharm. 2024;657:124197.
145. Luo Z-K, Qin H-M, Han J-M, Zhu J, Zeng Y-Y, Fan C-P, et al. Novel drug-drug co-amorphous systems of olaparib with nonsteroidal anti-inflammatory drugs with improved solubility, physical stability, antitumor activity and pharmacokinetics. J Drug Deliv Sci Technol. 2024;101:106232.
146. Song J, Bao R, Lin M, Li W, Zhao P, Liu X, et al. Investigation of the dissolution rate and oral bioavailability of atenolol-irbesartan co-amorphous systems. Int J Pharm. 2024;665:124704.
147. Yang Y, Ke Y, Xie W, Li Z, Tao L, Shen W, et al. Amphiphilic disodium glycyrrhizin as a co-former for ketoconazole co-amorphous systems: biopharmaceutical properties and underlying molecular mechanisms. Int J Pharm. 2024;665:124673.
148. Yarlagadda DL, Das S, Anand Vullendula SK, Manandhar S, Dengale SJ, Ranganath Pai KS, et al. Computational-based polyphenol therapy for nonsmall cell lung cancer: naringin coamorphous systems for solubility and bioavailability enhancement. Mol Pharm. 2024;21(8):3951–66.
149. Adams JS, Sutar Y, Mukkirwar S, Miglani C, Date AA. Sweetening the deal: sweetener-based ionic liquid of albendazole significantly enhances its solubility and oral bioavailability. Mol Pharm. 2025;22(5):2568–80.
150. Xie Z, Chen Y, Xie J, Lei Y, Jia C, Liang Y, et al. Mechanistic insight into the enhanced anti-pulmonary hypertension efficacy of wogonin co-amorphous. Pharmaceutics. 2025;17(6)
151. Guo Y, Wang H, Zhu Q, Mao Y, Wen X, Zhang X, et al. Exploration of enalapril-lacidipine co-amorphous system with superior dissolution, in vivo absorption and physical stability via incorporated into mesoporous silica. Eur J Pharm Sci. 2025;207:107033.
152. Huang P-f, Niu W-M, Liu L-l, Wang G-q, Chen J, Dong Z-x, et al. Enhanced anti-tumor efficacy of paclitaxel by the co-amorphous technology with resveratrol. J Pharm Sci. 2025;114(7):103844.
153. Liu J, Liu Q, Zhang Z, Yang W, Li H. Enhancement of solubility and bioavailability of bezafibrate based on hydrotropy and solid dispersion technique. J Pharm Innov. 2025;20(2):78.
154. Rode K, Maji I, Dhuri A, Manoharan B, Kanp T, Athavan E, et al. Enhancing oral bioavailability of pazopanib via co-amorphous system: formulation, characterization, pharmacokinetics and toxicity evaluation. Colloids Surf B: Biointerfaces. 2025;256:115034.
155. Hatanaka Y, Uchiyama H, Kadota K, Tozuka Y. Improved solubility and permeability of both nifedipine and ketoconazole based on coamorphous formation with simultaneous dissolution behavior. J Drug Deliv Sci Technol. 2021;65:102715.
156. Ruponen M, Kettunen K, Santiago Pires M, Laitinen R. Co-amorphous formulations of furosemide with arginine and P-glycoprotein inhibitor drugs. Pharmaceutics. 2021;13(2)
157. Shete S, Reddy SC, Lakshman YD, Vullendula SKA, Mehta CH, Nayak UY, et al. Implications of phase solubility/miscibility and drug-rich phase formation on the performance of co-amorphous materials: the case of Darunavir co-amorphous materials with ritonavir and indomethacin as co-formers. Int J Pharm. 2021;608:121119.
158. Prajapati B, Maji I, Kumar R, Tomar D, Khatri DK, Madan J, et al. Strategy to counteract the pyrazinamide induced hepatotoxicity by developing naringin based co-amorphous system with supplementary benefits. J Drug Deliv Sci Technol. 2022;69:103181.
159. Uppala S, Vullendula SKA, Yarlagadda DL, Dengale SJ. Exploring the utility of co-amorphous materials to concurrently improve the solubility and permeability of fexofenadine. J Drug Deliv Sci Technol. 2022;72:103431.
160. Garbiec E, Rosiak N, Zalewski P, Tajber L, Cielecka-Piontek J. Genistein co-amorphous systems with amino acids: an investigation into enhanced solubility and biological activity. Pharmaceutics. 2023;15(12)
161. Uchiyama H, Hanamoto Y, Hatanaka Y, Kadota K, Tozuka Y. The enhanced skin permeation of flavonoids via the application of a coamorphous in a microemulsion formulation. J Pharm Sci. 2023;112(12):3067–74.
162. Kanp T, Dhuri A, Aalhate M, Manoharan B, Rode K, Munagalasetty S, et al. Investigation of a palbociclib and naringin co-amorphous system to ameliorate anticancer potential: insights

on in silico modeling, physicochemical characterization, ex vivo permeation, and in vitro efficacy. Mol Pharm. 2025;22(5):2446–65.
163. Mallya P, Shenoy KM, Yarlagadda DL, Kini G, S, A Lewis S. An integrated computational and experimental framework for investigating flavonoid-driven bioenhancement of curcumin. J Pharm Innov. 2025;20(4):135.
164. Lenz E, Jensen KT, Blaabjerg LI, Knop K, Grohganz H, Löbmann K, et al. Solid-state properties and dissolution behaviour of tablets containing co-amorphous indomethacin–arginine. Eur J Pharm Biopharm. 2015;96:44–52.
165. Ojarinta R, Saarinen J, Strachan CJ, Korhonen O, Laitinen R. Preparation and characterization of multi-component tablets containing co-amorphous salts: combining multimodal non-linear optical imaging with established analytical methods. Eur J Pharm Biopharm. 2018;132:112–26.
166. Sørensen C-M, Rantanen J, Grohganz H. Compaction behavior of co-amorphous systems. Pharmaceutics. 2023;15(3)
167. da Costa NF, Daniels R, Fernandes AI, Pinto JF. Downstream processing of amorphous and co-amorphous olanzapine powder blends. Pharmaceutics. 2022;14(8)
168. Zhang Y, Gao Y, Du X, Guan R, He Z, Liu H. Combining co-amorphous-based spray drying with inert carriers to achieve improved bioavailability and excellent downstream manufacturability. Pharmaceutics. 2020;12(11)
169. Engelsing F, Buchart L, Grohganz H. Design of experiments approach on the compaction properties of co-amorphous tablets. Pharm Dev Technol. 2023;28(9):907–14.
170. Shi X-B, Wang Z-Q, Li H-T, Dai X-L, Long X-T, Huang Y-L, et al. Comparative evaluation of cocrystalline and coamorphous forms comprised of gefitinib and dasatinib for performance optimization. RSC Pharmaceutics. 2025;2(3):598–610.
171. Tanaka R, Ishihara S, Sasaki T, Hattori Y, Otsuka M. Injection-molded coamorphous tablets: analysis of intermolecular interaction and crystallization propensity. J Pharm Sci. 2021;110(9):3289–97.
172. da Costa NF, Azevedo RF, Lopes JA, Fernandes AI, Pinto JF. In situ co-amorphization of olanzapine in the matrix and on the coat of pellets. Pharmaceutics. 2022;14(12)
173. da Costa NF, Daniels R, Fernandes AI, Pinto JF. Amorphous and co-amorphous olanzapine stability in formulations intended for wet granulation and pelletization. Int J Mol Sci. 2022;23(18)
174. Hatwar P, Pathan IB, Chishti NAH, Ambekar W. Pellets containing quercetin amino acid co-amorphous mixture for the treatment of pain: formulation, optimization, in-vitro and in-vivo study. J Drug Deliv Sci Technol. 2021;62:102350.
175. Pathan İ, Chishti NA, Dehghan MH. Formulation, optimization and evaluation of pellets loaded glipizide co amorphous mixture using central composite design to enhance solubility. J Res Pharm. 2025;28(5):1536–49.
176. Schütz D, Timmerhaus A, Grohganz H. Wet granulation of co-amorphous indomethacin systems. Int J Pharm. 2023;644:123318.
177. Yin H, Jin W, Wang J, Ke J, Zhang W, Liu C, et al. Oral fast dissolving films for co-administration of breviscapine and matrine: formulation optimization and in vitro characterization. J Drug Deliv Sci Technol. 2024;95:105548.
178. Tian S, Ke J, Zhang T, Tan Q, Wang J, Liu C, et al. A co-amorphous oleogelator of curcumin and piperine for olive oil as multifunctional food additives. Curr Res Food Sci. 2025;11:101139.
179. Wang G, Li Y, Qin Z, Liu T. Nanosizing coamorphous drugs using top-down approach: the effect of particle size reduction on dissolution improvement. AAPS PharmSciTech. 2022;24(1):14.

# Chapter 8
# High Drug-Loading Amorphous Solid Dispersions

**Afroditi Kapourani and Panagiotis Barmpalexis**

**Abstract** Amorphous Solid Dispersions (ASDs) have become a cornerstone technology for enhancing the oral bioavailability of poorly water-soluble drugs (BCS Class II and IV). However, conventional ASDs are often limited by low drug loadings (typically 10–30% w/w), resulting in high excipient burdens and bulky dosage forms that hinder patient compliance. This chapter explores the critical transition from traditional ASDs to High Drug-Loaded (HDL) ASDs, which aim to increase drug fractions beyond 50% w/w. We review the fundamental thermodynamic and kinetic challenges associated with HDL systems, specifically the increased risk of phase separation and recrystallization during storage and dissolution (the "spring-and-parachute" phenomenon). To address these limitations, various innovative formulation and manufacturing strategies are evaluated, including polymer combinations, coprecipitation, chemical and thermal crosslinking, electrospinning, and Atomic Layer Coating (ALC). Additionally, the potential of biomacromolecules, such as whey proteins, as alternative stabilization carriers is discussed. These advanced techniques enable the design of physically stable, high-performance systems that reduce tablet mass while maintaining superior dissolution and therapeutic efficacy. Ultimately, HDL ASDs represent a pivotal advancement in overcoming the practical barriers of oral delivery for challenging new chemical entities.

A. Kapourani (✉)
Laboratory of Pharmaceutical Technology, School of Pharmacy, Faculty of Health Sciences, Aristotle University of Thessaloniki, Thessaloniki, Greece
e-mail: akapourag@pharm.auth.gr

P. Barmpalexis
Laboratory of Pharmaceutical Technology, School of Pharmacy, Faculty of Health Sciences, Aristotle University of Thessaloniki, Thessaloniki, Greece

Natural Products Research Centre of Excellence-AUTH (NatPro-AUTH), Center for Interdisciplinary Research and Innovation (CIRI-AUTH), Thessaloniki, Greece
e-mail: pbarmp@pharm.auth.gr



I. Nikolakakis, I. Partheniadis (eds.), *Amorphous Solid Dispersions*, AAPS Introductions in the Pharmaceutical Sciences 5,
https://doi.org/10.1007/978-3-032-24883-1_8

**Keywords** Amorphous solid dispersions (ASDs) · High drug loading (HDL) · Solubility enhancement · Physical stability · Polymer stabilization · Supersaturation

## 8.1 Introduction

Over the past few decades, the pharmaceutical landscape has undergone a dramatic transformation, driven by advances in combinatorial synthesis and high-throughput screening technologies [1]. While these innovations have significantly accelerated the identification of new chemical entities (NCEs) with targeted biological activity, they have also introduced a major formulation challenge: poor aqueous solubility. It is now widely recognized that a substantial proportion of candidate molecules entering preclinical and clinical development suffer from low solubility in water. According to data from industry pipelines, it is estimated that nearly 70–90% of small-molecule drug candidates exhibit either low aqueous solubility, low permeability, or both, placing them in Biopharmaceutics Classification System (BCS) Class II or IV [2, 3]. This has a significant impact on drug delivery via the oral route, resulting in reduced bioavailability of these poorly soluble compounds, which is a critical determinant of therapeutic efficacy [4]. Consequently, despite desirable pharmacological activity, many active pharmaceutical ingredients (APIs) fail to reach the market, a problem compounded by the high costs and low success rates of NCEs in clinical development [5].

The pharmaceutical industry has responded to this challenge by exploring various formulation strategies aimed at improving the solubility and dissolution behavior of hydrophobic drugs. These include particle size reduction (micronization and nanonization) [6, 7], salt formation [8, 9], prodrug approaches [10, 11], lipid-based drug delivery systems [12], solid dispersions [13, 14], and deep eutectic solvents [15]. Among these, the amorphous solid dispersion (ASD) approach has emerged as one of the most promising and widely applied technologies to enhance the dissolution rate and apparent solubility of poorly water-soluble drugs [2, 16].

By definition, such systems consist of an amorphous API molecularly dispersed in an excipient (or mixture of excipients) that acts as a matrix/carrier improving its physical stability and in vivo characteristics [17]. This molecular dispersion can be achieved by using a variety of manufacturing techniques, including spray drying, hot-melt extrusion, solvent evaporation, KinetiSol®, thin film freezing, etc. [18, 19].

### 8.1.1 Advantages of ASDs as a Formulation Approach

The rationale behind ASDs is rooted in the fundamental differences between crystalline and amorphous solids. Crystalline APIs are characterized by a highly ordered three-dimensional lattice, which provides thermodynamic stability but imposes an energetic barrier to dissolution. In contrast, amorphous solids lack long-range molecular order and instead exhibit only short-range molecular interactions, giving them liquid-like local properties while retaining the mechanical behavior of solids [20, 21]. This structural disorder results in higher free energy and enhanced molecular mobility, which together promote faster dissolution kinetics and enable the generation of supersaturated drug solutions in aqueous environments [22]. Such supersaturation is a key determinant of improved oral bioavailability, particularly for BCS Class II and IV compounds. Specifically, ASDs, like other supersaturated systems, can offer clear advantages compared with approaches that enhance the solubility of crystalline drugs but simultaneously reduce their thermodynamic activity. When crystallization is suppressed, highly supersaturated aqueous solutions of lipophilic drugs may undergo liquid–liquid phase separation (LLPS) once the concentration surpasses the amorphous solubility of the compound [23, 24]. At this point, the dissolved drug separates into two distinct phases: a water-rich continuous phase containing drug at its amorphous solubility limit, and a dispersed, colloidal phase composed of drug-rich amorphous particles. These phases remain in a metastable equilibrium. Experimental observations show that the amorphous solubility defines the upper limit of sustainable supersaturation. Characterization studies have shown that the drug-rich dispersed phase formed during LLPS generally consists of amorphous nanodroplets [25]. Their nanoscopic size and amorphous structure allow them to dissolve much more rapidly than crystalline solids. LLPS is thought to be beneficial in vivo, because the dissolved phase can maintain the highest possible flux across biological membranes, while the amorphous nanodroplets act as a readily dissolving reservoir that rapidly replenishes the drug depleted by absorption [26, 27]. The phenomenon described above is illustrated in Fig. 8.1.

Another critical advantage of ASDs lies in their high degree of formulation flexibility. The selection of polymeric carriers can be tailored to achieve specific functional outcomes by employing physicochemical parameters such as glass transition temperature ($T_g$), hygroscopicity, and hydrogen-bonding capacity [28–30]. Such customization enables fine control over drug dissolution profiles, stability, and even intestinal permeability after oral administration. This adaptability can be further expanded through the incorporation of various excipients like surfactants, plasticizers, or pH adjusters, which collectively broaden the performance potential of the system [31–33]. Importantly, ASDs also enable advanced formulation designs, including the codelivery of multiple APIs within a single dosage form. For this reason, they have played a pivotal role in developing bioavailable oral formulations for drugs in challenging therapeutic areas such as antiviral, oncology, and cystic fibrosis treatments [34]. Specifically, Fig. 8.2 illustrates the percentage distribution of drug product categories in which ASDs are employed [34]. Moreover, although

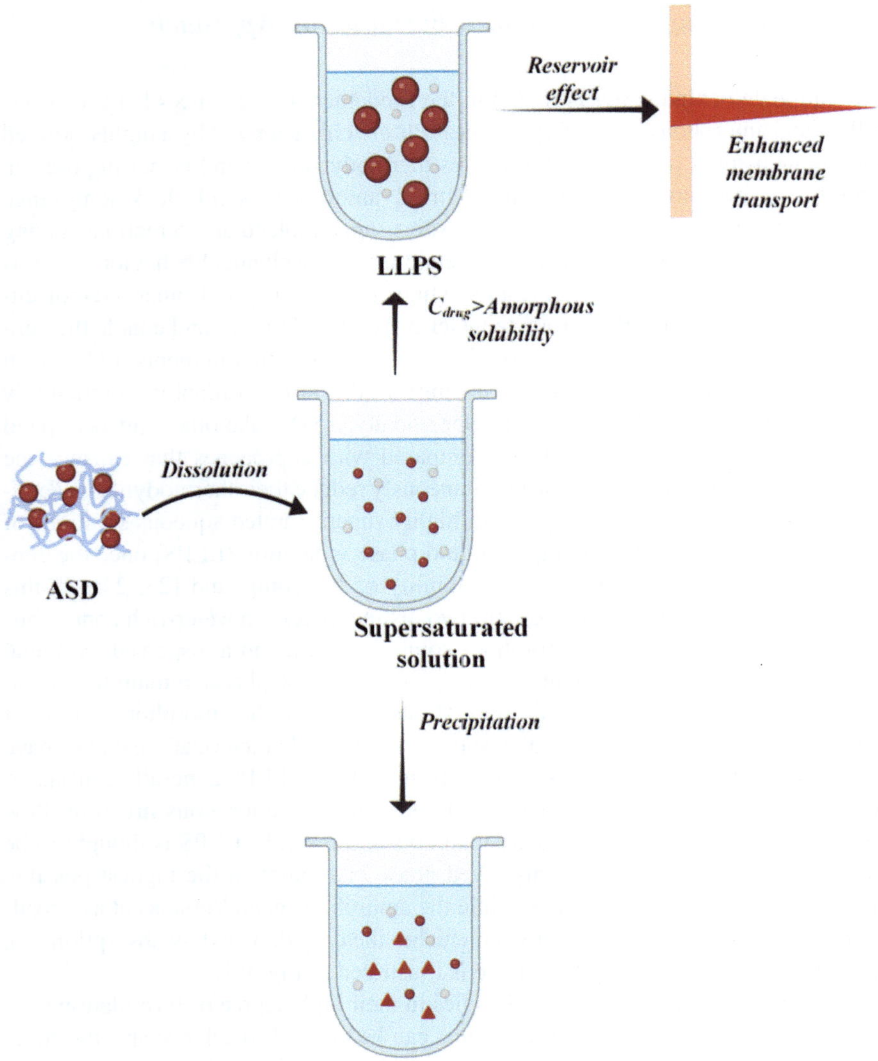


**Fig. 8.1** Possible phase transformations during ASDs' dissolution

most commercially available ASDs are formulated as tablets or capsules, several pediatric-friendly products have also been developed, including granule-based formulations such as Norvir (ritonavir), Kalydeco (ivacaftor), Orkambi (lumacaftor/ivacaftor) and Prograf (tacrolimus), which can be mixed with food or liquids to facilitate administration in children [22].

Another significant advantage of ASDs is their strong translational potential across the drug development pipeline. ASD formulations offer smooth progression from preclinical to clinical studies, often requiring only minor adjustments such as changes in polymer type or drug loading. This minimizes the risk of reformulation

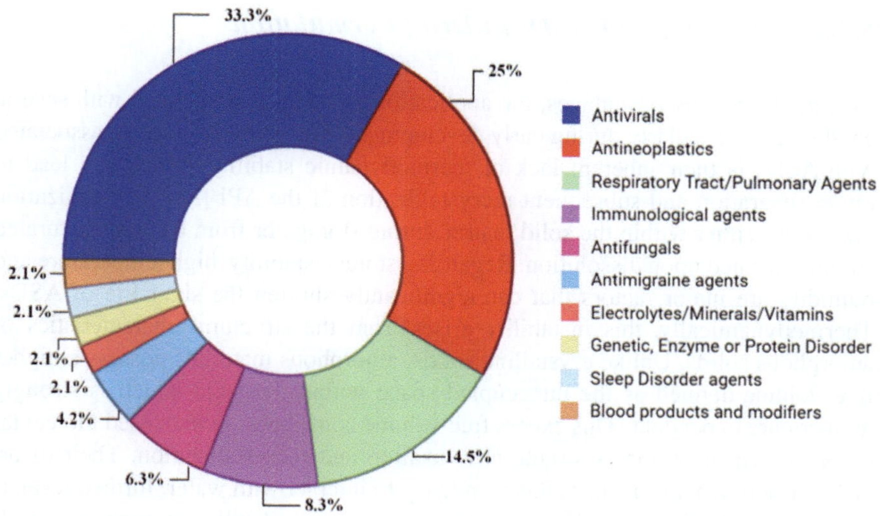


**Fig. 8.2** Drug product categories in which ASDs are employed

failure, streamlines development timelines, and avoids costly and time-consuming downstream optimization efforts.

From a manufacturing perspective, technologies for producing ASDs have matured significantly. Both spray drying and hot-melt extrusion are well established as robust, scalable, and regulatory-compliant continuous processes [35, 36]. Their adaptability allows them to accommodate both small-scale batches required during early development and large-scale production needed for commercial supply [36]. These technological advances have established ASDs as a mainstream formulation platform, applicable not only in early exploratory studies but also in late-stage development and full commercialization.

Safety and tolerability represent additional areas where ASDs offer clear advantages. During preclinical development, it is critical to verify that the excipients incorporated into formulations do not induce toxicity in rodents or higher animal species. The polymeric carriers employed in ASDs have been extensively characterized across species, and tolerability limits are well established. Many of these carriers are naturally derived, biocompatible, and nontoxic, with long-standing safety records and regulatory acceptance [37]. Nevertheless, only a select subset of polymers is commonly used for ASD development, reflecting a balance among performance, clinical precedence, regulatory approval history, and demonstrated safety in humans. Ultimately, the capacity of ASDs to enhance oral bioavailability arises from a combination of physical, chemical, and physiological mechanisms that work synergistically.

### 8.1.2 Challenges of ASDs in Drug Formulation

Despite their many advantages, the application of ASDs is associated with several challenges, particularly during early development. The primary concern associated with ASDs is their inherent lack of thermodynamic stability, which can lead to phase separation and subsequent recrystallization of the API [38]. Crystallization may occur either within the solid matrix during storage or from the supersaturated solution formed upon dissolution. Regarding storage stability, high temperature and humidity are major factors that can significantly shorten the shelf life of ASDs. Thermodynamically, this instability arises from the structural characteristics of amorphous solids. Unlike crystalline solids, amorphous materials possess a higher free volume-defined as the unoccupied space within the solid-which is strongly temperature dependent. This excess free volume contributes to increased molecular mobility, which promotes structural relaxation and recrystallization. Their disordered structure also enhances their tendency to interact with water, further accelerating crystallization and degradation processes. Specifically, amorphous solids exhibit both surface adsorption and bulk absorption of water, and they generally take up more moisture than crystalline solids [39]. The presence of water plasticizes the amorphous matrix, reducing its $T_g$, which in turn increases molecular mobility and can compromise the stability and performance of the system [40].

The instability of the amorphous state also affects the dissolution performance of ASDs. Under supersaturation conditions, the high free energy and molecular mobility of the amorphous drug can favor precipitation [41]. A typical dissolution profile of ASDs shows an initial rapid increase in drug concentration followed by a decline caused by recrystallization and precipitation of the API [42, 43]. This is known as the "spring-and-parachute" phenomenon [44]. The excess free energy in the amorphous form acts as a "spring," driving the concentration above the equilibrium solubility of the crystalline form to create a transient supersaturated solution [45]. As the system relaxes toward equilibrium, precipitation occurs. When an ASD contacts an aqueous medium, two main recrystallization pathways can occur: (1) solid-state recrystallization within the polymeric matrix or (2) solution-mediated recrystallization following rapid dissolution [46]. Often, both mechanisms act simultaneously. Solid-state recrystallization limits the extent of supersaturation, while solution--mediated crystallization reduces the time the drug remains in the supersaturated state. In either case, the loss of supersaturation negates the solubility advantage of the amorphous form. Therefore, an appropriate polymer must act as a "strong parachute," sustaining supersaturation and inhibiting drug precipitation during gastrointestinal transit. However, excipient selection for ASDs remains largely empirical. Despite advances in the field, most formulations are still developed using a trial--and-error approach with a small set of well-established polymers [47]. Experimental screening methods for identifying suitable carriers and optimizing drug-to-polymer ratios often require large quantities of API and extended timelines, which are rarely available during the early drug discovery phase [48]. As a result, early-stage data are frequently generated using suboptimal or nonoptimized formulations. Furthermore, in silico tools for ASD design are still limited. Computational models often oversimplify drug–polymer interactions, and the high molecular weight (MW) and structural complexity of polymers present additional challenges for accurate molecular-level simulations. In addition to selecting a suitable polymeric carrier, careful consideration must also be given to the choice of manufacturing technique, which is often constrained by the physicochemical properties of the API. For example, APIs with extremely low solubility in organic solvents cannot be processed using solvent-evaporation-based methods such as spray drying, while APIs with very high melting points or marked thermal sensitivity are unsuitable for melting--based approaches like hot-melt extrusion [49, 50].

While the above challenges have been widely discussed in literature and various strategies have been proposed to address them, the most critical obstacle to the broader adoption of ASDs remains the need to balance drug loading with physical stability [51, 52]. From a therapeutic and manufacturing perspective, achieving high drug loading is highly desirable, as it can reduce excipient content, enable smaller and more patient-friendly dosage forms, and improve compliance [53]. However, increasing the proportion of drug relative to polymer within the ASD matrix presents substantial challenges related to phase stability, miscibility, and the risk of crystallization.

## 8.2 Impact of Drug Loading on the Performance of ASDs and Their Tablet Formulations

Ongoing research on ASDs has shown that, to prevent drug recrystallization within the expected product shelf life (typically 1–5 years), it is almost always necessary to incorporate relatively large amounts of the selected polymeric carrier [54]. The main rationale behind this approach is that as the drug-to-polymer ratio increases, the polymer's capacity to molecularly stabilize the amorphous drug diminishes, thereby increasing the likelihood of forming drug-rich domains [55, 56]. These domains are highly susceptible to phase separation and recrystallization. In essence, when the polymer content is insufficient, large intermolecular "gaps" remain between polymer chains, providing sites that facilitate nucleation and crystal growth. This crystallization not only undermines the solubility advantage of the amorphous state but can also introduce batch variability and lower both bioavailability and overall product performance. This risk is especially critical for compounds with poor miscibility in commonly used polymers. Although there are no universal rules defining the optimal polymer content, recent studies have highlighted the pivotal role of the relationship between polymer concentration and the critical polymer overlap concentration (C*) [57–59]. ASDs have been conceptualized as solid polymer-in-drug solutions, and when a polymer interacts favorably with a solvent, there exists a critical concentration (C*) below which polymer chains behave independently, and above which they overlap to form a continuous network.

When the polymer concentration equals or exceeds C*, intermolecular gaps are minimized. In this set framework, it has been shown that when C > C*, kinetic stabilization of ASDs can be maintained at room temperature for up to 12 months. Therefore, the extent of crystallization inhibition is governed by the polymer concentration (C), polymer molecular weight (MW), and the strength of drug–polymer interactions. The latter two factors are embedded within the C* value, which can be experimentally determined using dynamic light scattering or rheological analysis of polymers dissolved in molten drug [60]. Because C* decreases with increasing polymer MW, using higher-MW polymers allows for lower polymer loadings. However, the use of high-MW polymers can be limited by processing challenges.

Beyond the physical stability concerns associated with higher drug loadings, a decline in dissolution performance is also frequently observed [61]. At high drug contents, the drug and polymer initially dissolve independently (with the polymer dissolving more rapidly), followed by the formation of a drug-rich barrier layer at the dissolution front, which subsequently controls the release rate. At low drug loadings, the amount of drug present is insufficient to form a continuous drug-rich layer, and drug release is then governed by polymer dissolution. In other words, at low loadings, both components dissolve congruently, that is, at the same rate [62]. Because polymer dissolution rates are generally several orders of magnitude higher than those of poorly soluble drugs, an effective strategy is to formulate ASDs at loadings that allow congruent dissolution of drug and polymer, enabling rapid and complete drug release [63]. A notable outcome of this behavior is that the apparent drug concentration in solution can exceed the amorphous solubility, representing the maximum achievable supersaturation. The drug loading at which release behavior shifts from congruent to incongruent has been termed the "limit of congruency" (LoC) [64, 65]. However, this congruent release pattern has primarily been observed only at relatively low drug loadings (≤ 25% w/w), especially in ASDs based on polyvinylpyrrolidone/vinyl acetate (PVPVA). For these reasons, most ASDs are formulated within the range of approximately 10–30% w/w drug content, with about 25% w/w frequently targeted, whereas loadings approaching 50% are rare and considered exceptions [66].

In this context, to avoid issues related to the physical stability of ASDs-and consequently to prevent loss of their therapeutic effectiveness-relatively high amounts of polymeric carrier are often incorporated [67]. However, this approach results in very low drug content, which may be insufficient to achieve the required therapeutic dose [68]. Notably, approximately 24% of marketed oral drug products worldwide are administered at doses of 200 mg or higher [69], whereas the typical drug content in an ASD formulation is 10–30% w/w, and the ASD portion of the final dosage form generally constitutes 40–80% w/w [70, 71]. Based on these values, the total mass of an oral dosage form required to deliver a 100 mg dose can range from 300 to 2500 mg. Such large dosage forms frequently present swallowing difficulties, which can reduce patient adherence, particularly in populations that already face swallowing challenges, such as pediatric or geriatric patients. In these cases, ASDs are at a clear disadvantage compared with dosage forms containing higher proportions of drug substance.

Additionally, when the proportion of ASD within an oral tablet exceeds 30% w/w, disintegration time tends to increase [70–72]. This effect is especially evident in ASD-based tablets formulated with polymers that are soluble across the entire gastrointestinal pH range, such as PVP and HPMC. These polymers hydrate rapidly upon contact with gastrointestinal fluids, leading to a higher tendency to form gels compared with polymers insoluble at low pH, such as Eudragit L100 and HPMCAS. Because of this gelling tendency, relatively high amounts of additional excipients-such as inorganic salts-are often incorporated to promote faster disintegration [73, 74].

Therefore, incorporating higher amounts of drug within ASDs (≥50% w/w) could offer several potential benefits, including reduced dosage form size, fewer units per dose, and improved patient adherence [75]. Nonetheless, such high drug-loaded (HDL) systems face substantial challenges in maintaining physical stability, which continues to be one of the major obstacles limiting their broader application.

## 8.3 Formulation Strategies for Developing HDL ASDs

### *8.3.1 Polymer Combination*

Owing to the significant challenge of using a single polymer to simultaneously ensure physical stability in HDL ASDs and support both the generation and maintenance of drug supersaturation during dissolution, recent years have seen a growing interest in ternary ASD systems, particularly those comprising a drug and a binary polymer blend. In these ternary systems, the overall performance is largely determined by the individual characteristics of each polymer. Typically, one polymer is selected for its ability to promote rapid drug release, while the other is chosen for its superior capacity to inhibit drug recrystallization, both within the solid matrix and during in vitro dissolution. Consequently, the strategic combination of polymers with varying hydrophilicity, drug release profiles, and crystallization inhibition capacities has been proposed and investigated as a promising approach to optimize both the performance and physical stability of HDL ASDs.

In greater detail, in an effort to develop HDL ASDs, the influence of various polymer combinations on the stability of Celecoxib ASDs with drug content up to 50% w/w was explored [76]. The polymers assessed included different grades of polyvinylpyrrolidone (PVP K12 and K29/32), as well as HPMC and HPMCAS. Both single-polymer and binary-polymer systems were prepared using solvent evaporation. The results showed that binary ASDs containing PVP, regardless of grade, exhibited faster dissolution rates compared to those with HPMC or HPMCAS. However, PVP was the least effective at maintaining drug supersaturation, with recrystallization occurring after approximately 2 h in solution. Interestingly, replacing 20% of the PVP content with HPMCAS led to a markedly different dissolution profile. Although the overall dissolution rate decreased, the supersaturated state was preserved for a longer duration. This suggested that the ternary ASD offered superior physical stability compared to the binary formulation based solely on PVP. Hence, this work introduced a forward-looking formulation strategy, demonstrating that combining two polymers with distinct functional roles-one providing crystallization inhibition and the other enabling rapid drug release-can enable the successful development of physically stable, HDL ASDs.

In a subsequent study [77], HDL ASDs of Celecoxib formulated with poly(acrylic acid) (PAA) in combination with HPMC or HPMCAS as polymeric carriers were investigated. The results revealed that both HPMCAS and HPMC were effective in inhibiting drug recrystallization, not only in the solid state but also during the dissolution process. However, the drug release from these ASDs, particularly those with moderate to high drug loads (30–50% w/w), was relatively slow and could only be improved by reducing the drug content. In contrast, PAA was found to be a relatively weak crystallization inhibitor, but it facilitated very rapid drug release, even at HDL, albeit with incomplete release due to recrystallization.

Furthermore, in an effort to develop HDL ASDs of Pretomanid, a poor glass former with a strong tendency to recrystallize from both the amorphous and supersaturated states, various combinations of polymers were investigated. These included the hydrophilic HPMC and PVP, the amphiphilic polymer Soluplus®, and more hydrophobic carriers such as Eudragit S100, HPMCAS, and HPMCP, at varying weight ratios [78]. Polymer selection based on saturation solubility and drug–polymer miscibility studies. Both binary and ternary ASDs, incorporating either a single polymer or a binary polymer blend, were prepared using solvent evaporation method. The findings highlighted the limitations of using a single polymer, as none of the individual carriers was capable of fully converting crystalline Pretomanid into the amorphous state at drug loadings exceeding 20% w/w. This limitation necessitated the addition of a second polymer to improve system performance. Specifically, the combination of HPMCAS and PVP demonstrated a synergistic effect on drug–polymer miscibility, enabling the successful development of ASDs with drug loadings as high as 30% w/w. Moreover, the ternary ASDs outperformed the binary ones in terms of apparent solubility and in vitro dissolution performance, and they also showed improved outcomes in in vivo pharmacokinetic studies compared to the crystalline API and binary ASDs, respectively. The results support the strategy of using complementary polymer pairs to overcome the formulation challenges associated with poorly glass-forming, high-recrystallization-tendency drugs in HDL ASD systems.

In the same context, the role of amphiphilic polymer combinations in the development of ternary Griseofulvin ASDs has been systematically explored [79]. By combining polymers with distinct hydrophilic-hydrophobic characteristics, such as HPC-SL, Soluplus®, copovidone, and HPMCAS, it was demonstrated that ternary matrices can surpass the performance of conventional binary systems, particularly at elevated drug loadings. Among the investigated formulations, the HPMCAS-HPC-SL ternary system proved to be the most effective, exhibiting strong polymer-drug interactions that resulted in enhanced physical stability and extended supersaturation. These findings underscore the potential of pairing hydrophobic and hydrophilic amphiphilic polymers to generate complementary mechanisms of solubilization and precipitation inhibition, thereby providing a pathway for achieving high drug loadings while preserving stability and dissolution performance.

### 8.3.2 *Coprecipitation*

Coprecipitation has also been explored as a promising approach to modulate dissolution behavior and increase the drug-loading capacity of ASDs, owing to the markedly different physical properties that can be achieved through this technique [80, 81]. The process involves dissolving both the API and the polymeric carrier in a common solvent, followed by their simultaneous precipitation in a nonsolvent (antisolvent) in which both components are insoluble. Schenck et al. [82] investigated this method as an alternative to spray drying for the production of HDL ASDs. In their study, amorphous dispersions of a proprietary compound (API-A) were prepared with HPMCAS as the polymeric carrier, using both spray drying and coprecipitation for comparison. Because API-A exhibited slow recrystallization kinetics, the coprecipitation process was performed under both high- and low-shear conditions. The most striking differences between the spray-dried and coprecipitated dispersions were related to the physical properties of the resulting solids. The spray-dried material consisted mainly of a mixture of collapsed and intact spherical particles, whereas the coprecipitated ASD particles displayed a thin, paper-like morphology with smaller particle size and significantly larger specific surface area. Interestingly, the coprecipitated dispersions exhibited slower dissolution rates than their spray-dried counterparts, a finding attributed to the reduced wettability of the coprecipitated particles. Dry granulation by roller compaction provided only a modest improvement in dissolution performance, which remained inferior to that of the spray-dried dispersions. To overcome this limitation, the authors incorporated the coprecipitated dispersions into a matrix of highly water-soluble excipients. This additional step substantially enhanced the surface properties of the particles and ultimately resulted in dissolution behavior comparable to that of the spray-dried formulations. Consequently, a major advantage of coprecipitation when prepared HDL ASDs is its flexibility as a platform process. When followed by evaporative isolation, it enables the continuous incorporation of multiple components into the amorphous matrix during formation, facilitating the production of engineered particles. In contrast, techniques such as spray drying and hot-melt extrusion would require additional downstream processing to create layered or compositionally complex particles, thereby complicating manufacturing. With coprecipitation, supplementary components can simply be added to the antisolvent or to the precipitated suspension before evaporation, either dissolved or as a slurry (as demonstrated in this work), and subsequently recovered together with the amorphous material during isolation and drying. Importantly, this study showed that the coprecipitation approach could increase the maximum achievable drug loading by up to 20% w/w, underscoring its potential as a robust and versatile manufacturing strategy for high-drug-content amorphous solid dispersions.

Recently, Hiew et al. [83] demonstrated that coprecipitated ASDs of Grazoprevir with HPMCAS can achieve drug loadings as high as 75% w/w while maintaining physical stability. Specifically, the coprecipitation approach produced particles with a considerably larger specific surface area compared with conventional solvent-evaporated ASDs, which can be advantageous for enhancing dissolution performance. However, HDL coprecipitated ASDs often exhibit poor wettability, which may impair drug release. To address this, the researchers developed a hierarchical particle design, wherein a secondary hydrophilic polymer (such as HPMC or HPC) was introduced into the antisolvent during precipitation. This process deposited a thin hydrophilic layer on the surface of the primary particles, effectively reducing surface drug enrichment and improving wetting properties. Remarkably, hierarchical coprecipitated ASDs exhibited superior dissolution and even achieved greater in vivo oral bioavailability in dogs than binary coprecipitated ASDs, showing approximately 1.8-fold higher *AUC* values. These findings confirm that coprecipitation, particularly when combined with hierarchical surface engineering, provides a viable platform to develop HDL ASDs without compromising dissolution performance or bioavailability.

The coprecipitation method has been shown to be valuable not only for enabling the development of HDL ASDs but in the same time producing systems with improved mechanical properties. In particular, the influence of the preparation method on the mechanical characteristics of ASDs was thoroughly investigated using the development compound GDC-0810. GDC-0810 is an orally administered anticancer agent designed for the treatment of estrogen receptor–positive breast cancer and is characterized as a weak acid with very low intrinsic solubility. In this study, ASDs containing 50% w/w GDC-0810 with HPMCAS were prepared using three different techniques: (1) spray drying, (2) coprecipitation with overhead mixing, and (3) coprecipitation with resonant acoustic mixing. All three approaches produced ASDs exhibiting good phase homogeneity and comparable glass transition temperatures. However, the coprecipitated powders (prepared via both overhead mixing and resonant acoustic mixing) displayed markedly superior tabletability and flow properties compared with the spray-dried material, highlighting the potential of coprecipitation to enhance both drug loading and downstream manufacturability of ASD formulations [84].

### *8.3.3 Chemical Crosslinking*

It is well recognized that the establishment of strong molecular interactions between the drug and the polymer is expected to enhance the physical stability of ASDs [85]. Among these interactions, ionic bonding is considered particularly advantageous, as it has the potential to not only increase the achievable drug loading but also to suppress molecular mobility and delay phase separation. When sufficiently strong, such interactions can lower the thermodynamic driving force for recrystallization, thereby contributing to stabilization of the amorphous state.

Polyethylenimine (PEI), a highly basic polymer commonly used as a nonviral vector in gene delivery applications, has recently gained attention as a potential carrier for ASDs [86]. Its remarkably high density of basic amine groups (approximately 24 mmol N/g PEI) offers a strong capacity for electrostatic interactions with drugs, which can both enhance drug incorporation and improve physical stability. For comparison, this nitrogen content is roughly 60 times higher than that of Eudragit E100 (~0.4 mmol N/g). This high density of basic sites facilitates the incorporation of larger amounts of acidic drugs and improves their physical stability compared with conventional basic polymers [86]. Notably, chemically crosslinked PEI (cPEI) has been shown to be more effective for gene delivery than high-molecular-weight branched PEI (bPEI), while exhibiting lower cytotoxicity [87]. This finding provided an initial indication that such crosslinked polymeric systems could also be exploited as oral drug delivery carriers, benefiting from the fact that crosslinking increases the $T_g$ of the polymer, thereby enhancing the physical stability of the amorphous state.

Based on this rationale, Kelsall et al. [88] investigated the use of cPEI as a novel carrier for ASD development, specifically for the oral delivery of Flufenamic acid, a poorly soluble nonsteroidal anti-inflammatory drug (BCS Class II). In their study, chemical crosslinking was achieved using terephthaloyl chloride as the crosslinking agent. The influence of several formulation parameters was systematically evaluated, including solvent type, crosslinking density, reactant concentration, solution viscosity, reaction temperature, and the molecular weight of the parent bPEI. The resulting ASDs contained amorphous, deprotonated Flufenamic acid at high drug contents (50–60% w/w) and exhibited markedly improved drug release under acidic conditions compared with crystalline Flufenamic acid. Furthermore, stability studies revealed no evidence of recrystallization even after 29 weeks of storage under various relative humidity conditions.

In addition to cPEI-based systems, crosslinked poly(2-hydroxyethyl methacrylate) (PHEMA) hydrogels have also been examined as carriers for high drug-loaded ASDs. PHEMA is a water-insoluble, biocompatible polymer capable of absorbing substantial amounts of drug solution within its hydrophilic network, allowing the generation of amorphous dispersions at elevated drug contents [89]. In the study conducted by Sun et al. [90], Indomethacin-loaded PHEMA beads were successfully prepared at loadings up to 48% w/w, with approximately 34% w/w identified as the threshold beyond which crystallization initiated. These formulations exhibited a single $T_g$ across the drug-loading range, confirming the formation of a single amorphous phase without drug–polymer phase separation. Dissolution testing under nonsink conditions revealed that PHEMA-based ASDs displayed a slower, diffusion-controlled release mechanism compared to conventional water-soluble polymer carriers such as PVP and HPMCAS. This controlled release profile minimized the abrupt surge in supersaturation typically observed with soluble carriers, thereby reducing the risk of nucleation and recrystallization events. As a result, PHEMA-based systems were able to maintain elevated drug concentrations for longer periods, particularly at high drug loadings, ultimately outperforming their spray-dried counterparts in sustaining supersaturation over 24 h. The kinetic solubility advantage was most evident at drug loadings above 30%, where the performance of conventional carriers typically declined due to partial crystallization.

### 8.3.4 Thermal Crosslinking

Thermal crosslinking has recently emerged as a promising strategy for the development of HDL ASDs, effectively addressing the long-standing challenge of balancing drug loading with physical stability. This approach relies on the formation of covalent or physical bonds between polymer chains upon thermal treatment, generating a three-dimensional crosslinked network that restricts molecular mobility and thereby mitigates the risk of recrystallization. The resulting structural confinement substantially enhances the ability of the polymeric carrier to stabilize large amounts of amorphous drug while maintaining the functional performance of the formulation.

Sahoo et al. [91] were the first to investigate the development of crosslinked ASDs containing up to 95% w/w Ketoconazole using poly(acrylic acid) (PAA) as the primary polymer and poly(vinyl alcohol) (PVA) as the crosslinking agent. Ketoconazole is known for its high crystallization tendency, which makes its amorphous stabilization particularly challenging. In this study, thermal treatment promoted esterification between the carboxyl groups of PAA and hydroxyl groups of PVA, thereby generating a crosslinked polymeric network. Increasing the PVA concentration (0.25–1.0% w/w) and applying higher thermal crosslinking temperatures significantly increased both system viscosity and α-relaxation times, which are indicative of reduced molecular mobility. These structural changes translated to markedly enhanced physical stability, with the crosslinked ASDs resisting recrystallization even under stress conditions of elevated temperature and humidity. In addition to enhanced stability, dissolution studies revealed a notable increase in the degree of supersaturation in the dissolution medium, which was sustained for a longer duration compared to non–crosslinked formulations. These findings suggest that thermal crosslinking not only mitigates physical instability but also contributes to improved release profiles and extended supersaturation, even at drug loadings as high as 95% w/w.

In this set framework, Kapourani et al. [56, 92, 93] conducted a systematic series of investigations further investigating the concept by demonstrating the feasibility of in situ thermal crosslinking to produce HDL ASDs of Indomethacin at loadings up to 90% w/w using PAA/PVA blends. This work further elucidated critical factors affecting the performance of crosslinked ASDs, such as the degree of PVA hydrolysis, crosslinking temperature and duration, and reactant concentrations. Structural analysis confirmed extensive crosslink formation, which increased the glass transition temperature ($T_g$), slowed molecular α-relaxation dynamics, and stabilized the amorphous drug even under ICH long-term storage conditions (25 °C/60% RH) for over 3 months. Complementary rheological studies and molecular dynamics simulations further revealed that the elastic and viscous properties of the resulting crosslinked matrices contributed directly to drug stabilization by restricting molecular mobility. Dissolution studies showed that the crosslinked systems provided a diffusion-controlled release and sustained supersaturation for more than 5 h, whereas non–crosslinked Indomethacin rapidly recrystallized following an initial burst release.

Consequently, thermal crosslinking offers a powerful formulation platform to overcome the traditional trade-off between high drug loading and physical stability in ASDs. By creating a crosslinked polymer network that immobilizes drug molecules and retards both solid-state and solution-mediated crystallization, this approach enables the design of stable and effective HDL ASD systems for poorly water-soluble compounds.

### *8.3.5 Spray Drying–Granulation Hybrid Approach*

An alternative strategy proposed to increase drug loading in ASDs while simultaneously reducing the overall tablet mass of ASD-based dosage forms was presented by Mudie et al. [69]. Specifically, in this work, the authors developed an innovative high-loaded dosage form platform that strategically decouples the two principal functions of ASD polymers, which are physical stabilization and concentration sustainment. In this approach, the drug is first spray-dried with a high-$T_g$ polymer to produce a physically stable HDL ASD. This spray-dried intermediate is then granulated with a second polymer, termed the concentration-sustaining polymer (CSP), before compression into tablets. Within this system, the dispersion polymer embedded "inside" the ASD provides solid-state stabilization by limiting molecular mobility and suppressing recrystallization, whereas the external CSP, distributed "outside" the ASD, prevents precipitation and sustains supersaturation during dissolution.

In their proof-of-concept study, Erlotinib was employed as a model compound, and Eudragit® L100 was selected as the dispersion polymer owing to its high $T_g$ (≈187 °C)-approximately 30–70 °C higher than that of widely used ASD polymers such as HPMCAS, PVP, or HPMC-which facilitates the incorporation of higher drug contents while preserving physical stability. HPMCAS was chosen as the CSP because of its amphiphilic nature upon ionization at intestinal pH. For comparison, a conventional lower drug-loaded ASD was also prepared using HPMCAS alone ($T_g$ = 119 °C). The HLDF approach resulted in a reduction of tablet mass by approximately 40% compared with conventional single-polymer ASDs, while maintaining similar in vitro dissolution profiles and physical stability.

Building on these findings, the same research group [94] further demonstrated the in vivo efficacy of their strategy using Posaconazole, a BCS Class II drug with a low $T_g$ (59 °C) and a strong tendency to crystallize. In this study, Posaconazole was spray-dried with Eudragit® L100 to produce an HDL ASD, which was then granulated with HPMCAS serving as the CSP. Pharmacokinetic evaluation in beagle dogs showed that the new ASD tablets, despite being up to 40% smaller in mass than conventional ASD benchmark tablets, achieved comparable plasma exposure and approximately doubled the area under the plasma concentration–time curve (*AUC*) relative to the marketed crystalline suspension (Noxafil® OS). These results confirmed that external placement of the CSP, rather than its incorporation within the ASD matrix, can effectively sustain supersaturation while allowing for significantly higher drug loadings.

## 8.3.6 Electrospun HDL ASDs

Electrospinning is an emerging and highly versatile technique for the fabrication of nanofibrous polymer-based systems, in which the application of a high-voltage electric field to a polymer solution produces continuous fibers with diameters typically below 500 nm [95]. During electrospinning, the solvent evaporates extremely rapidly as the charged jet travels toward the collector, leading to the rapid solidification of the polymeric matrix. This ultrafast drying process kinetically entraps drug molecules in a highly dispersed manner within the polymer network, markedly reducing their molecular mobility. The resulting reduction in molecular mobility is a critical factor in suppressing the thermodynamic tendency for recrystallization, thereby enhancing the physical stability of ASDs. In addition to this kinetic stabilization, the electrospun nanofibers possess a high specific surface area and interconnected porosity, both of which facilitate rapid wetting and dissolution [96, 97]. These morphological characteristics accelerate drug release and improve the apparent solubility and dissolution rate of poorly water-soluble compounds. Furthermore, electrospinning offers notable practical advantages as it operates at relatively low temperatures, minimizing thermal stress on heat-sensitive drugs, and it is readily scalable to industrial levels. Furthermore, it provides significant formulation flexibility, as the release profile can be tailored by selecting polymers with different physicochemical properties. These attributes set electrospinning as a highly promising platform for the development of ASDs, and particularly for producing HDL ASDs.

Becelaere et al. [98] developed stable nanofibrous ASDs of the poorly water-soluble anthelmintic Flubendazole using solvent electrospinning with poly(2-ethyl-2-oxazoline) (PEtOx) as the polymeric carrier. They achieved drug loadings up to 55% w/w, while maintaining a fully amorphous state and long-term stability for at least one year. Crucially, lowering the PEtOx concentration below its normal electrospinnability threshold was necessary, as strong Flubendazole-PEtOx hydrogen bonding increased solution viscosity and enabled nanofiber formation at higher drug fractions. ATR-FTIR analysis and molecular dynamics simulations confirmed that these drug–polymer hydrogen bonds predominated over drug-drug interactions, thereby suppressing nucleation and crystallization. As a result, the electrospun ASDs showed a fourfold enhancement in dissolution rate compared with crystalline Flubendazole, achieving complete drug release within 1 h, whereas the crystalline API reached only about 30% release after 5 h. The combination of kinetic entrapment of amorphous drug molecules, high miscibility with the polymer, and the presence of stable drug–polymer hydrogen bonding interactions contributed to the exceptional stability of these systems, as the nanofibrous ASDs remained fully amorphous even after 2.5 years of storage.

Following this work, the same group later demonstrated the broader applicability of electrospinning for other BCS class II drugs, including Itraconazole, Mebendazole, Celecoxib, and Fenofibrate [99]. Using electrospinning approach, researchers successfully fabricated nanofibrous ASDs with exceptionally high drug loadings, reaching 80–85% w/w, by combining Itraconazole, Mebendazole, and Ciclopirox with PEtOx, as well as Flufenamic acid with PVP. Achieving drug contents above 40% w/w required reducing the polymer concentration in the electrospinning solutions below their conventional spinnability threshold. This approach is thought to be feasible because strong drug–polymer interactions make up for the reduced polymer content, enabling sufficient molecular entanglement for fiber formation. Importantly, for Itraconazole and Flufenamic acid, these stabilizing interactions were shown not to be limited to hydrogen bonding; van der Waals forces also contributed effectively to polymer-drug association and chain entanglement. Dissolution studies revealed up to a 50-fold increase in apparent solubility compared with the corresponding crystalline drugs. Furthermore, the experimentally measured $T_g$ values were positively shifted relative to theoretical predictions, providing additional evidence of strong intermolecular interactions that kinetically stabilized the amorphous state.

Consequently, these findings demonstrate that electrospinning can reproducibly generate physically stable, HDL ASDs for a broad range of APIs by leveraging strong drug–polymer miscibility combined with the ultrafast solidification inherent to the electrospinning process.

### *8.3.7 Atomic Layer Coating (ALC)*

Engineering the surface properties of drug or ASD particles has emerged as a novel approach to improve the physical stability of HDL systems. It is well established that crystal nucleation and growth occur more rapidly at the surface of amorphous solids than within their bulk, primarily due to the substantially higher molecular mobility of surface molecules, which possess fewer neighboring molecules and thus greater freedom of movement [100, 101]. Consequently, coating the free surface of ASD particles with appropriate crystallization inhibitors can effectively suppress surface-induced recrystallization and enhance physical stability. However, conventional particle-coating techniques, such as electrospraying, are typically solution based and offer limited control over coating thickness, and uniformity, particularly at the nanometer or molecular scale. To overcome these limitations, ALC has been proposed as a vapor-phase deposition technique capable of depositing ultrathin, conformal films on complex three-dimensional surfaces with angstrom-level precision.

Briefly, ALC involves the alternate exposure of a substrate, bearing reactive functional groups such as hydroxyl or amine groups, to vapor-phase metal precursors and coreactants. Each exposure results in a self-limiting surface reaction that forms a single molecular layer, after which excess precursor and reaction byproducts are purged using an inert gas (e.g., nitrogen or argon). This cycle is repeated multiple times to achieve the desired coating thickness. Owing to its precise thickness control and excellent surface conformity, ALC is widely employed in the semiconductor, catalysis, and biomedical device fields for depositing functional thin films on inorganic substrates.

Recognizing the potential of this approach, Van Duong et al. [102] applied ALC to stabilize spray-dried Ezetimibe-HPMCAS ASDs containing 50% and 70% w/w drug. Aluminum oxide ($Al_2O_3$) was deposited on the surface of ASD particles, and the resulting coating thickness was characterized using electron microscopy combined with focused ion beam (FIB) milling. The physical state of the systems was evaluated after storage at 40 °C/75% RH for up to 2 years. The $Al_2O_3$ nanocoating effectively prevented recrystallization over the entire storage period and did not negatively affect the drug release profile. The authors attributed the marked improvement in physical stability to two main factors: (i) the effective isolation of the amorphous drug from ambient moisture and (ii) the suppression of molecular mobility at the particle surface. In addition, the ALC treatment reduced hygroscopicity, particle agglomeration, and improved powder-handling properties such as flowability and compressibility.

Following these findings, Moseson et al. [103] systematically evaluated ALC as a platform technology across various drug–polymer ASD systems prepared by both spray drying and hot-melt extrusion, including Erlotinib/HPMCAS (60%w/w drug loading), Erlotinib/PVPVA (50% w/w drug loading), Naproxen/HPMCAS (10% w/w drug loading), and Lumefantrine/PVPVA (40% w/w drug loading). They applied $Al_2O_3$ or zinc oxide (ZnO) layers of 5–25 nm thickness using a rotary powder ALC reactor. Transmission electron microscopy confirmed uniform nanoscale coatings, and powder rheology measurements showed improved flowability and reduced cohesiveness. Under accelerated stability conditions (40 °C and 31–75% RH), uncoated ASDs exhibited surface crystallization within 1–4 weeks, whereas ALC-coated counterparts showed delayed or completely inhibited crystallization for up to 48 weeks, even when stored close to or slightly above their glass transition temperatures. Importantly, this crystallization suppression was independent of drug/polymer chemistry (presence or absence of hydroxyl groups), particle size, or manufacturing method. The authors concluded that the coating primarily acts by reducing surface molecular mobility, possibly also limiting surface water adsorption, thereby preventing surface-initiated nucleation.

### 8.3.8 Protein-Based HDL ASDs

A review of the relevant literature on ASDs clearly indicates that most carriers investigated to date are hydrophilic, semisynthetic polymers or their combinations. By contrast, biomacromolecules have received little attention as potential ASD carriers, despite their theoretical suitability arising from their exceptional ability to establish reversible interactions with drug molecules. Their capacity to form hydrogen bonds, as well as electrostatic or hydrophobic interactions, could facilitate the maintenance of a supersaturated drug state in the gastrointestinal tract, thereby potentially enhancing oral bioavailability.

In this context, Pas et al. [104] examined Gelatin as a matrix/carrier for ASDs of 12 poorly water-soluble APIs (BCS Class II) at drug loadings in the range 5–40% w/w. The ASDs were produced via freeze-drying, and the results revealed complete or partial drug amorphization depending on the drug-to-gelatin ratio. Notably, complete amorphization and markedly enhanced dissolution rates compared with the crystalline forms were achieved at lower drug contents (<20–30% w/w). However, and most importantly, gelatin at this stage can be regarded as a promising alternative to the commonly used (semi-)synthetic polymers in ASD formulations, warranting further investigation.

In this context, Mishra et al. [105] explored the potential of dairy-derived whey proteins as ASD carriers. Whey proteins are mainly composed of β-lactoglobulin (BLG), α-lactalbumin (ALA), casein glycomacropeptides (cGMP), bovine serum albumin (BSA), immunoglobulin G (IgG), and lactoferrin, and are widely available as whey protein isolate (WPI) or whey protein hydrolysate (WPH). In their study, Indomethacin, Carvedilol, and Furosemide, all poorly water-soluble drugs, were formulated into ASDs at 50% w/w drug loading using ball milling technique. Solid-state analysis confirmed that both WPI and WPH successfully stabilized the drugs in an amorphous state, as evidenced by single glass transition events and the absence of diffraction peaks. Compared with the respective crystalline drugs, the WPI- and WPH-based ASDs exhibited substantially increased $T_g$ values, 3–10-fold higher intrinsic dissolution rates, and higher apparent solubilities. Importantly, WPI-based ASDs generally outperformed the WPH systems, which was attributed to the more intact and hydrophilic nature of WPI. The ASDs also showed remarkable physical stability, with Indomethacin-based ASDs remaining fully amorphous for over 27 months, and Carvedilol- and Furosemide-based systems being stable for approximately 8 and 17 months, respectively. But, on the contrary, pure amorphous drugs recrystallized within 1 week. These findings demonstrate that whey proteins can serve as highly effective amorphous stabilizers, enabling the production of stable, HDL ASDs.

In a more recent study, Leng et al. [106] further investigated the role of individual whey protein components, BLG, ALA, and cGMP, in ASD formation and performance. Using ball milling, they prepared ASDs of two poorly water-soluble model compounds (referred to as Compound A and Compound B) with BLG, ALA, cGMP, and WPI at 50%, 60%, and 70% w/w drug loading. All formulations were initially amorphous and showed accelerated dissolution relative to the crystalline drugs. However, differences emerged in terms of stabilization capacity. For Compound A, BLG- and ALA-based ASDs at 50% and 60% loading remained fully amorphous after 5 weeks of storage at 40 °C/75% RH, whereas WPI- and cGMP-based systems showed signs of recrystallization. At 70% w/w loading, all formulations showed partial recrystallization, highlighting the challenges at very high drug fractions. For Compound B, all protein-based ASDs remained amorphous and displayed rapid dissolution (80–95% drug release within 20 min in acidic medium).

**Table 8.1** Formulation strategies and key findings in the development of HDL ASDs

| Reference | Formulation strategy | ASD systems | Drug loading | Findings |
|---|---|---|---|---|
| Xie et al. [76] | Polymer combination | Celecoxib with PVP, HPMC, HPMCAS | Up to 50% w/w | Binary systems with PVP showed improved dissolution performance but had poor supersaturation maintenance; ternary systems with HPMCAS showed improved stability and prolonged supersaturation. Combining polymers with complementary roles enhanced stability at higher loadings |
| Xie et al. [77] | Polymer combination | Celecoxib with PAA and HPMC/HPMCAS | 30–50% w/w | HPMC/HPMCAS effectively inhibited recrystallization but slowed release at higher loads; PAA promoted rapid release but incomplete recovery due to precipitation--supporting dual-polymer designs |
| Juneja et al. [78] | Polymer combination | Pretomanid with HPMC, PVP, Soluplus®, Eudragit S100, HPMCAS, HPMCP | Up to 30% w/w | Individual polymers were insufficient to achieve complete amorphization of the drug, whereas binary and ternary blends provided enhanced stabilization |
| Kapourani et al. [79] | Polymer combination | Griseofulvin with HPC-SL, Soluplus®, PVPVA, HPMCAS | Up to 50% w/w | Ternary amphiphilic matrices outperformed binary systems. HPMCAS–HPC/SL showed strongest drug–polymer interactions, superior stability, and prolonged supersaturation |
| Schenck et al. [82] | Coprecipitation vs. spray dry | API-A with HPMCAS | Increased by up to 20% | Coprecipitation gave thin, high surface area particles but slower dissolution; layering with excipients improved release |

(continued)

**Table 8.1** (continued)

| Reference | Formulation strategy | ASD systems | Drug loading | Findings |
|---|---|---|---|---|
| Hiew et al. [83] | Coprecipitation | Grazoprevir with HPMCAS | Up to 75% w/w | Secondary hydrophilic polymer layer improved wettability, dissolution, and resulted in 1.8-fold higher oral bioavailability in dogs |
| Hou et al. [84] | Coprecipitation | GDC-0810 with HPMCAS | 50% w/w | All ASDs were successfully developed with similar $T_g$; coprecipitated powders displayed significantly superior tabletability and flowability vs. spray-dried material |
| Kelsall et al. [88] | Chemical crosslinking | Flufenamic acid with cPEI | 50–60% w/w | Improved drug release compared with crystalline Flufenamic acid. No evidence of recrystallization even after 29 weeks of storage |
| Sun et al. [90] | Chemical crosslinking | Indomethacin with crosslinked PHEMA | Up to 48% | Elevated supersaturation extent for longer period |
| Sahoo et al. [91] | Thermal crosslinking | Ketoconazole with PAA/PVA | Up to 95% w/w | Crosslinked networks reduced mobility, enhanced stability, and sustained supersaturation |
| Kapourani et al. [56, 92, 93] | Thermal crosslinking | Indomethacin with PAA/PVA | Up to 90% w/w | In situ crosslinking increased $T_g$, slowed α-relaxation, and stabilized amorphous drug; ICH long-term stability (25 °C/60% RH) >3 months; diffusion-controlled release sustained supersaturation >5 h |
| Mudie et al. [69] | Spray drying-granulation hybrid approach | Erlotinib with Eudragit L100 and HPMCAS | 29% w/w | Reduced tablet mass by approximately 40% while preserving physical stability and dissolution performance |
| Mudie et al. [94] | Spray drying-granulation hybrid | Posaconazole with Eudragit L100 and HPMCAS | Up to 40% w/w | Tablets were ~40% smaller but achieved comparable exposure and two-fold increased *AUC* in dogs |
| Becelaere et al. [98] | Electrospinning | Flubendazole with PEtOx | Up to 55% w/w | Long-term stability for at least 1 year. Four-fold enhancement in dissolution rate compared with crystalline Flubendazole |
| Becelaere et al. [99] | Electrospinning | Itraconazole, Mebendazole, Celecoxib, and Fenofibrate with PEtOx | Up to 80–85% w/w | Strong drug–polymer interactions. Improved dissolution performance compared with the corresponding crystalline drugs |

(continued)

**Table 8.1** (continued)

| Reference | Formulation strategy | ASD systems | Drug loading | Findings |
|---|---|---|---|---|
| Van Duong et al. [102] | Atomic layer coating | Ezetimibe--HPMCAS | Up to 70% w/w | $Al_2O_3$ nanocoating stabilized ASD for 2 years, no recrystallization, maintained release |
| Moseson et al. [103] | Atomic layer coating | Erlotinib, Naproxen, Lumefantrine ASDs | 10–60% w/w | ALC delayed/inhibited crystallization up to 48 weeks and improved flowability and stability |
| Pas et al. [104] | Protein carrier | Gelatin-based ASDs with 12 APIs | 5–40% w/w | Complete amorphization and improved dissolution, but stability not assesse |
| Mishra et al. [105] | Protein carrier | WPI/WPH with Indomethacin, Carvedilol, Furosemide | Up to 50% w/w | Higher $T_g$, faster dissolution, and improved solubility vs. crystalline drugs; long-term stability |
| Leng et al. [106] | Protein carrier | BLG, ALA, cGMP, WPI with compounds A and B | 50–70% w/w | All systems enhanced dissolution; BLG/ALA best stabilized amorphous Compound A at 50–60%; Compound B remained amorphous up to 70% with rapid release (80–95% in 20 min) |

Dissolution tests revealed that BLG provided the most consistent improvement in both the extent and maintenance of supersaturation, particularly for Compound A. Overall, this study confirmed that whey proteins, especially BLG and ALA, can produce physically stable ASDs even at drug loadings up to 70% w/w, significantly enhancing dissolution performance compared to their crystalline counterparts.

Therefore, whey proteins represent a promising new class of excipients for the development of HDL ASDs. Their ability to form strong molecular interactions with drug molecules, combined with their capacity to enhance dissolution, maintain supersaturation, and confer significant physical stability, renders them as valuable alternatives to conventional polymeric carriers in ASD technology.

To provide a comprehensive overview of the current advances in the field, Table 8.1 summarizes the aforementioned published studies investigating various formulation strategies for the development of HDL ASDs.

## 8.4 Conclusion

ASDs have emerged as a particularly powerful approach for enhancing the apparent solubility and oral absorption of poorly water-soluble drugs. Nevertheless, despite the substantial progress made in this field, conventional ASDs are frequently constrained by low drug-loading capacities, typically within the range of 10–30% w/w, which leads to several downstream challenges, including large dosage forms, high excipient burdens, and compromised patient compliance.

Consequently, the development of HDL ASDs has emerged as a critical objective in contemporary pharmaceutical formulation science. Increasing the drug fraction within ASDs offers several interrelated advantages. Most notably, it allows for the design of smaller, more patient-friendly dosage forms by reducing the overall mass and volume required to deliver a therapeutic dose. This is especially important for drugs administered at high daily doses or for patient populations, such as pediatric and geriatric patients, who frequently encounter difficulties swallowing large tablets or capsules. Additionally, reducing excipient content not only improves patient compliance but can also mitigate excipient-related adverse effects and lower manufacturing costs.

However, the pursuit of HDL ASDs is inherently challenging due to the delicate balance between drug loading and physical stability. Increasing the proportion of drug relative to polymer compromises the polymer's ability to molecularly stabilize the amorphous drug, heightening the risk of phase separation, crystallization, and loss of supersaturation. This necessitates innovative formulation strategies that extend beyond traditional polymer selection and process optimization. This chapter has highlighted several promising approaches, including coprecipitation, electrospinning, thermal and chemical crosslinking, spray drying-granulation hybrid techniques, ALC, and the use of alternative carriers such as proteins or novel polymeric systems. These techniques aim to enhance drug–polymer interactions, tailor the microstructure to restrict molecular mobility, or introduce external stabilizing barriers, enabling the production of ASDs with substantially higher drug loadings without compromising stability or performance.

In summary, the transition from conventional to HDL ASDs marks a pivotal step toward overcoming long-standing limitations in oral drug delivery. Continued innovation in formulation design, materials engineering, and processing technologies is expected to unlock the full therapeutic potential of poorly soluble compounds while reducing the size, cost, and complexity of final dosage forms. HDL ASDs, by combining enhanced bioavailability with practical manufacturability, are poised to play an increasingly central role in the future development of oral pharmaceutical products.

## References


1. Bergström CAS, Charman WN, Porter CJH. Computational prediction of formulation strategies for beyond-rule-of-5 compounds. Adv Drug Deliv Rev. 2016;101:6–21.
2. Anane-Adjei AB, Jacobs E, Nash SC, Askin S, Soundararajan R, Kyobula M, et al. Amorphous solid dispersions: utilization and challenges in preclinical drug development within AstraZeneca. Int J Pharm. 2022;614:121387.
3. Coty J-B, Martin C, Telò I, Spitzer D. Use of Spray Flash Evaporation (SFE) technology to improve dissolution of poorly soluble drugs: case study on furosemide nanocrystals. Int J Pharm. 2020;589:119827.
4. Boyd BJ, Bergström CAS, Vinarov Z, Kuentz M, Brouwers J, Augustijns P, et al. Successful oral delivery of poorly water-soluble drugs both depends on the intraluminal behavior of drugs and of appropriate advanced drug delivery systems. Eur J Pharm Sci. 2019;137:104967.
5. Zhong H, Chan G, Hu Y, Hu H, Ouyang D. A comprehensive map of FDA-approved pharmaceutical products. Pharmaceutics. 2018;10(4):263.
6. Khadka P, Ro J, Kim H, Kim I, Kim JT, Kim H, et al. Pharmaceutical particle technologies: an approach to improve drug solubility, dissolution and bioavailability. Asian J Pharm Sci. 2014;9(6):304–16.
7. Kumar R, Thakur AK, Chaudhari P, Banerjee N. Particle size reduction techniques of pharmaceutical compounds for the enhancement of their dissolution rate and bioavailability. J Pharm Innov. 2022;17(2):333–52.
8. Serajuddin ATM. Salt formation to improve drug solubility. Adv Drug Deliv Rev. 2007;59(7):603–16.
9. Bharate SS. Recent developments in pharmaceutical salts: FDA approvals from 2015 to 2019. Drug Discov Today. 2021;26(2):384–98.
10. de Souza MM, Gini ALR, Moura JA, Scarim CB, Chin CM, dos Santos JL. Prodrug approach as a strategy to enhance drug permeability. Pharmaceuticals. 2025;18(3):297.
11. Cheng Y, Zhong C, Yan S, Chen C, Gao X. Structure modification: a successful tool for prodrug design. Future Med Chem. 2023;15(4):379–93.
12. Kesharwani R, Jaiswal P, Patel DK, Yadav PK. Lipid-based drug delivery system (LBDDS): an emerging paradigm to enhance Oral bioavailability of poorly soluble drugs. Biomed Mater Devices. 2023;1(2):648–63.
13. Chaudhary S, Nair AB, Shah J, Gorain B, Jacob S, Shah H, et al. Enhanced solubility and bioavailability of dolutegravir by solid dispersion method: in vitro and in vivo evaluation—a potential approach for HIV therapy. AAPS PharmSciTech. 2021;22(3):127.
14. Patel K, Shah S, Patel J. Solid dispersion technology as a formulation strategy for the fabrication of modified release dosage forms: A comprehensive review. DARU J Pharm Sci. 2022;30(1):165–89.
15. Pedro SN, Freire CSR, Silvestre AJD, Freire MG. Deep eutectic solvents and pharmaceuticals. Encyclopedia. 2021;1(3):942–63.
16. Van den Mooter G. The use of amorphous solid dispersions: A formulation strategy to overcome poor solubility and dissolution rate. Drug Discov Today Technol. 2012;9(2):e79–85.
17. Baghel S, Cathcart H, O'Reilly NJ. Polymeric amorphous solid dispersions: A review of amorphization, crystallization, stabilization, solid-state characterization, and aqueous solubilization of biopharmaceutical classification system class II drugs. J Pharm Sci. 2016;105(9):2527–44.
18. Hughey JR, DiNunzio JC, Bennett RC, Brough C, Miller DA, Ma H, et al. Dissolution enhancement of a drug exhibiting thermal and acidic decomposition characteristics by fusion processing: A comparative study of hot melt extrusion and KinetiSol® dispersing. AAPS PharmSciTech. 2010;11(2):760–74.
19. Tan DK, Davis DA Jr, Miller DA, Williams RO 3rd, Nokhodchi A. Innovations in thermal processing: hot-melt extrusion and KinetiSol® dispersing. AAPS PharmSciTech. 2020;21(8):312.

20. Liu X, Feng X, Williams RO, Zhang F. Characterization of amorphous solid dispersions. J Pharm Investig. 2018;48(1):19–41.
21. Grohganz H, Priemel PA, Löbmann K, Nielsen LH, Laitinen R, Mullertz A, et al. Refining stability and dissolution rate of amorphous drug formulations. Expert Opin Drug Deliv. 2014;11(6):977–89.
22. Taylor LS, Zografi G. From unwanted annoyances to Oral delivery Saviors: the rollercoaster journey of amorphous drugs. Mol Pharm. 2025;22(9):5125–38.
23. Ilevbare GA, Taylor LS. Liquid–liquid phase separation in highly supersaturated aqueous solutions of poorly water-soluble drugs: implications for solubility enhancing formulations. Cryst Growth Des. 2013;13(4):1497–509.
24. Deac A, Qi Q, Indulkar AS, Purohit HS, Gao Y, Zhang GGZ, et al. Dissolution mechanisms of amorphous solid dispersions: role of drug load and molecular interactions. Mol Pharm. 2023;20(1):722–37.
25. Yoshikawa E, Ueda K, Hakata R, Higashi K, Moribe K. Quantitative investigation of intestinal drug absorption enhancement by drug-rich nanodroplets generated via liquid–liquid phase separation. Mol Pharm. 2024;21(4):1745–55.
26. Kong Y, Wang W, Wang C, Li L, Peng D, Tian B. Supersaturation and phase behavior during dissolution of amorphous solid dispersions. Int J Pharm. 2023;631:122524.
27. Kapourani A, Pantazos I, Valkanioti V, Chatzitheodoridou M, Kalogeri C, Barmpalexis P. Unveiling the impact of preparation methods, matrix/carrier type selection and drug loading on the supersaturation performance of amorphous solid dispersions. Int J Pharm. 2025;671:125242.
28. Li N, Taylor LS. Tailoring supersaturation from amorphous solid dispersions. J Control Release. 2018;279:114–25.
29. Lin X, Hu Y, Liu L, Su L, Li N, Yu J, et al. Physical stability of amorphous solid dispersions: a physicochemical perspective with thermodynamic, kinetic and environmental aspects. Pharm Res. 2018;35(6):125.
30. Konno H, Taylor LS. Influence of different polymers on the crystallization tendency of molecularly dispersed amorphous felodipine. J Pharm Sci. 2006;95(12):2692–705.
31. Halder S, Tabata A, Seto Y, Sato H, Onoue S. Amorphous solid dispersions of carvedilol along with pH-modifiers improved pharmacokinetic properties under hypochlorhydoria. Biopharm Drug Dispos. 2018;39(4):232–42.
32. Iyer R, Petrovska Jovanovska V, Berginc K, Jaklič M, Fabiani F, Harlacher C, et al. Amorphous solid dispersions (ASDs): the influence of material properties, manufacturing processes and analytical technologies in drug product development. Pharmaceutics. 2021;13(10):1682.
33. Patel K, Shah S, Patel J. Solid dispersion technology as a formulation strategy for the fabrication of modified release dosage forms: a comprehensive review. Daru. 2022;30(1):165–89.
34. Moseson DE, Tran TB, Karunakaran B, Ambardekar R, Hiew TN. Trends in amorphous solid dispersion drug products approved by the U.S. Food and Drug Administration between 2012 and 2023. Int J Pharm X. 2024;7:100259.
35. Kshirsagar MM, Chatale BC, Dyawanapelly S, Vora LK, Amin PD. Continuous processing strategies for amorphous solid dispersions of itraconazole: impact of polymer selection and manufacturing techniques. Pharmaceutics. 2025;17(9):1090.
36. Tambe S, Jain D, Agarwal Y, Amin P. Hot-melt extrusion: highlighting recent advances in pharmaceutical applications. J Drug Deliv Sci Technol. 2021;63:102452.
37. He Y, Ho C. Amorphous solid dispersions: utilization and challenges in drug discovery and development. J Pharm Sci. 2015;104(10):3237–58.
38. Theil F, Anantharaman S, Kyeremateng SO, van Lishaut H, Dreis-Kühne SH, Rosenberg J, et al. Frozen in time: kinetically stabilized amorphous solid dispersions of nifedipine stable after a quarter century of storage. Mol Pharm. 2017;14(1):183–92.
39. Wu T, Yu L. Surface crystallization of indomethacin below Tg. Pharm Res. 2006;23(10):2350–5.
40. Grönniger B, Kimpe K, Singh A, Sadowski G. Simultaneous water sorption and crystallization in ASDs 1: stability studies lasting for two years. Mol Pharm. 2024;21(2):957–69.

41. Schneider R, Kerkhoff J, Danzer A, Mattusch A, Ohmann A, Thommes M, et al. The interplay of dissolution, solution crystallization and solid-state transformation of amorphous indomethacin in aqueous solution. Int J Pharm: X. 2020;2:100063.
42. Sun DD, Lee PI. Evolution of supersaturation of amorphous pharmaceuticals: the effect of rate of supersaturation generation. Mol Pharm. 2013;10(11):4330–46.
43. Chen Y, Wang S, Wang S, Liu C, Su C, Hageman M, et al. Initial drug dissolution from amorphous solid dispersions controlled by polymer dissolution and drug-polymer interaction. Pharm Res. 2016;33(10):2445–58.
44. Guzmán HR, Tawa M, Zhang Z, Ratanabanangkoon P, Shaw P, Gardner CR, et al. Combined use of crystalline salt forms and precipitation inhibitors to improve oral absorption of celecoxib from solid oral formulations. J Pharm Sci. 2007;96(10):2686–702.
45. Jørgensen JR, Mohr W, Rischer M, Sauer A, Mistry S, Rades T, et al. In vitro-in vivo relationship for amorphous solid dispersions using a double membrane dissolution-permeation setup. Eur J Pharm Biopharm. 2023;188:26–32.
46. Alonzo DE, Zhang GGZ, Zhou D, Gao Y, Taylor LS. Understanding the behavior of amorphous pharmaceutical systems during dissolution. Pharm Res. 2010;27(4):608–18.
47. Lu T, Wu Y, Xiong P, Zhong H, Ding Y, Li H, et al. Combining high-throughput screening and machine learning to predict the formation of both binary and ternary amorphous solid dispersion formulations for early drug discovery and development. Pharm Res. 2025;42(4):697–709.
48. Block M, Sieger P, Truenkle C, Saal C, Simon R, Truebenbach I. Miniaturized screening and performance prediction of tailored subcutaneous extended-release formulations for preclinical in vivo studies. Eur J Pharm Sci. 2024;196:106733.
49. Bhujbal SV, Pathak V, Zemlyanov DY, Taylor LS, Zhou Q. Physical stability and dissolution of lumefantrine amorphous solid dispersions produced by spray anti-solvent precipitation. J Pharm Sci. 2021;110(6):2423–31.
50. Tomar D, Singh PK, Hoque S, Modani S, Sriram A, Kumar R, et al. Amorphous systems for delivery of nutraceuticals: challenges opportunities. Crit Rev Food Sci Nutr. 2022;62(5):1204–21.
51. Pendam D, Tomake P, Debaje S, Guleria K, Saha A, Thakran P, et al. Advances in formulation strategies and stability considerations of amorphous solid dispersions. J Drug Deliv Sci Technol. 2025;108:106922.
52. Bookwala M, Wildfong PLD. The implications of drug-polymer interactions on the physical stability of amorphous solid dispersions. Pharm Res. 2023;40(12):2963–81.
53. Correa-Soto CE, Gao Y, Indulkar AS, Zhang GGZ, Taylor LS. Role of surfactants in improving release from higher drug loading amorphous solid dispersions. Int J Pharm. 2022;625:122120.
54. Tian Y, Jacobs E, Jones DS, McCoy CP, Wu H, Andrews GP. The design and development of high drug loading amorphous solid dispersion for hot-melt extrusion platform. Int J Pharm. 2020;586:119545.
55. Mosquera-Giraldo LI, Li N, Wilson VR, Nichols BLB, Edgar KJ, Taylor LS. Influence of polymer and drug loading on the release profile and membrane transport of telaprevir. Mol Pharm. 2018;15(4):1700–13.
56. Kapourani A, Andriotis EG, Chachlioutaki K, Kontogiannopoulos KN, Klonos PA, Kyritsis A, et al. High-drug-loading amorphous solid dispersions via in situ thermal cross-linking: Unraveling the mechanisms of stabilization. Mol Pharm. 2021;18(12):4393–414.
57. Song S, Wang C, Zhang B, Sun CC, Lodge TP, Siegel RA. A rheological approach for predicting physical stability of amorphous solid dispersions. J Pharm Sci. 2023;112(1):204–12.
58. Song S, Xu J, Chen Z, Sun CC, Munson EJ, Siegel RA. Miscibility of amorphous solid dispersions: a rheological and solid-state NMR spectroscopy study. J Pharm Sci. 2025;114(1):119–26.
59. Xiang T, Song S, Siegel RA, Sun CC. Efficient development of high drug loaded posaconazole tablets enabled by amorphous solid dispersion. RSC Pharm. 2025;2(1):178–85.

60. Sahoo A, Suryanarayanan R, Siegel RA. Stabilization of amorphous drugs by polymers: the role of overlap concentration (C*). Mol Pharm. 2020;17(11):4401–6.
61. Yang R, Mann AKP, Van Duong T, Ormes JD, Okoh GA, Hermans A, et al. Drug release and nanodroplet formation from amorphous solid dispersions: insight into the roles of drug physicochemical properties and polymer selection. Mol Pharm. 2021;18(5):2066–81.
62. Saboo S, Moseson DE, Kestur US, Taylor LS. Patterns of drug release as a function of drug loading from amorphous solid dispersions: a comparison of five different polymers. Eur J Pharm Sci. 2020;155:105514.
63. Saboo S, Mugheirbi NA, Zemlyanov DY, Kestur US, Taylor LS. Congruent release of drug and polymer: a "sweet spot" in the dissolution of amorphous solid dispersions. J Control Release. 2019;298:68–82.
64. Deac A, Qi Q, Indulkar AS, Gao Y, Zhang GGZ, Taylor LS. Dissolution mechanisms of amorphous solid dispersions: a close look at the dissolution interface. Mol Pharm. 2023;20(4):2217–34.
65. B S, Ghosh A. Mechanistic insights into amorphous solid dispersions: bridging theory and practice in drug delivery. Pharm Res. 2025;42(1):1–23.
66. Oberoi HS, Arce F, Purohit HS, Yu M, Fowler CA, Zhou D, et al. Design of a re-dispersible high drug load amorphous formulation. J Pharm Sci. 2023;112(1):250–63.
67. Vo CL-N, Park C, Lee B-J. Current trends and future perspectives of solid dispersions containing poorly water-soluble drugs. Eur J Pharm Biopharm. 2013;85(3, Part B):799–813.
68. Ojo AT, Lee PI. A mechanistic model for predicting the physical stability of amorphous solid dispersions. J Pharm Sci. 2021;110(4):1495–512.
69. Mudie DM, Buchanan S, Stewart AM, Smith A, Shepard KB, Biswas N, et al. A novel architecture for achieving high drug loading in amorphous spray dried dispersion tablets. Int J Pharm: X. 2020;2:100042.
70. Démuth B, Nagy ZK, Balogh A, Vigh T, Marosi G, Verreck G, et al. Downstream processing of polymer-based amorphous solid dispersions to generate tablet formulations. Int J Pharm. 2015;486(1):268–86.
71. Agrawal A, Dudhedia M, Deng W, Shepard K, Zhong L, Povilaitis E, et al. Development of tablet formulation of amorphous solid dispersions prepared by hot melt extrusion using quality by design approach. AAPS PharmSciTech. 2016;17(1):214–32.
72. Pantazos I, Poimenidou M, Kouskouridas D, Tzaferas E, Karava V, Cholevas C, et al. Influence of drug properties, formulation composition, and processing parameters on the stability and dissolution performance of amorphous solid dispersions-based tablets. Polymers. 2025;17(18):2484.
73. Takano R, Maurer R, Jacob L, Stowasser F, Stillhart C, Page S. Formulating amorphous solid dispersions: impact of inorganic salts on drug release from tablets containing itraconazole-HPMC extrudate. Mol Pharm. 2020;17(8):2768–78.
74. Kapourani A, Chatzitheodoridou M, Valkanioti V, Manioudaki A-E, Bikiaris ND, Barmpalexis P. Evaluating the effect of kosmotropic inorganic salts in the in vitro dissolution behavior of tablets containing amorphous indomethacin-polyvinylpyrrolidone solid dispersions. J Drug Deliv Sci Technol. 2022;72:103421.
75. Dedroog S, Huygens C, Van den Mooter G. Chemically identical but physically different: a comparison of spray drying, hot melt extrusion and cryo-milling for the formulation of high drug loaded amorphous solid dispersions of naproxen. Eur J Pharm Biopharm. 2019;135:1–12.
76. Xie T, Taylor LS. Dissolution performance of high drug loading celecoxib amorphous solid dispersions formulated with polymer combinations. Pharm Res. 2016;33(3):739–50.
77. Xie T, Taylor LS. Improved release of celecoxib from high drug loading amorphous solid dispersions formulated with polyacrylic acid and cellulose derivatives. Mol Pharm. 2016;13(3):873–84.

78. Juneja M, Mehtre K, Saini V, Singh R, Amate P, Kashyap M, et al. Synergistic effect of polymers in stabilizing amorphous pretomanid through high drug loaded amorphous solid dispersion. Drug Deliv Transl Res. 2025;15(2):717–31.
79. Kapourani A, Proaki L-E, Anagnostaki M-E, Gkougkourelas I, Barmpalexis P. Deciphering the role of amphiphilic polymer combinations in the formation of ternary griseofulvin amorphous solid dispersions. J Pharm Sci. 2025;114(11):104000.
80. Dong Z, Chatterji A, Sandhu H, Choi DS, Chokshi H, Shah N. Evaluation of solid state properties of solid dispersions prepared by hot-melt extrusion and solvent co-precipitation. Int J Pharm. 2008;355(1):141–9.
81. Shah N, Sandhu H, Phuapradit W, Pinal R, Iyer R, Albano A, et al. Development of novel microprecipitated bulk powder (MBP) technology for manufacturing stable amorphous formulations of poorly soluble drugs. Int J Pharm. 2012;438(1):53–60.
82. Schenck L, Mann AKP, Liu Z, Milewski M, Zhang S, Ren J, et al. Building a better particle: leveraging physicochemical understanding of amorphous solid dispersions and a hierarchical particle approach for improved delivery at high drug loadings. Int J Pharm. 2019;559:147–55.
83. Hiew TN, Saboo S, Zemlyanov DY, Punia A, Wang M, Smith D, et al. Improving dissolution performance and drug loading of amorphous dispersions through a hierarchical particle approach. J Pharm Sci. 2023;112(8):2057–68.
84. Hou HH, Rajesh A, Pandya KM, Lubach JW, Muliadi A, Yost E, et al. Impact of method of preparation of amorphous solid dispersions on mechanical properties: comparison of coprecipitation and spray drying. J Pharm Sci. 2019;108(2):870–9.
85. Wegiel LA, Mauer LJ, Edgar KJ, Taylor LS. Crystallization of amorphous solid dispersions of resveratrol during preparation and storage—impact of different polymers. J Pharm Sci. 2013;102(1):171–84.
86. Kelsall KN, Schubiner RO, Schenck L, Frank DS, Matzger AJ. Enhancing the acidity of polymers for improved stabilization of amorphous solid dispersions: protonation of weakly basic compounds. ACS Appl Polym Mater. 2024;6(3):1592–8.
87. Ismail A, Chou SF. Polyethylenimine carriers for drug and gene delivery. Polymers (Basel). 2025;17(15):2150.
88. Kelsall KN, Foroughi LM, Frank DS, Schenck L, LaBuda A, Matzger AJ. Structural modifications of polyethylenimine to control drug loading and release characteristics of amorphous solid dispersions. Mol Pharm. 2023;20(3):1779–87.
89. Zahedi P, Lee PI. Solid molecular dispersions of poorly water-soluble drugs in poly(2-hydroxyethyl methacrylate) hydrogels. Eur J Pharm Biopharm. 2007;65(3):320–8.
90. Sun DD, T-cR J, Lee PI. Enhanced kinetic solubility profiles of indomethacin amorphous solid dispersions in poly(2-hydroxyethyl methacrylate) hydrogels. Eur J Pharm Biopharm. 2012;81(1):149–58.
91. Sahoo A, Kumar NSK, Suryanarayanan R. Crosslinking: an avenue to develop stable amorphous solid dispersion with high drug loading and tailored physical stability. J Control Release. 2019;311-312:212–24.
92. Kapourani A, Chachlioutaki K, Andriotis EG, Fatouros DG, Barmpalexis P. Evaluating PAA/PVA thermal crosslinking process during the preparation of in-situ high-drug loading amorphous solid dispersions. J Drug Deliv Sci Technol. 2023;79:104030.
93. Kapourani A, Chatzitaki A-T, Vizirianakis IS, Fatouros DG, Barmpalexis P. Assessing the performance of thermally crosslinked amorphous solid dispersions with high drug loadings. Int J Pharm. 2023;640:123004.
94. Mudie DM, Stewart AM, Biswas N, Brodeur TJ, Shepard KB, Smith A, et al. Novel high-drug-loaded amorphous dispersion tablets of posaconazole; in vivo and in vitro assessment. Mol Pharm. 2020;17(12):4463–72.
95. Kumar Sharma G, Rachel James N. Electrospinning: the technique and applications. In: Khan M, Chelladurai SJS, editors. Recent developments in nanofibers research. London: IntechOpen; 2022.

96. Agarwal S, Wendorff JH, Greiner A. Use of electrospinning technique for biomedical applications. Polymer. 2008;49(26):5603–21.
97. Chachlioutaki K, Karavasili C, Adamoudi E, Tsitsos A, Economou V, Beltes C, et al. Electrospun nanofiber films suppress inflammation in vitro and eradicate endodontic bacterial infection in an E. Faecalis-infected ex vivo human tooth culture model. ACS Biomater Sci Eng. 2022;8(5):2096–110.
98. Becelaere J, Van Den Broeck E, Schoolaert E, Vanhoorne V, Van Guyse JFR, Vergaelen M, et al. Stable amorphous solid dispersion of flubendazole with high loading via electrospinning. J Control Release. 2022;351:123–36.
99. Becelaere J, Frateur O, Schoolaert E, Vanhoorne V, D'Hooge DR, Vervaet C, et al. Solvent electrospinning amorphous solid dispersions with high itraconazole, celecoxib, mebendazole and fenofibrate drug loading and release potential. J Control Release. 2023;362:268–77.
100. Yu L. Surface mobility of molecular glasses and its importance in physical stability. Adv Drug Deliv Rev. 2016;100:3–9.
101. Zhu L, Wong L, Yu L. Surface-enhanced crystallization of amorphous nifedipine. Mol Pharm. 2008;5(6):921–6.
102. Van Duong T, Nguyen HT, Wang F, Wang M, Narwankar PK, Taylor LS. Surface nanocoating of high drug-loading spray-dried amorphous solid dispersions by atomic layer coating: excellent physical stability under accelerated storage conditions for two years. Int J Pharm. 2022;620:121747.
103. Moseson DE, Benson EG, Nguyen HT, Wang F, Wang M, Zheng K, et al. Atomic layer coating to inhibit surface crystallization of amorphous pharmaceutical powders. ACS Appl Mater Interfaces. 2022;14(36):40698–710.
104. Pas T, Vergauwen B, Van den Mooter G. Exploring the feasibility of the use of biopolymers as a carrier in the formulation of amorphous solid dispersions—part I: Gelatin. Int J Pharm. 2018;535(1–2):47–58.
105. Mishra J, Bohr A, Rades T, Grohganz H, Löbmann K. Whey proteins as stabilizers in amorphous solid dispersions. Eur J Pharm Sci. 2019;128:144–51.
106. Leng D, Bulduk B, Widmer T, Wiborg O, Sanchez-Felix M, Löbmann K. Protein based amorphous solid dispersion: a case study investigating different whey proteins at high drug loading. Pharm Res. 2023;40(7):1865–72.

# Part IV
# Downstream Processing and Patient-Centric Design

# Chapter 9
# Downstream Processing of Amorphous Solid Dispersions of Drugs

**Ioannis Partheniadis, Margarita Maria Gkantanidou, and Ioannis Nikolakakis**

**Abstract** The superior solubility and absorption of amorphous drugs have fueled massive research on Amorphous Solid Dispersions (ASDs), where "Generally Recognized as Safe" (GRAS) polymers stabilize the drug form. To meet clinical demand, industry has adopted continuous processing methods such as spray drying, hot-melt extrusion, and electrospraying. These technological advances, supported by predictive modeling and rigorous characterization, have enabled so far the development of over 1000 publications and 58 marketed products. Critically, ASDs have allowed high-molecular-weight, highly lipophilic drugs ($\log P > 5$) to be formulated into life-saving therapies. This chapter reviews 25 years of ASD development, beginning with a longitudinal analysis of preparation trends and publication rates. It provides an extensive discussion on downstream processing, comparing the processability and tableting feasibility of ASDs produced via different methods. Key variables are examined, including polymer grade, formulation composition, and the impact of compression pressure, dwell time, and drug load on chemical and physical stability. The chapter concludes with a comprehensive survey of FDA- and EMA-approved ASD products. The increasing number of marketing licenses granted over the last 2.5 decades confirms that ASD-based products will remain a cornerstone of pharmaceutical research and formulation for years to come.

**Keywords** Hot-melt extrusion · Spray drying · Downstream processing · Tableting · Physical stability · Marketed pharmaceutical products

I. Partheniadis
Department of Pharmacy, Faculty of Health and Medical Sciences, University of Copenhagen, Copenhagen, Denmark
e-mail: ioannis.partheniadis@sund.ku.dk

M. M. Gkantanidou · I. Nikolakakis (✉)
Laboratory of Pharmaceutical Technology, School of Pharmacy, Faculty of Health Sciences, Aristotle University of Thessaloniki, Thessaloniki, Greece
e-mail: mgkanta@pharm.auth.gr; yannikos@pharm.auth.gr

 
I. Nikolakakis, I. Partheniadis (eds.), *Amorphous Solid Dispersions*, AAPS Introductions in the Pharmaceutical Sciences 5,
https://doi.org/10.1007/978-3-032-24883-1_9

## 9.1 Introduction

Solid dispersions have been originally defined by Chiou and Riegelman as "dispersions of one or more active ingredients in an inert carrier at the solid state, prepared by the melting, the solvent or the melting-solvent method" [1]. Due to solubility issues, the research interest is mainly focused on solid dispersions where the drug is present in its amorphous form instead of micro- or nanocrystalline, stabilized by inert auxiliaries. As a result, a number of drug molecules have completed the journey to the market, and there is a large number of polymeric amorphous solid drug dispersions (ASDs) in the development pipelines of pharmaceutical companies.

Today, amorphous drug forms are mostly stabilized by dispersing the amorphous drug in polymers. However, despite the solubility and dissolution improvements achieved, polymeric ASDs have inherent problems, such as oversized dosage units due to the high polymer-to-drug ratio [2], poor tablet compressibility [3], slow disintegration due to polymer gelation [4] and pressure-induced recrystallization [5]. Also, the processing steps toward the final product, comprising milling, blending, granulation, tableting, and coating, may seriously challenge stability, since they can trigger recrystallization [6, 7]. In particular, the high compressional forces applied during industrial tablet production and the resulting high mechanical stresses, together with the produced frictional heat, may inflict solid-state changes to the drug. Due to the above reasons, it is apparent that downstream processing constitutes a major challenge to formulation. The problems can be solved by adding special additives and proper selection of the ASDs preparation conditions, depending on the drug and the type of dosage form pursued (enteric, orodispersible, instant release). These problems have been addressed in the papers described in Sect. 9.3 of the present Chapter.

Recently, co-amorphous drug systems (CAMS) were introduced as a polymer-free approach to conventional ASDs, to overcome problems of oversized dosage forms, disintegration, and gelation. However, their downstream processing into dosage forms is understudied and there are no yet marketed products [8–12]. For this reason, the ASD formulation platforms and downstream processing into dosage forms and brand name products reported in this Chapter concern only polymeric amorphous drug dispersions (ASDs).

## 9.2 Trends for ASD Preparation Methods in the Last Decades

The literature survey of publications during the last decades has shown that the preferred methods for the laboratory-scale and large-scale ASD manufacturing are divided into three main categories: (i) thermomechanical methods represented by hot melt extrusion (HME) and lately by KinetiSol® technology (fusion/mixing); (ii) solvent evaporation (SE) represented by rotary evaporation, spray drying (SD, heat assisted evaporation), freeze drying (FD), and electrospinning (ES), and (iii) mechanochemical methods represented by nanomilling. Pure heating methods such as quenching and microwaves (for in situ ASD formation), and granulation (wet, dry, and melt) are reported but to a smaller extent. Regarding ASD manufacturability and drug loading capacity, Dedroog et al. compared SD with HME and milling. They concluded that HME with suitable polymeric carriers is preferable for high loading capacities, specifically highlighting the critical role of thermal energy input [13].

### *9.2.1 Literature Survey for ASD Preparation Methods in the Last Decades*

To provide an overview of ASD preparation methods, we conducted a systematic literature review of articles published between 1987 and 2026 using the Scopus® (Elsevier) database. We excluded review papers and categorized the retrieved information based on the underlying scientific principles of these methods (e.g., spray drying and solvent evaporation, or hot-melt extrusion and KinetiSol® technology). Subsequently, a further survey was conducted (Sect. 9.3) to identify drug molecules that successfully progressed to a final solid dosage form.

Table 9.1 presents the results of the identification of ASD preparation studies via the Scopus® database. The retrieved data are summarized as the annual number of publications over the past four decades, categorized by preparation methods with industrial scale-up application. In this context, simple laboratory methods such as quenching are excluded from the analysis. The complete search algorithm for the used database is presented separately for each ASD method. Notably, the first paper on ASDs was published by Takeuchi et al. (1987), who reported the use of spray drying for ASD preparation [14].

Figure 9.1 presents the frequency distribution of the data on preparation ASD methods listed in Table 9.1. From Figure 9.1, it appears that the greatest volume of published research is based on the application of the two continuous manufacturing processes of SD (34.1% of totally retrieved articles) and HME (33.5%), followed by the more traditional common laboratory methods of SE (14.2%), milling (6.8%), FD (5.2%), ES (2.2%), KinetiSol® technology (fusion-based 1.9%), granulation (1.2%), and microwaves (0.8%).

To illustrate the growing interest in ASD preparation, Figure 9.2 was constructed presenting the evolution of annual publication volume for various methods since 2000. Research activity remained relatively low during the first decade of the century, dominated primarily by SE and FD. However, the landscape shifted radically after 2010, marked by a surge in publications focusing on SD and HME. While these continuous manufacturing methods were already established in the food industry, they only gained prominence in the pharmaceutical sector once the clinical benefits of ASDs were fully recognized. Meanwhile, SE and milling remain viable

**Table 9.1** Identification of ASD preparation studies via Scopus database and retrieved results

| Preparation method | Scopus search algorithm | Results |
|---|---|---|
| Electrospinning | TITLE: "Amorphous solid dispersions" TITLE-ABS-KEY "Electrospinning" AND NOT "Review" | Years 2015–2025 ($n = 15$) Excluded ($n = 0$) |
| Freeze drying | TITLE: "Amorphous solid dispersions" TITLE-ABS-KEY "Freeze drying" OR "Lyophilization" AND NOT "Review" | Years 2004–2025 ($n = 35$) Excluded ($n = 0$) |
| Granulation | TITLE: "Amorphous solid dispersions" TITLE-ABS-KEY "Granulation" AND NOT "Review" AND NOT "Spray drying" | Years 2010–2025 (n = 10) Excluded conference ($n = 2$) and book chapter (n = 1) |
| Hot melt extrusion | TITLE: "Amorphous solid dispersions" TITLE-ABS-KEY "Hot melt extrusion" AND NOT "Review" AND NOT '3-D PRINTING" | Years 2008–2025 ($n = 227$) Excluded ($n = 0$) |
| KinetiSol® | TITLE: "Amorphous solid dispersions" TITLE-ABS-KEY "Kinetisol" AND NOT "Review" | Years 2010–2023 ($n = 13$) Excluded conference proceedings ($n = 1$) |
| Microwaves | TITLE: "Amorphous solid dispersions" TITLE-ABS-KEY "Microwaves" AND NOT "Review" | Years 2019–2025 ($n = 6$) Excluded ($n = 0$) |
| Milling | TITLE: "Amorphous solid dispersions" TITLE-ABS-KEY "Milling" AND NOT "Review" | Years 2010–2025 ($n = 46$) Excluded ($n = 0$) |
| Solvent evaporation | TITLE: "Amorphous solid dispersions" TITLE-ABS-KEY "Solvent evaporation" AND NOT "Review" AND NOT "Spray drying" | Years 2008–2025 ($n = 96$) Excluded conference proceedings ($n = 1$) |
| Spray drying | TITLE: "Amorphous solid dispersions" TITLE-ABS-KEY "Spray drying" AND NOT "Review" | Years 1987–2025 ($n = 231$) Excluded conference proceedings ($n = 9$) |

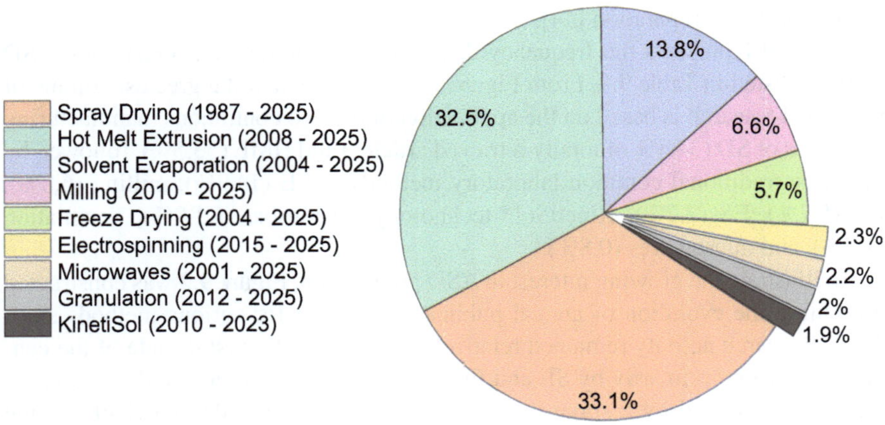


**Fig. 9.1** Frequency distribution of the preparation ASD methods derived from Table 9.1

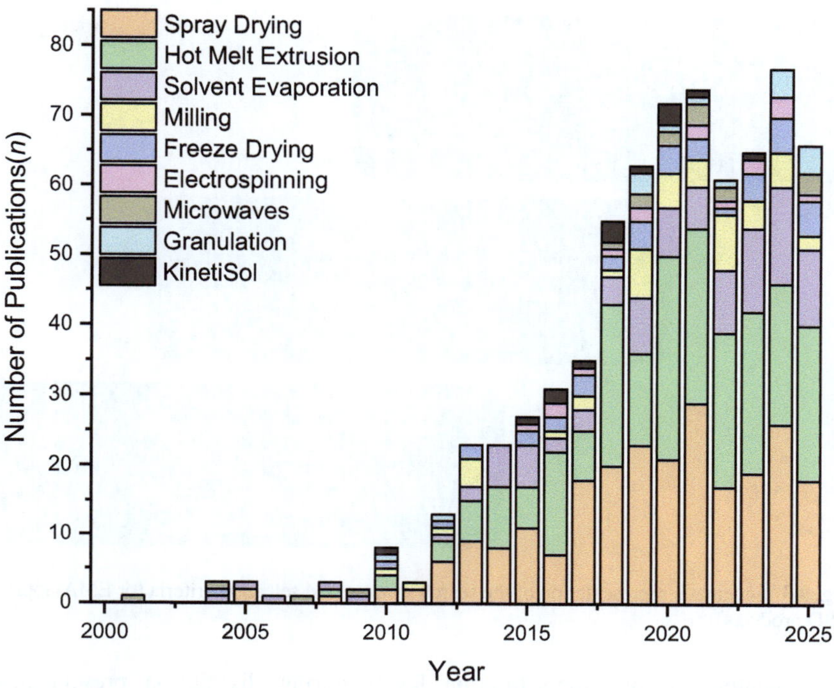


**Fig. 9.2** Number of papers published annually in the last 2.5 decades reporting different ASD preparation methods

laboratory-scale techniques, and interest in electrospinning is steadily emerging. Additionally, KinetiSol™ high-rate infusion technology has been introduced as an alternative to HME for thermolabile drugs and high-viscosity polymers, offering processing times under 1 min [15]. Finally, while microwave irradiation was pioneered for ASDs as early as 1998 [16], it has seen a resurgence lately for in situ ASD formation [17].

## 9.3 Downstream Processing of ASDs

Once the solid dispersion is made available, it follows the journey toward the final formulated dosage form and market product. The technological advances in characterization methods, the predictive models applied during the preformulation phase, and the knowledge gained have aided in understanding the potential challenges that may be encountered during the downstream processing of ASDs. For example, direct feeding of ASDs into processing machines is successful only if the ASD powder is within a certain range of particle size, bulk density, flowability, compressibility, and compactibility, which are rarely found in unprocessed ASD intermediate products. Furthermore, physicochemical instability, processing of high drug loads,

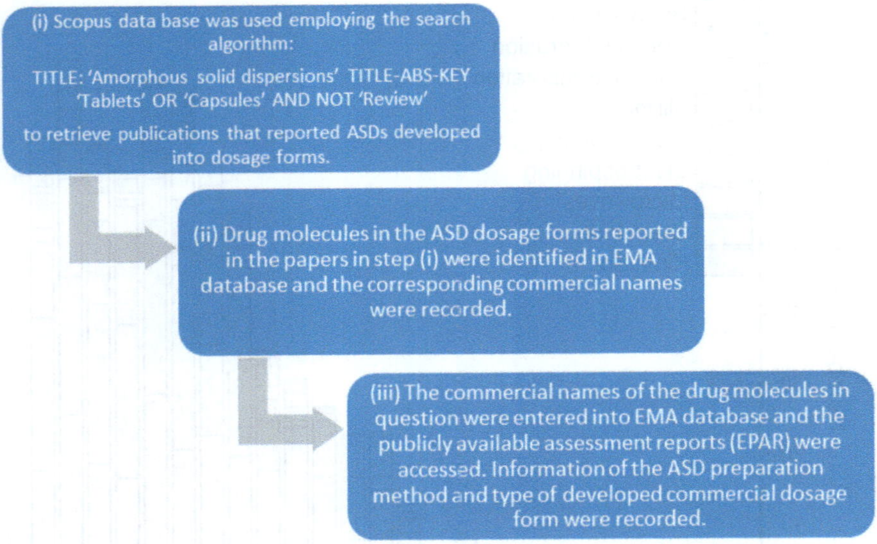


**Fig. 9.3** Schematic representation of the search strategy and selection criteria for EMA-approved ASD products

and maintenance of supersaturation levels during dissolution present major challenges.

## *9.3.1 Downstream Processibility of ASDs*

As demonstrated earlier in Figure 9.3 by the number of annually published papers reporting ASD preparation, the continuous manufacturing methods of spray drying (SD) and hot melt extrusion (HME) constitute the lion's share of the investigated ASD preparation methods and have been mostly explored for the downstream processing of ASDs. Among the other methods with good ASD stabilization ability but smaller scale-up potential, freeze drying, solvent evaporation, granulation, and quenching, as well as the newer methods of electrospinning, KinetiSol® (mixing/fusion) technology, and microwaves, also appear in studies of ASD development into dosage forms.

### 9.3.1.1 Compression Behavior of ASDs

One of the early studies by Grymonpré et al. proposed HME as a robust process for tableting ASDs [18]. Extrudate and tablet properties were mainly influenced by the drug load and less by the process parameters. On the other hand, a systematic study of the milling and tableting of a HME extruded itraconazole/hypromellose acetate succinate (HPMCAS) ASD showed that there is dependence of the critical quality attributes (CQAs) of ASD powder on the processing of extrudate strand. Among the techniques of hand-cutting, pelletizing, and chill rolling/flaking, applied before milling, chill rolling/flaking showed the highest percentage yield, lowest particle size ($D_{50}$), highest compactibility, and highest drug dissolution rate [19].

Doktorovova et al. published results on the compression mechanisms of ASDs [20]. They examined the strain rate sensitivity under compression of spray-dried ASDs with HPMC, HPMCAS, PVP-VA at varying drug content up to 50% w/w, using a compaction simulator. The ASDs exhibited plastic flow and deformation and formed robust compacts at slow compression speeds, but at high speeds, tablet defects were noticed. The study of Zhang et al. emphasized the need for thorough characterization of the compaction behavior of itraconazole ASDs prior to tablet development, especially at high drug loadings [21]. Tabletability and compactability of the ASD powders differed significantly, while their compressibility was similar. At compressions from 50 to 150 MPa, tensile strength of ASD tablets increased with drug content, reaching maximum at 40–60% drug and then decreased. However, at 200 MPa pressure, a monotonic decrease of tensile strength with drug loading was observed.

Ekdahl et al. prepared ASDs with 20% w/w felodipine dispersed in PVP-VA, by the SD method [22]. Processing conditions were tuned to obtain powders with different physical properties. Powder flowability measured by shear cell ranged from poor to easy flowing, but all powders produced strong tablets at similar solid fractions and compression pressures. The work demonstrated that ASD particle morphology can be engineered to achieve desired powder flow and mechanical properties to reduce downstream processing risks and increase process throughput.

Using celecoxib/PVP/VA64 ASD powder with a narrow particle size range, Osei-Yeboa and Sun found that increasing the relative humidity (RH) from 11% to 67% led to an improvement in ASD tablet tensile strength [23]. The increase in strength depended on the tablet hardness and elasticity. Due to the plastic deformation mechanism, compression at high pressures and exposure to higher RH had small effects on tabletability, because the particles were already deformed and the bonding areas were developed.

In summary, these studies identify plastic deformation as the predominant compaction mechanism for ASD powders, though this is highly sensitive to the "thermal and mechanical history" of the material. The comminution method for extruded strands (e.g., chill rolling vs. pelletizing) and the particle engineering of spray-dried intermediates are critical, as they dictate the initial powder morphology and subsequent compactibility. Furthermore, while relative humidity (RH) can enhance tensile strength by plasticizing the polymer matrix, this effect diminishes at high compression pressures where the inherent plasticity of the ASD already maximizes the available bonding area. Consequently, successful tableting requires a precise balance among drug loading, compression force, and environmental variables to avoid defects such as capping or lamination.

#### 9.3.1.2 Comparison of the Processibility and Tableting of ASDs Prepared by Different Methods

Results of the work of Mahmah et al. on felodipine indicated that ASDs produced by SD exhibited faster drug release than the corresponding HME formulations, although the latter were more stable [24]. Furthermore, the study by Davis et al. showed that the preparation of itraconazole amorphous solid dispersion by either SD or HME led to significantly different powder properties [25]. Although the SD product exhibited poorer flow and reduced sensitivity to aeration relative to milled extrudate, it could be compressed into stronger compacts than HME and released the drug faster and completely.

Ritonavir/copovidone/sorbitan monolaurate ASDs and respective tablet blends (filler, glidant, lubricant) were investigated by Schönfeld et al. [26]. Milled HME extrudate showed superior powder properties compared to vacuum drum drying (VDD) and SD. The VDD product was directly compressible, whereas the SD product was not, resulting in tablets with defects due to elastic recovery. Compared to HME, VDD showed superior tabletability resulting in stronger compacts at lower solid fractions. Despite the differences in compressibility, tablets showed similar disintegration and dissolution, independent of the ASD origin.

Since poor ASD flowability is an obstacle to automatic direct tablet compression, roller compaction is the method of choice to assist downstream processing of SD products. Compaction force, gap and mesh aperture of the roller compactor have major impact. The results of the work of Henriques et al. showed that a large gap between the rollers and low force was an optimal combination of roller compaction parameters leading to powder flow improvement and hence low tablet weight variation and good ASD compressibility [27].

Dry granulation was also applied by Song et al. to overcome poor flowability of the coprecipitation prepared ASD of sorafenib/HPMC-AS with 40% drug loading [28]. The ASD was compressed into tablet, and the tablet formulation was evaluated using a material-sparing, expedited approach to optimize compaction conditions for low friability and rapid disintegration. The ASD tablets exhibited better bioavailability than the marketed tablet.

Jia et al. evaluated the interrelationship of microstructure, properties, and dissolution performance of ASDs produced by SD and by co-precipitation, as well as the corresponding tablet forms [29]. Distinct differences in microstructure, particle size distribution, specific surface area, bulk and tapped density, were observed. The coprecipitated ASD powders demonstrated superior compactibility, while the compressibility of the two ASDs was comparable. SD powder and tablets showed higher in vitro dissolution than the respective co-precipitated.

Trenkenschuh et al. compared ASD tablets of an API produced by HME and SD and compared their in vitro dissolution [30]. They found that drug release can be controlled by adjusting the particle size of melt-extruded ASDs, leading to a similar release profile with the tablets from SD. This confirmed the possibility of technology shift from SD to HME during product development.

Overall, the choice of ASD preparation method fundamentally dictates the downstream strategy. While SD products generally offer superior compressibility and faster dissolution, which is likely due to higher intraparticle porosity and surface area [31], their poor flowability often necessitate intermediate steps like roller compaction. In contrast, HME extrudates typically exhibit superior flow and bulk density but may require precise control over milling and particle size distribution to match the rapid release profiles of SD counterparts. These findings suggest that while the ASD preparation method results in inherent mechanical differences, downstream unit operations (e.g., dry granulation) and engineering of the solid state can be leveraged to harmonize the performance of the final dosage form across different production platforms.

#### 9.3.1.3 Influence of Polymer Type/Grade on Tableting and ASD Chemical Stability

Casian et al. studied the influence of polymer type and processing of electrospun meloxicam ASDs on the properties of resulting orodispersible tablets [32]. The high surface area of the fibers enabled tableting at low pressures with PVP K30 grade making stronger compacts. Eudragit® E formulation showed lower sticking tendency during tableting, as reflected in lower detachment and ejection forces. Tableting was influenced by polymer mechanical properties. PVP K30 fibers could be delivered in the form of orodispersible webs and conventional tablets, whereas Eudragit® E fibers as orodispersible tablets only.

Browne et al. explored the ability of different polyvinyl-based polymer grades (PVA, PVP, PVP-VA, polyvinyl acetate, and polyvinyl acetate phthalate) to produce ASDs with ketoprofen with desired stability, dissolution, and tablet strength [33]. Polymer's molecular weight and the vinyl pyrrolidone-to-vinyl acetate ratio were related to relative humidity (RH) level inducing glass transition, and also to ketoprofen's supersaturation achieved during solubility testing. ASD tablets with vinylpyrrolidone functional group exhibited higher tensile strength, due to the binding ability of vinylpyrrolidone.

Zhang et al. prepared tablets of indomethacin (IND) ASDs with various polymer matrices utilizing a microwave dielectric method [34]. Comparative studies between the ASD tablets and directly compressed (DC) conventional tablets showed that tablets of greater hardness were obtained using Soluplus® and HPC HF than HPMC, HPMCAS, Eudragit® PO, and PEO. Supersaturation during dissolution of the ASD tablets was stabilized by the polymeric matrix and drug release exhibited mostly Fickian diffusion.

Butreddy et al. compared the impact of peroxides in Plasdone™ S630 Ultra grade and in regular Plasdone™ S630 grade on the oxidative degradation of quetiapine fumarate/Plasdone™ ASDs (from HME) compressed into tablets with the aid of microcrystalline cellulose and croscarmellose sodium [35]. Stability studies monitoring N-oxide degradation impurity showed that its levels in the ASD extrudates prepared with Plasdone™ S630 Ultra grade and tablets were reduced by two- and three-fold, compared to regular Plasdone™ S630. The reduced oxidative degradation and improved HME processability using grade S630 Ultra is a better choice for oxidation-labile drugs.

On a similar investigation line, Saraf et al. examined the impact of three copovidone grades, together with the HME barrel temperature, and presence of colloidal silicon dioxide ($SiO_2$) on the impurity levels, stability, and drug release from ASD tablets of a pyrrolopyridine type drug [36]. ASD tablets exhibited greater physical stability than the corresponding milled extrudates, due to the reduced environmental exposure. $SiO_2$ in the extrudate composition improved the physical stability of the "in-tablet" ASD, but had an adverse effect on chemical stability, promoting oxidative degradation and hydroxylation of the drug.

In conclusion, the selection of the polymeric carrier and its specific grade is a dual-purpose decision, governing both the mechanical performance and the physicochemical stability of the dosage form. While vinyl-pyrrolidone-based polymers generally facilitate superior bonding and tablet strength, their moisture sorption tendency and the presence of residual peroxides in standard grades can compromise the formulation of oxidation-labile drugs. Furthermore, the use of stabilizers like $SiO_2$ presents a trade-off, that is, while they may enhance physical stability by protecting the amorphous state within the tablet matrix, they can promote chemical degradation. Therefore, optimizing the downstream performance of ASDs requires a holistic approach that matches the polymer's mechanical profile (e.g., plasticity vs. brittleness) with its chemical purity and compatibility with processing aids.

#### 9.3.1.4 Influence of ASD-Tablet Formulation Factors on Drug Release

Studies on the dissolution of uncompacted and compacted ASDs of felodipine with the water soluble polymer copovidone revealed that compaction significantly decreased the rate and extent of dissolution of felodipine and there was a strong dependence on drug loading, especially for the uncompacted samples. Low drug content tablets (5% and 15% w/w) eroded with linear kinetics at identical rates, pointing to matrix-controlled release, whereas dissolution of compacts containing higher drug loads (≥30%, w/w), was dominated by the drug properties [37].

The work by Démuth et al. investigated the influence of excipients on the dissolution and scale-up feasibility of electrospun itraconazole/PVP-VA ASDs [38]. Dissolution from tablets was mainly composition dependent. Magnesium stearate added as lubricant interacted and promoted nucleation and crystallization, which hindered drug release. This was avoided when the water-soluble lubricant, sodium stearyl fumarate was used. The optimized formulation for the final tablet was scaled up using a rotary tablet press.

Lu et al. also studied the dissolution of felodipine from high drug load ASDs with PVP-VA with added sodium dodecyl sulphate (SDS) surfactant aiming for dissolution enhancement [39]. The binary drug/PVP-VA (1:1 and 1:3) and ternary drug/PVP-VA/SDS (1:1:0.02–0.16 and 1:3:0.02–0.32) combinations formed ASDSs. The surfactant enhanced wetting, erosion and dissolution of the

PVP-VA-rich ASDs, but induced drug recrystallization and incomplete dissolution. For drug/surfactant ratio of 1:0.08–0.12, the SDS significantly increased dissolution in both extrudates and compressed tablets. Chiang et al. also added surfactant in order to improve dissolution of ASD tablets of drug candidate [40]. In vitro evaluations showed that the added surfactant improved both the disintegration and the dissolution of the drug in the ASD tablets.

Chae et al. incorporated alkalinizing agents (NaOH) into ASDs of telmisartan with PEG 6000, PVP K30, Poloxamer 407, and Soluplus® polymers in order to improve drug's dissolution [41]. The ASD prepared with Soluplus® exhibited enhanced dissolution with extended supersaturation and was formulated into a direct compression tablet with the aid of Vivapur® 105 and croscarmellose, as a binder and superdisintegrant, respectively. The developed tablets demonstrated longer supersaturation maintenance than the commercial product (Micardis®).

Gelation of polymeric carriers in ASDs is the main cause of slow disintegration. To mitigate polymer gelation on the surface of tablets of itraconazole/HPMC ASDs (prepared by HME), Takano et al. used inorganic salts [42]. A kosmotropic effect on HPMC was observed with KCl, $KH_2PO_4$, and $KHCO_3$, but a chaotropic effect was observed with KI. ASD/croscarmellose sodium/MCC tablets containing $KHCO_3$, $KH_2PO_4$, KCl, and KI compressed at solid fraction of 0.83–0.85 showed faster release compared to tablets without salts. Also, tablets without salt stored under dry or humid conditions exhibited large difference in dissolution, whereas tablets with kosmotropic salts did not. On similar lines, Xi et al. used fast-dissolving kosmotropic salts in concentrated solutions, added inside restrained tablet matrices to inhibit gelation of copovidone [43]. The outcome of this study has enabled the design of a higher ASD loading platform formulation for copovidone based ASD.

Mudie et al. studied ASDs prepared with the SD method, comprising 50% acalabrutinib and 50% hydroxy-propyl methylcellulose acetate succinate (HPMCAS, H grade) formulated as an immediate-release (IR) tablet [44]. The ASD tablets outperformed the bioavailability of commercial Calquence® capsules by 2.4-fold. An in vitro multicompartment model successfully predicted the dissolution performance, and the ASD tablets presented good physicochemical stability and successful laboratory-scale manufacturability.

HME produced granules of itraconazole/HPMC ASD were efficiently absorbed into mesoporous silica. The ternary system containing high-viscosity HPMC (AF4M) maintained drug supersaturation in neutral pH dissolution media in which crystalline drug solubility was below 1 μg/mL. The granules were compressed into tablets (containing 80% w/w of ASD granules and 40% w/w drug), with acceptable physical size, hardness and disintegration. The dissolution of the ternary ASD tablet exhibited supersaturation maintenance similar to that of the ASD granules [45].

Furthermore, Giri et al. prepared telmisartan/Soluplus® ASD with a pH-modulator (sodium carbonate 0–10%) with drug content 10–60% w/w, Soluplus® at 30–90% w/w, and sodium carbonate at 0–10% w/w [46]. Optimized tablets showed strikingly improved (≈30-fold) in vitro dissolution profile compared to free drug tablets. Studies with itraconazol/HPMCAS ASD (SD method) tablets showed that tablet surface characterization (contact angle, surface free energy, and surface polar fraction) can be used to identify the factors affecting drug dissolution rate [47]. HPMCAS was already proved to be a good release and supersaturation maintenance improver due to its amphiphilic nature allowing insoluble drug molecules to interact with the hydrophobic regions of the polymer, whereas the hydrophilic regions allow the ASD structures to remain as stable colloids in aqueous solution [48].

Zhang et al. investigated the effects of ASD loading, filler type and ratio, disintegrant type and level on the performance of ASD (from SD) tablets prepared by direct compression [49]. Disintegration time increased when the ASD content exceeded 50%. Lactose gave faster tablet disintegration compared to mannitol when they were used as water-soluble fillers. Croscarmellose sodium (5%) performed the best in facilitating the ASD tablet disintegration, followed by sodium starch glycolate. Crospovidone was the worst disintegrant, indicating wicking as the major contributing disintegration step.

Bochman et al. introduced a novel method requiring low material- and time-consumption for the microscopic erosion time test (METT), to predict the API-specific drug load threshold between erosion and non-erosion mechanisms of ASD in tablets [50]. Even a drug load as small as 2.5% above the threshold led to a non-erosion delayed release of the API from the ASD tablets and impeded tablet disintegration. Nineteen APIs in copovidone-based ASDs were characterized via the METT test. From these, indomethacin, celecoxib, dipyridamole, fenofibrate, naproxen, and ritonavir were investigated in more detail.

The results of the above studies point to the importance of the correct choice of polymeric carrier in the ASD, as well as of disintegrant and lubricant in order to avoid interactions with the drug in the dosage form and recrystallization. While the addition of surfactants can improve wetting and disintegration, it is not without the risk of inducing recrystallization and impairing release. Furthermore, the incorporation of inorganic salts or pH-modulators is shown to be an effective strategy to reduce polymer gelation and maintain supersaturation. Ultimately, the successful formulation of ASD tablets requires drug content below an API-specific load threshold to ensure an erosion-controlled release mechanism and rapid disintegration.

### 9.3.2 ASDs in Pediatric and Enteric Release Formulations

Layering ASDs onto inert beads has been useful in the formulation of low dose drugs. Boel et al. investigated the influence of solvent type (acetonitrile, methanol, ethanol, acetone, 2-propanol, dichloromethane, and ethyl acetate) on bead coating process and on the solid-state characteristics of the resulting ASDs of felodipine/PVP-VA coated onto MCC beads [51]. The coated beads were further dried under vacuum at room temperature. Different morphologies were observed for the PVP-VA and FEL-PVP-VA amorphous dispersions after deposition onto beads from the various solvents, which was attributed to differences in solvent evaporation rates.

Spray dried ASDs of indomethacin with PVP or HPMCAS carriers were prepared by Autzen Virtanen et al. [52]. In the next step the ASD powder was blended with MCC and lactose and compressed into instant release paediatric minitablets with 6.22% drug content. Kokkot et al. prepared ASDs of ritonavir/PVP-VA and lopinavir/HPMCAS by HME [53]. The extrudates were milled, sieved, and blended with mannitol/cellulose/disintegrant based co-processed excipients (CPE) and compressed to paediatric tablets. The selected ASD particle size for compression was the key parameter for fast disintegration and high mechanical strength.

Nguyen et al. applied enteric coating in order to minimize contact of the delamanid/HPMCP ASD particles with the acidic stomach fluids and achieve drug release at higher pH conditions [54]. To this end, they formulated an ASD tablet that was then coated with a methacrylic acid copolymer.

Ellenberger et al. minimized the size of dosage form and reduced patient pill burden of ritonavir, using KinetiSol® technology [55]. ASD of the drug with 45% reduction in dosage form mass was developed and was compared to the commercial product (Norvir®). The developed ASD tablet presented approximately 20% $C_{max}$ reduction compared to the reference tablet (Norvir®).

The results of these studies point to the feasibility to formulate ASDs into tablets for specific needs, such as paediatric minitablets and pellets (beads) and enteric-coated formulations. For low-dose drugs, spray coating of ASDs on a core tablet or pellet surface is a possibility for the preparation of dosage units with uniform low dose drug content. Furthermore, the selection of ASD particle size and the use of specialized technologies like KinetiSol® could enable the development of dosage forms with reduced mass and improved mechanical strength, effectively addressing patient pill burden and disintegration requirements.

### 9.3.3 Stability of ASDs During Processing

Démuth et al. confirmed the feasibility of processing of electrospun itraconazole/PVP-VA64 ASD fibers into direct compression tablets of good physical stability, adequate strength and acceptable disintegration time [56]. This was achieved by adding 76.25% fillers in the 9.5 mm flat-faced tablet formulation and compressing with 6 kN force. Electron microscopy confirmed the maintenance of fibrous tablet surface structure. Moisture sorption experiments suggested good physical stability (25 °C/60% RH).

Bērziņš and Suryanarayanan prepared tablets of amorphous sucrose and sucrose/PVP ASDs [7]. Tablets were compressed at 25, 75, or 150 MPa and dwell times were between 5 and 900 s. Compression induced physical destabilization of the ASDs. At the highest applied pressure, sucrose crystallization was evident immediately after compression. Addition of PVP, even at a low concentration (1% w/w), inhibited crystallization. Nucleation was completely prevented at higher PVP concentrations (≥15% w/w) and at a standard 150 MPa compression pressure (5 s dwell time). However, increase in dwell time (e.g., to 60 s) prompted nucleation, and nucleation density increased with dwell time.

Other workers examined the sensitivity to compression of ASDs of paracetamol with different PVP grades containing up to 30% drug [57]. Even high compression forces did not affect ASD's stability. By applying the concept of polymer overlap concentration ($c^*$) (concentration where polymer coils interpenetrate, creating a hypothetically single homogeneous phase in ASD), Xiang et al. identified the maximum posaconazole content in ASDs with HPMCAS, that allowed physical stability of "in-tablet" ASDs for 1 month at 40 °C [58].

Yu and Huang compared the performance of common diluents on the tableting behavior, dissolution, and physical stability of itraconazole/HPMCAS ASDs [59]. Crystalline diluents such as lactose and mannitol facilitated "in-tablet" ASD recrystallization, signifying that physical stability of the ASD is not solely dependent on moisture sorption. However, the effect of RH on stability cannot be overlooked. One week of exposure to 76% RH resulted in weight increase of tadalafil/Soluplus® ASDs to about 7%, and diffraction patterns showed the onset of recrystallization of the amorphous drug. Hence, packaging with high resistance to moisture was recommended for these formulations [60].

The importance of correct solvent selection in order to exclude solvent residues is emphasized in the work of Dohrn et al. [61]. The same group investigated phase changes during the drying process of API/polymer/solvent systems from a thermodynamic perspective, aiming for prediction and selection of solvents that do not cause phase separation during drying.

Neuwirth et al. studied the impact of film coating on the physical stability of ASDs layered onto pellets [62]. Stability was found to be related to the composition of film coat material, and to coating thickness. On the same lines, the work of Punia et al. reported that an increase in coating thickness led to reduction of physical stability of the "in-tablet" ASD [63]. Therefore, coating of ASD sprayed on tablets or incorporated in the tablet matrix is not without risk, due to the possible leaching induced by the presence of plasticizers in film coatings [64].

Neilly et al. applied X-ray microscopy to trace crystalline fenofibrate residues as well as their concentration (down to 0.2% w/w of tablet), particle size (down to 10 μm), and distribution in ASD model tablets spiked with known concentrations of drug crystals [65]. This technique is nondestructive and provided quantitative measures of crystallinity below standard X-ray powder diffraction methods, as well as three-dimensional information regarding size, shape, and distribution of the API crystals.

The results of the above studies point to the critical role of compression pressure, dwell time, drug load, and the crystallinity of tableting excipients on the physical stability of processed ASDs. Crystalline diluents and high RH levels can facilitate "in-tablet" recrystallization, signifying that stability is not solely dependent on moisture sorption. When solvent evaporation or SD is used for ASD preparation, solvent choice should ensure there are no remaining residues in order to avoid instability and phase separation. Furthermore, when coating is applied to tablets or pellets having ASD spray-coated on the surface or inside the matrix, the choice of film coat composition and the amount of plasticizer are critical in order to avoid possible leaching into the ASD and subsequent reduction of stability.

### *9.3.4 General Comments on the Results of Presented Published Works*

From the account of the presented papers above that were published in the last 26 years, it appears that the success of downstream processing of the ASDs into dosage forms depends to a large extent on the physical and mechanical properties of the ASD itself and on the conditions during manufacturing into the final dosage form. In particular, the interplay among ASD composition (drug type and loading, type and grade of polymer, presence of auxiliaries such as surfactants, presence of solvent residues), ASD production method, and ASD characteristics (bulk density, flowability, compressibility, compactibility) will define the downstream strategy. Furthermore, the selection of tablet formulation parameters and manufacturing conditions (compression pressure, dwell time, humidity, and excipient crystallinity) are critical to maintaining the amorphous state and ensuring an erosion-controlled release mechanism. This holistic understanding provides the basis for an experimental design (QbD) that can effectively convert an ASD intermediate into a stable and high-performing pharmaceutical product.

## 9.4 ASD Products in the Pharmaceutical Market

In their review paper, Tambe et al. presented a list of ASD products in the market [66]. More recently Moseson et al. published a review paper listing ASD products approved by the FDA for marketing during the period 2012–2023 [67]. Analysis of the data in the last paper showed that 42 products containing ASDs have been approved representing 46 drug molecules, in 5 different dosage forms: tablets (71.4%), capsules (10.2%), granules for oral suspensions (6.1%), pellets (6.1%), and powder for oral suspension (4.1%). Most of the ASDs in the approved products are prepared by spray drying (52.9%) and hot melt extrusion (33.3%), and the rest of them (13.7%) by wet or melt granulation and electrospraying.

Since no similarly published data for approved products for marketing by the European Medicines Agency (EMA) is available in the literature, the task was undertaken in this Chapter to present them. The data covers the time-period from 2008 until today and is presented in Table 9.2. They provide an overview of drug molecules and related therapeutic areas, their progress from ASD intermediates to dosage forms, and ultimately to market products. The information in Table 9.2 was gathered following the search line shown in Fig. 9.3.

More specifically, the information in Table 9.2 concerns only marketed ASD products: that is, active substances and formulations that are already approved and commercially authorized by the EMA. This ensures that Table 9.2 lists regulatory confirmed ASD products rather than experimental or preclinical formulations. For fixed-dose combinations (FDCs), the table presents them if at least one of the components is formulated as an amorphous solid dispersion. In these cases, the ASD

**Table 9.2** Drug molecules that have been investigated for downstream processing to solid dosage forms and corresponding approved products by the European Medicine Agency European (EMA) as they appear in the EMA public assessment reports

| | Drug/API | Therapeutic area | ASD preparation method | Trade name | Dosage form | Public EMA EPAR |
|---|---|---|---|---|---|---|
| 01 | Deucravacitinib | Treatment of psoriasis | SD | Sotyktu | Film-coated tablets | [68] |
| 02 | Enzalutamide | Anticancer | SD | Xtandi | Film-coated tablets | [69] |
| 03 | Etravirine | Antiviral (HIV) | SD | Intelence | Tablets | [70] |
| 04 | Everolimus | Anticancer/Immunosuppressant | SD[b] | Afinitor | Tablets | [71] |
| 05 | Glecaprevir[a]/Pibrentasvir[a] | Antiviral (HCV) | HME[b] | Maviret | Film-coated tablets | [72] |
| 06 | Lonafarnib | Hutchinson-Gilford progeria syndrome | SD | Zokinvy | Capsules | [73] |
| 07 | Ivacaftor | Cystic fibrosis | SD | Kalydeco | Tablets/granules | [74] |
| 08 | Lumacaftor/Ivacaftor[a] | Cystic fibrosis | SD | Orkambi | Tablets/granules | [75] |
| 09 | Ivacaftor[a]/Tezacaftor[a] | Cystic fibrosis | SD | Symkevi | Film-coated tablets | [76] |
| 10 | Ivacaftor[a]/Tezacaftor[a]/ Elexacaftor | Cystic fibrosis | SD | Kaftrio | Tablets/granules | [77] |
| 11 | Ledipasvir[a]/Sofosbuvir | Antiviral (HCV) | SD | Harvoni | Film-coated tablets | [78] |
| 12 | Pirtobrutinib | Anticancer | SD | Jaypirca | Film-coated tablets | [79] |
| 13 | Paclitaxel | Anticancer | HSH/NAB™ | Abraxane | Powder for dispersion | [80] |
| 14 | Posaconazole | Antifungal | HME | Noxafil | Tablets | [81] |
| 15 | Tolvaptan | Congestive Heart Failure | SD | Samsca | Tablets/capsules | [82] |
| 16 | Ritonavir | Antiviral (HIV) | HME | Norvir | Tablets | [83] |

*ASD* amorphous solid dispersion, *SD* spray drying, *HSH* high pressure homogenization, *NAB*™ nanoparticle albumin bound, *HME* hot melt extrusion *EPAR* European public assessment report

[a]In fixed dose combinations, asterisks denote the amorphous drug component

[b]ASD preparation method not reported but deducted based on the context

component is the only one requiring amorphization to improve solubility or bioavailability, which is the rationale for classifying the entire combination as an ASD product. The other components may remain crystalline, but they do not affect the classification of the amorphous drug in the ASD. From the 171 retrieved records of drug molecules in ASDs found in the Scopus database, 81 were filtered out, because the term “tablets” was not supported by experimental data (e.g., comparator’s marketed products, or reduntant use of the term ‘tablets’). From the 90 records left, 15 drug molecules were traced in EPARs that were progressed to commercial ASDs products.

From the analysis of the results presented in Table 9.2, it appears that EMA has granted license to 16 ASDs products representing 15 drug molecules. The products comprise different dosage forms including tablets (plain and film coated), capsules, powder for dispersion, and granules. The predominant form is tablets comprising 70% of the total final dosage forms. Most of the ASDs in the approved products are prepared by spray drying (75.0%).

The collected EMA data together with those of the FDA (from Moseson et al. [67]) in the last 2.5 decades are summarized in Table 9.3. In both sets of the collected results from the two agencies, unambiguously, spray drying and hot-melt extrusion make up the majority of ASD preparation methods, indicating the great role of solvent evaporation and heating/mechanical stresses, respectively, as the most effective mechanisms for ASD formation. Additionally, the vast majority of ASD product dosage forms are in tablet form, indicating that despite the low or moderate compactibility of the ASDs, the tablet remains their preferred final dosage form. This is due to the intrinsic compactibility of certain stabilizing polymers containing vinyl pyrrolidone group in their molecular structure, or to the functionality of the added tableting excipients, provided there are no stability and incompatibility issues [33].

Furthermore, from Table 9.3, it appears that the ASD products contain life-saving drugs, since they target life-threatening diseases such as cancer (25.0% and 41.3% for EMA and FDA respectively), viral infections (25.0% and 28.3%), or cystic fibrosis (25.0% and 10.9%). Therefore, most of the ASD products contain either new drug molecules or reformulated anticancer, antiviral, and immunosuppressant drugs, aiming for greater bioavailability, better plasma level control, and minimal losses of expensive materials by enhancement of the drugs’ solubility and dissolution profile. Regarding the cost of these medicines, many antiviral medications are considered high-cost specialized drugs, particularly the newer branded versions. While some older, generic antivirals are affordable, others, especially those for chronic infections like hepatitis C and HIV, have historically high and sometimes prohibitive costs. In this context, it is encouraging to point out that the leading manufacturing methods of ASDs, namely spray drying and hot melt extrusion, may contribute to the reduction of the high cost of these drug products, being continuous affordable methods.

**Table 9.3** Statistics of the selected drug molecules, ASD preparation method, final dosage form, and therapeutic areas of products that were approved by the EMA and FDA in the last two and a half decades

| Health authority | Drug molecules in ASDs (*n*) | ASD products (*n*) | Dosage forms (%) | | | | ASD method (%) | | | Therapeutic area (%) | | |
|---|---|---|---|---|---|---|---|---|---|---|---|---|
| | | | Tablets | Capsules | Granules | Other[a] | Spray drying | Hot melt extrusion | Other[b] | | EMA | FDA |
| EMA | 15 | 16 | 70.0 | 5 | 20.0 | 5.0 | 75.0 | 18.8 | 6.2 | Anticancer | 25.0 | 41.3 |
| | | | | | | | | | | Antiviral | 25.0 | 28.3 |
| FDA | 46 | 42 | 71.4 | 10.2 | 6.1 | 12.3 | 52.9 | 33.3 | 13.7 | Cystic fibrosis | 25.0 | 10.9 |
| | | | | | | | | | | Others[c] | 25.0 | 19.5 |

[a]EMA: powder for dispersion; FDA: pellets, powder for dispersion, gel
[b]EMA: high shear homogenization; FDA: wet or melt granulation, electrospraying
[c]EMA: treatment of psoriasis, Hutchinson-Gilford progeria syndrome, antifungal, congestive heart failure; FDA: fungal infections, insomnia, migraine, psoriasis, hyponatremia

## 9.5 Conclusions

ASDs have proved to be a vital formulation tool for reviving important and life-saving drugs, which, in the past, were discarded due to their very low solubility. This is accomplished by the mediating role of the polymer ASD coformers. They enhance dissolution performance and maintain drug supersaturation in vitro and in vivo, thus increasing absorption and bioavailability. Additionally, ASD structure provides physical stability of the drug amorphous form, which is a prerequisite for downstream processing of the amorphous drug into stable solid or occasionally semisolid dosage forms. Correct choice of polymeric carrier, suitable adjustment of the applied temperature and solvent type during spray drying and solvent evaporation methods, as well as correct temperature range and residence time during HME processing, are important for obtaining ASDs with desired features for processing into stable dosage forms. Similarly, control of operation parameters during the manufacturing of dosage forms (such as tensile and shear strength of the ASD powder, level of gap length between cylinders during roller compaction, flowability, bulk density, compression pressure, and dwell time during tableting) have to be adjusted for best results. The rapidly increasing number of marketing licenses granted by the EMA and the FDA guarantees that ASD-based products will be the future research topic for many years, transforming poorly soluble drugs into vital tools for lifesaving and health-promoting therapies.

## References

1. Chiou WL, Riegelman S. Pharmaceutical applications of solid dispersion systems. J Pharm Sci. 1971;60(9):1281–302.
2. Démuth B, Nagy ZK, Balogh A, Vigh T, Marosi G, Verreck G, et al. Downstream processing of polymer-based amorphous solid dispersions to generate tablet formulations. Int J Pharm. 2015;486(1–2):268–86.
3. Triboandas H, Pitt K, Bezerra M, Ach-Hubert D, Schlindwein W. Itraconazole amorphous solid dispersion tablets: formulation and compaction process optimization using quality by design principles and tools. Pharmaceutics. 2022;14(11):2398.
4. Agrawal A, Dudhedia M, Deng W, Shepard K, Zhong L, Povilaitis E, et al. Development of tablet formulation of amorphous solid dispersions prepared by hot melt extrusion using quality by design approach. AAPS PharmSciTech. 2016;17(1):214–32.
5. Patel S, Kou X, Hou HH, Huang YB, Strong JC, Zhang GGZ, et al. Mechanical properties and tableting behavior of amorphous solid dispersions. J Pharm Sci. 2017;106(1):217–23.
6. Ayenew Z, Paudel A, Rombaut P, Van den Mooter G. Effect of compression on non-isothermal crystallization behaviour of amorphous indomethacin. Pharm Res. 2012;29(9):2489–98.
7. Bērziņš K, Suryanarayanan R. Compression-induced crystallization in sucrose-polyvinylpyrrolidone amorphous solid dispersions. Cryst Growth Des. 2018;18(2):839–48.
8. Engelsing F, Buchart L, Grohganz H. Design of experiments approach on the compaction properties of co-amorphous tablets. Pharm Dev Technol. 2023;28(9):907–14.
9. Lenz E, Jensen KT, Blaabjerg LI, Knop K, Grohganz H, Löbmann K, et al. Solid-state properties and dissolution behaviour of tablets containing co-amorphous indomethacin-arginine. Eur J Pharm Biopharm. 2015;96:44–52.

10. Ojarinta R, Saarinen J, Strachan CJ, Korhonen O, Laitinen R. Preparation and characterization of multi-component tablets containing co-amorphous salts: combining multimodal non-linear optical imaging with established analytical methods. Eur J Pharm Biopharm. 2018;132:112–26.
11. Shi Q, Wang Y, Moinuddin SM, Feng X, Ahsan F. Co-amorphous drug delivery systems: a review of physical stability, in vitro and in vivo performance. AAPS PharmSciTech. 2022;23(7):259.
12. Sørensen C-M, Rantanen J, Grohganz H. Compaction behavior of co-amorphous systems. Pharmaceutics. 2023;15(3):858.
13. Dedroog S, Huygens C, Van den Mooter G. Chemically identical but physically different: a comparison of spray drying, hot melt extrusion and cryo-milling for the formulation of high drug loaded amorphous solid dispersions of naproxen. Eur J Pharm Biopharm. 2019;135:1–12.
14. Takeuchi H, Handa T, Kawashima Y. Spherical solid dispersion containing amorphous tolbutamide embedded in enteric coating polymers or colloidal silica prepared by spray-drying technique. Chem Pharm Bull (Tokyo). 1987;35(9):3800–6.
15. Miller DA, DiNunzio JC, Hughey JR, Williams R, McGinity JW. KinetiSol: a new processing paradigm for amorphous solid dispersion systems. Drug Dev Deliv. 2012;12(9):1749–68.
16. Kerc J, Srcic S, Kofler B. Alternative solvent-free preparation methods for felodipine surface solid dispersions. Drug Dev Ind Pharm. 1998;24(4):359–63.
17. Qiang W, Löbmann K, McCoy CP, Andrews GP, Zhao M. Microwave-induced in situ amorphization: a new strategy for tackling the stability issue of amorphous solid dispersions. Pharmaceutics. 2020;12(7):655.
18. Grymonpré W, Bostijn N, Herck SV, Verstraete G, Vanhoorne V, Nuhn L, et al. Downstream processing from hot-melt extrusion towards tablets: a quality by design approach. Int J Pharm. 2017;531(1):235–45.
19. Mishra SM, Richter M, Mejia L, Sauer A. Downstream processing of itraconazole:HPMCAS amorphous solid dispersion: from hot-melt extrudate to tablet using a quality by design approach. Pharmaceutics. 2022;14(7):1429.
20. Doktorovová S, Stone EH, Henriques J. A fundamental study on compression properties and strain rate sensitivity of spray-dried amorphous solid dispersions. AAPS PharmSciTech. 2022;23(4):96.
21. Zhang W, Sluga KK, Yost E, Phan J, Nagapudi K, Helen HH. Impact of drug loading on the compaction properties of itraconazole-PVPVA amorphous solid dispersions. Int J Pharm. 2022;629:122366.
22. Ekdahl A, Mudie D, Malewski D, Amidon G, Goodwin A. Effect of spray-dried particle morphology on mechanical and flow properties of felodipine in PVP VA amorphous solid dispersions. J Pharm Sci. 2019;108(11):3657–66.
23. Osei-Yeboah F, Sun CC. Effect of drug loading and relative humidity on the mechanical properties and tableting performance of celecoxib–PVP/VA 64 amorphous solid dispersions. Int J Pharm. 2023;644:123337.
24. Mahmah O, Tabbakh R, Kelly A, Paradkar A. A comparative study of the effect of spray drying and hot-melt extrusion on the properties of amorphous solid dispersions containing felodipine. J Pharm Pharmacol. 2014;66(2):275–84.
25. Davis MT, Potter CB, Walker GM. Downstream processing of a ternary amorphous solid dispersion: the impacts of spray drying and hot melt extrusion on powder flow, compression and dissolution. Int J Pharm. 2018;544(1):242–53.
26. Schönfeld BV, Westedt U, Wagner KG. Compression of amorphous solid dispersions prepared by hot-melt extrusion, spray drying and vacuum drum drying. Int J Pharm: X. 2021;3:100102.
27. Henriques J, Moreira J, Doktorovová S. QbD approach to downstream processing of spray-dried amorphous solid dispersions — a case study. Pharm Dev Technol. 2021;26(3):269–77.
28. Song S, Wang C, Wang S, Siegel RA, Sun CC. Efficient development of sorafenib tablets with improved oral bioavailability enabled by coprecipitated amorphous solid dispersion. Int J Pharm. 2021;610:121216.

29. Jia W, Yawman PD, Pandya KM, Sluga K, Ng T, Kou D, et al. Assessing the interrelationship of microstructure, properties, drug release performance, and preparation process for amorphous solid dispersions via noninvasive imaging analytics and material characterization. Pharm Res. 2022;39(12):3137–54.
30. Trenkenschuh E, Blattner SM, Hirsh D, Hoffmann R, Luebbert C, Schaefer K. Development of ternary amorphous solid dispersions manufactured by hot-melt extrusion and spray-drying—comparison of in vitro and in vivo performance. Mol Pharm. 2024;21(3):1309–20.
31. Rassu G, Eissens AC, Bolhuis GK. Tableting properties of an improved spray-dried lactose. J Drug Deliv Sci Technol. 2006;16(6):455–9.
32. Casian T, Borbás E, Ilyés K, Démuth B, Farkas A, Rapi Z, et al. Electrospun amorphous solid dispersions of meloxicam: influence of polymer type and downstream processing to orodispersible dosage forms. Int J Pharm. 2019;569:118593.
33. Browne E, Worku ZA, Healy AM. Physicochemical properties of poly-vinyl polymers and their influence on ketoprofen amorphous solid dispersion performance: a polymer selection case study. Pharmaceutics. 2020;12(5):433.
34. Zhang J, Thakkar R, Zhang Y, Maniruzzaman M. Microwave induced dielectric heating for the on-demand development of indomethacin amorphous solid dispersion tablets. J Drug Deliv Sci Technol. 2021;61:102109.
35. Butreddy A, Sarabu S, Bandari S, Batra A, Lawal K, Chen NN, et al. Influence of Plasdone(TM) S630 ultra-an improved Copovidone on the processability and oxidative degradation of quetiapine fumarate amorphous solid dispersions prepared via hot-melt extrusion technique. AAPS PharmSciTech. 2021;22(5):196.
36. Saraf I, Jakasanovski O, Stanić T, Kralj E, Petek B, Williams JD, et al. Investigation of the influence of copovidone properties and hot-melt extrusion process on level of impurities, in vitro release, and stability of an amorphous solid dispersion product. Mol Pharm. 2024;21(11):5703–15.
37. Langham ZA, Booth J, Hughes LP, Reynolds GK, Wren SAC. Mechanistic insights into the dissolution of spray-dried amorphous solid dispersions. J Pharm Sci. 2012;101(8):2798–810.
38. Démuth B, Farkas A, Balogh A, Bartosiewicz K, Kállai-Szabó B, Bertels J, et al. Lubricant-induced crystallization of itraconazole from tablets made of electrospun amorphous solid dispersion. J Pharm Sci. 2016;105(9):2982–8.
39. Lu Y, Chen J, Yi S, Xiong S. Enhanced felodipine dissolution from high drug loading amorphous solid dispersions with PVP/VA and sodium dodecyl sulfate. J Drug Deliv Sci Technol. 2019;53:101151.
40. Chiang CW, Lubach JW, Chen T, Chin S, Ly J, Zhang W, et al. Development of an amorphous solid dispersion formulation for mitigating mechanical instability of crystalline form and improving bioavailability for early phase clinical studies. Mol Pharm. 2023;20(5):2452–64.
41. Chae JS, Chae BR, Shin DJ, Goo YT, Lee ES, Yoon HY, et al. Tablet formulation of a polymeric solid dispersion containing amorphous alkalinized telmisartan. AAPS PharmSciTech. 2018;19(7):2990–9.
42. Takano R, Maurer R, Jacob L, Stowasser F, Stillhart C, Page S. Formulating amorphous solid dispersions: impact of inorganic salts on drug release from tablets containing itraconazole-HPMC extrudate. Mol Pharm. 2020;17(8):2768–78.
43. Xi H, Ren J, Novak JM, Kemp E, Johnson G, Klinzing G, et al. The effect of inorganic salt on disintegration of tablets with high loading of amorphous solid dispersion containing copovidone. Pharm Res. 2020;37(4):70.
44. Mudie DM, Stewart AM, Rosales JA, Biswas N, Adam MS, Smith A, et al. Amorphous solid dispersion tablets overcome acalabrutinib pH effect in dogs. Pharmaceutics. 2021;13(4):557.
45. Hanada M, Jermain SV, Thompson SA, Furuta H, Fukuda M, Williams RO III. Ternary amorphous solid dispersions containing a high-viscosity polymer and mesoporous silica enhance dissolution performance. Mol Pharm. 2021;18(1):198–213.

46. Giri BR, Kwon J, Vo AQ, Bhagurkar AM, Bandari S, Kim DW. Hot-melt extruded amorphous solid dispersion for solubility, stability, and bioavailability enhancement of telmisartan. Pharmaceuticals (Basel). 2021;14(1)
47. Yu D, Fiddler F, Ibrahim A, Sanedrin R, Tremblay H, Hoag SW. Surface characterization as a tool for identifying the factors affecting the dissolution rate of amorphous solid dispersion tablets. AAPS PharmSciTech. 2022;23(8):282.
48. Friesen DT, Shanker R, Crew M, Smithey DT, Curatolo WJ, Nightingale JA. Hydroxypropyl methylcellulose acetate succinate-based spray-dried dispersions: an overview. Mol Pharm. 2008;5(6):1003–19.
49. Zhang W, Thool P, Weitz BW, Hou HH. Investigating the effects of formulation variables on the disintegration of spray dried amorphous solid dispersion tablets. J Pharm Sci. 2025;114(1):304–12.
50. Bochmann ES, Steidel A, Rosenblatt KM, Gessner D, Liepold B. Assessment of the amorphous solid dispersion erosion behavior following a novel small-scale predictive approach. Eur J Pharm Sci. 2021;158:105682.
51. Boel E, Giacomini F, Van den Mooter G. Solvent influence on manufacturability, phase behavior and morphology of amorphous solid dispersions prepared via bead coating. Eur J Pharm Biopharm. 2021;167:175–88.
52. Autzen Virtanen A, Myślińska M, Healy AM, Power E, Madi A, Sivén M. The challenge of downstream processing of spray dried amorphous solid dispersions into minitablets designed for the paediatric population — a sustainable product development approach. Eur J Pharm Sci. 2024;196:106752.
53. Kokott M, Lura A, Breitkreutz J, Wiedey R. Evaluation of two novel co-processed excipients for direct compression of orodispersible tablets and mini-tablets. Eur J Pharm Biopharm. 2021;168:122–30.
54. Nguyen HT, Van Duong T, Taylor LS. Enteric coating of tablets containing an amorphous solid dispersion of an enteric polymer and a weakly basic drug: a strategy to enhance in vitro release. Int J Pharm. 2023;642:123139.
55. Ellenberger DJ, Miller DA, Kucera SU, Williams RO 3rd. Generation of a weakly acidic amorphous solid dispersion of the weak base ritonavir with equivalent in vitro and in vivo performance to Norvir tablet. AAPS PharmSciTech. 2018;19(5):1985–97.
56. Démuth B, Farkas A, Szabó B, Balogh A, Nagy B, Vágó E, et al. Development and tableting of directly compressible powder from electrospun nanofibrous amorphous solid dispersion. Adv Powder Technol. 2017;28(6):1554–63.
57. Ritters L, Tian Y, Reichl S. Spray-dried paracetamol/polyvinylpyrrolidone amorphous solid dispersions: part I — stability of powders and tablets. Pharmaceutics. 2021;13(11):1938.
58. Xiang T, Song S, Siegel RA, Sun CC. Efficient development of high drug loaded posaconazole tablets enabled by amorphous solid dispersion. RSC Pharmaceutics. 2025;2(1):178–85.
59. Yu D, Hoag SW. The impact of diluents on the compaction, dissolution, and physical stability of amorphous solid dispersion tablets. Int J Pharm. 2024;654:123924.
60. Nowak P, Krupa A, Kubat K, Węgrzyn A, Harańczyk H, Ciułkowska A, et al. Water vapour sorption in tadalafil-Soluplus co-milled amorphous solid dispersions. Powder Technol. 2019;346:373–84.
61. Dohrn S, Luebbert C, Lehmkemper K, Kyeremateng SO, Degenhardt M, Sadowski G. Phase behavior of pharmaceutically relevant polymer/solvent mixtures. Int J Pharm. 2020;577:119065.
62. Neuwirth M, Kappes SK, Hartig MU, Wagner KG. Amorphous solid dispersions layered onto pellets — an alternative to spray drying? Pharmaceutics. 2023;15(3):764.
63. Punia A, Biyyala V, Faassen F, Ash J, Lamm MS. Detrimental effect of the film coat chemistry and thickness on the physical stability of amorphous solid dispersions in tablet formulations. J Pharm Sci. 2023;112(3):708–17.

64. Chatterjee K, Punia A, Confer AM, Lamm MS. Understanding the effect of plasticizers in film coat materials on the physical stability of amorphous solid dispersions. J Pharm Sci. 2025;114(1):468–76.
65. Neilly JP, Yin L, Leonard S-E, Kenis PJA, Danzer GD, Pawate AS. Quantitative measures of crystalline fenofibrate in amorphous solid dispersion formulations by X-ray microscopy. J Pharm Sci. 2020;109(10):3078–85.
66. Tambe S, Jain D, Meruva SK, Rongala G, Juluri A, Nihalani G, et al. Recent advances in amorphous solid dispersions: preformulation, formulation strategies, technological advancements and characterization. Pharmaceutics. 2022;14(10):2203.
67. Moseson DE, Tran TB, Karunakaran B, Ambardekar R, Hiew TN. Trends in amorphous solid dispersion drug products approved by the U.S. Food and Drug Administration between 2012 and 2023. Int J Pharm X. 2024;7:100259.
68. EMA/68815/2023, (CHMP) CfMPfHU. Assessment report, Sotyktu, International non-proprietary name: deucravacitinib, Procedure No. EMEA/H/C/005755/0000 2023. Available from: https://www.ema.europa.eu/en/documents/assessment-report/sotyktu-epar-public-assessment-report_en.pdf
69. EMA/CHMP/432249/2017,(CHMP)CfMPfHU.CHMPassessmentreportonextension(s)ofmarketing authorisation, Xtandi, International non-proprietary name: enzalutamide, Procedure No. EMEA/H/C/002639/X/0029 2017. Available from: https://www.ema.europa.eu/en/documents/variation-report/xtandi-h-c-002639-x-0029-epar-assessment-report-variation_en.pdf
70. EMEA/CHMP/43952/2008, (CHMP) CfMPfHU. CHMP ASSESSMENT REPORT FOR INTELENCE, International Nonproprietary Name: etravirine, Procedure No. EMEA/H/C/000900 2008. Available from: https://www.ema.europa.eu/en/documents/assessment-report/intelence-epar-public-assessment-report_en.pdf
71. EMEA/533232/2009, (CHMP) CfMPfHU. CHMP ASSESSMENT REPORT FOR Afinitor, International Nonproprietary Name: everolimus Procedure No. EMEA/H/C/001038 2009. Available from: https://www.ema.europa.eu/en/documents/assessment-report/afinitor-epar-public-assessment-report_en.pdf
72. EMA/449689/2017, (CHMP) CfMPfHU. Assessment report Maviret, International non-proprietary name: glecaprevir / pibrentasvir, Procedure No. EMEA/H/C/004430/0000 2017. Available from: https://www.ema.europa.eu/en/documents/assessment-report/maviret-epar-public-assessment-report_en.pdf.
73. Rev.1 E, (CHMP) CfMPfHU. Assessment report Zokinvy, International non-proprietary name: lonafarnib, Procedure No. EMEA/H/C/005271/0000 2022. Available from: https://www.ema.europa.eu/en/documents/assessment-report/zokinvy-epar-public-assessment-report_en.pdf
74. EMA/473279/2012, (CHMP) CfMPfHU. Assessment report Kalydeco, ivacaftor, Procedure No.: EMEA/H/C/002494//0000 2012. Available from: https://www.ema.europa.eu/en/documents/assessment-report/kalydeco-epar-public-assessment-report_en.pdf
75. EMA/843650/2018, (CHMP) CfMPfHU. Assessment report Orkambi, International non-proprietary name: lumacaftor / ivacaftor, Procedure No. EMEA/H/C/003954/X/0034/G 2018. Available from: https://www.ema.europa.eu/en/documents/variation-report/orkambi-h-c-3954-x-0034-g-epar-assessment-report-variation_en.pdf
76. EMA/CHMP/567306/2018, (CHMP) CfMPfHU. Assessment report Symkevi, International non-proprietary name: tezacaftor / ivacaftor, Procedure No. EMEA/H/C/004682/0000 2018. Available from: https://www.ema.europa.eu/en/documents/assessment-report/symkevi-epar-public-assessment-report_en.pdf
77. Rev.1 E, (CHMP) CfMPfHU. Assessment report Kaftrio, International non-proprietary name: ivacaftor / tezacaftor / elexacaftor, Procedure No. EMEA/H/C/005269/0000 2020. Available from: https://www.ema.europa.eu/en/documents/assessment-report/kaftrio-epar-public-assessment-report_en.pdf
78. EMA/702742/2014, (CHMP) CfMPfHU. Assessment report Harvoni, International non-proprietary name: ledipasvir / sofosbuvir, Procedure No. EMEA/H/C/003850/0000

2014. Available from: https://www.ema.europa.eu/en/documents/assessment-report/harvoni-epar-public-assessment-report_en.pdf
79. EMA/308908/2023, (CHMP) CfMPfHU. Assessment report Jaypirca, International non-proprietary name: pirtobrutinib, Procedure No. EMEA/H/C/005863/0000 2023. Available from: https://www.ema.europa.eu/en/documents/assessment-report/jaypirca-epar-public-assessment-report_en.pdf
80. EMA/116095/2020, (CHMP) CfMPfHU. Assessment report Abraxane, International non-proprietary name: paclitaxel, Procedure No. EMEA/H/C/000778/II/0097 2020. Available from: https://www.ema.europa.eu/en/documents/variation-report/abraxane-h-c-778-ii-0097-epar-assessment-report-variation_en.pdf
81. EMA/159150/2014, (CHMP) CfMPfHU. Assessment report Noxafil, International non-proprietary name: POSACONAZOLE, Procedure No. EMEA/H/C/000610/X/0028 2014. Available from: https://www.ema.europa.eu/en/documents/variation-report/noxafil-h-c-610-x-0028-epar-scientific-discussion-extension_en.pdf
82. EMEA/502935/2009, (CHMP) CfMPfHU. CHMP ASSESSMENT REPORT FOR Samsca, International Nonproprietary Name: tolvaptan, Procedure No. EMEA/H/C/000980 2009. Available from: https://www.ema.europa.eu/en/documents/assessment-report/samsca-epar-public-assessment-report_en.pdf
83. EMADOC-1700519818-2433345, (CHMP) CfMPfHU. Assessment report, Invented name: Norvir, International non-proprietary name: Ritonavir, Procedure No. EMA/VR/0000249795 2025. Available from: https://www.ema.europa.eu/en/documents/variation-report/norvir-vr-0000249795-epar-assessment-report-variation_en.pdf

# Chapter 10
# Improving Paediatric Drug Delivery Through Amorphous Solid Dispersions (ASDs)

**Konstantina Chachlioutaki, Melpomeni Liogka, and Dimitrios G. Fatouros**

**Abstract** Amorphous solid dispersions (ASDs) are widely used to improve the solubility and oral bioavailability of poorly water-soluble drugs. Although this approach has been successfully applied in many adult medicines, its potential in paediatric drug development has only recently attracted greater attention. Formulating medicines for children presents additional challenges related to dose flexibility, physiological and pharmacokinetic differences across age groups, excipient safety, and patient acceptability. In clinical practice, the limited availability of licensed paediatric products often leads to the manipulation of adult dosage forms, which may compromise dose accuracy, stability, and palatability. These factors frequently limit the performance of conventional dosage forms and highlight the need for formulation strategies capable of enhancing drug performance and therapeutic efficacy whilst enabling age-appropriate administration. The role of ASDs in addressing these challenges is therefore of increasing interest in paediatric pharmacotherapy. The chapter first outlines the key constraints associated with paediatric drug development and subsequently discusses the contribution of amorphous solid dispersions to advancing formulation strategies for paediatric medicines. Manufacturing approaches commonly used to produce ASDs for paediatric drug formulations, including spray drying and hot melt extrusion, are reviewed together with emerging technologies that may further expand formulation possibilities. Formulation aspects particularly relevant to paediatric medicines, including polymer selection, stability of the amorphous form, and strategies to improve palatability, are also discussed. Finally, clinically relevant ASD formulations and current regulatory considerations are presented, highlighting the growing influence of this technology in the development of safer and more acceptable medicines for children.

K. Chachlioutaki (✉) · M. Liogka · D. G. Fatouros
Laboratory of Pharmaceutical Technology, School of Pharmacy, Faculty of Health Sciences, Aristotle University of Thessaloniki, Thessaloniki, Greece

Center for Interdisciplinary Research and Innovation (CIRI-AUTH), Thessaloniki, Greece
e-mail: kchachlio@pharm.auth.gr; mliogka@pharm.auth.gr; dfatouro@pharm.auth.gr



I. Nikolakakis, I. Partheniadis (eds.), *Amorphous Solid Dispersions*, AAPS Introductions in the Pharmaceutical Sciences 5,
https://doi.org/10.1007/978-3-032-24883-1_10

**Keywords** Poorly water-soluble drugs · Child-appropriate dosage forms · Palatability · Acceptability · Dose accuracy

## 10.1 Introduction

Amorphous solid dispersions (ASDs) were first introduced in the late 1960s as a strategy to improve the dissolution and oral bioavailability of poorly water-soluble drugs [1]. The concept involves dispersing an active pharmaceutical ingredient, most of the times a poorly water-soluble drug, in an inert polymeric carrier in its amorphous (non-crystalline) state. In an ASD, the drug and excipient form a homogeneous single-phase system, meaning the drug's molecules are "dissolved" in a solid matrix rather than arranged in a crystal lattice [2]. This lack of crystalline structure endows the drug with higher apparent solubility and faster dissolution rates than its crystalline form [2]. In ASDs, the polymer/carrier not only converts the drug to its amorphous form but also improves wettability and maintains supersaturation, ultimately leading to better oral absorption [2].

Over the past two decades, the concept of solid dispersions has acquired significant importance as a strategy for enhancing the bioavailability of poorly water-soluble drugs [2, 3]. It is estimated that 70–90% of new chemical entities are classified as poorly soluble, specifically categorised as BCS Class II or IV [3], which presents considerable challenges for the development of oral formulations [2]. ASDs have emerged as one of the most effective approaches to overcome poor drug solubility, with numerous marketed drug products now utilising ASD technology. Between 2012 and 2023, the U.S. Food and Drug Administration (FDA) approved 48 drug products incorporating ASDs, encompassing 36 unique drugs across various therapeutic categories [4]. This corresponds to an average of four approvals per year, highlighting the steady acceptance of ASDs within the pharmaceutical industry [4]. Notably, the most frequently used polymers in these products are copovidone (polyvinylpyrrolidone–vinyl acetate copolymer, PVP/VA) and hypromellose acetate succinate (HPMCAS), whilst spray drying (~54%) and hot melt extrusion (~35%) dominate as the manufacturing methods [4]. Several ASD-enabled products, including itraconazole, posaconazole, and lumacaftor/ivacaftor, have demonstrated enhanced clinical efficacy as a direct result of improved solubility and absorption.

Although most initial ASD products were developed for adults, there is a growing interest in extending this technology to paediatric drug formulation. Recent approvals have included child-friendly dosage forms, such as granules or powders for suspension, as line extensions to adult tablets, reflecting a move towards patient-centric dosing strategies for children [4]. Building on their proven success in adults, ASDs are increasingly being explored as a promising solution to the specific challenges of paediatric drug formulation. Children, like adults, are affected by conditions requiring poorly soluble drugs (e.g. fungal infections, HIV, epilepsy, transplant rejection, cystic fibrosis), but their treatment is complicated by additional formulation barriers. ASDs address these issues by enhancing solubility and dissolution without the need for potentially harmful solubilising excipients or extreme pH modifications, an approach of particular value in paediatric therapy. The following sections will outline the special considerations of paediatric formulations, the role of ASDs in paediatric therapy, the selection of appropriate preparation methods, the selection of child-appropriate excipients, regulatory aspects concerning ASDs in paediatric medicine, current applications, and future directions.

## 10.2 Challenges of Paediatric Drug Development

Formulating medicines for paediatric patients requires careful attention to their unique needs and limitations [5, 6]. It is often stated that "children are not just small adults." This principle underscores the imperative that children should benefit from advances in pharmaceutical science through age-appropriate formulations rather than being treated with scaled-down versions of adult medicines. Extemporaneous formulations, also known as compounding, remain common in clinical practice, particularly when licensed paediatric products are unavailable. These are typically prepared by modifying adult dosage forms in pharmacies or hospitals, for instance, by crushing tablets, dispersing them in water, and adding excipients to create liquids. However, such preparations are often suboptimal [7]. They may lack validated stability data, suffer from poor palatability, and pose risks of inaccurate dosing. Additionally, liquid formulations derived from adult products frequently require antimicrobial preservatives and viscosity modifiers to maintain usability, which may introduce excipients that are inappropriate or potentially harmful in paediatric use [8].These limitations highlight the urgent need for medicines specifically designed and tested for paediatric use.

Differences in physiology, pharmacokinetics, and patient acceptability necessitate the design of paediatric formulations tailored to specific age groups [9] (Fig. 10.1, Table 10.1). A further major challenge is dose flexibility and accuracy. Paediatric dosing typically must be individualised according to age, body weight, or surface area [5]. Achieving this precision often requires multiple strengths or adaptable dosage forms, unlike the "one-size" adult tablet [9]. Formulations such as oral liquids [10], orodispersible tablets [10, 11], mini-tablets [12–15], and multiparticulates [14] allow caregivers to measure and administer doses tailored to a child's needs. For example oral syringes can be used with liquids to deliver small, accurate doses, whilst mini-tablets can be counted out to match a specific weight band. This inherent complexity in dose adjustment distinguishes paediatric therapy fundamentally from adult treatment [9].

Swallowability and palatability represent additional, equally important barriers [17]. Children, especially infants and toddlers, are often unable to swallow conventional tablets or capsules and may even choke on large solid dosage forms. Therefore, age-appropriate dosage forms [5], such as oral liquids, orally dispersible tablets (ODTs that dissolve in the mouth), powders or granules, chewable tablets are

**Fig. 10.1** Age classification system due to ontogenic changes that occur during paediatric development [6, 16]

**Table 10.1** Age classification system illustrating key ontogenic changes during paediatric development, including fluid volumes, enzyme activity, gastrointestinal motility, intestinal permeability, and diet composition [6, 16]

| Parameters | 0–1 months | 12–24 months | 2–4 years | 5–11 years | 11–18 years |
|---|---|---|---|---|---|
| Fluid volumes | Low volume, high pH | Increasing volume & acidity | Near adult volumes/ acidity | Adult like | Adult like |
| Enzyme activity | Limited (immature liver, pancreas) | Increasing but still low | Approaching adult levels | Nearly adult | Adult-level activity |
| GI motility | Slow transit | Increasing transit speed | Near adult motility | Adult like | Adult like |
| Intestinal permeability | High ("leaky gut") | Decreasing permeability | Approaching adult | Adult like | Adult like |
| Food components | Milk based | Milk transitioning to solids | Increasingly varied | Varied, diverse | Adult-like diet & variability |

required. Even amongst older children capable of swallowing pills, taste plays a pivotal role in compliance. Bitter or unpleasant-tasting medicines are often refused or expelled by children, leading to poor adherence and treatment failure [18]. In one survey, taste was identified by paediatric care providers as one of the biggest hurdles to medication compliance [19]. Consequently, taste-masking strategies, such as the use of sweeteners, flavours, coatings, or microencapsulation, are integral to paediatric formulation design.

Another critical issue is the safety of excipients. Many excipients commonly used in adult formulations (such as certain preservatives, solvents, and surfactants) may be unsuitable or even harmful for children, especially neonates and infants, due to immature metabolic and detoxification pathways. For instance excipients like benzyl alcohol, propylene glycol, or certain surfactants have known toxicities in neonates (e.g. "gasping syndrome" from benzyl alcohol) and must be avoided or strictly limited [20–23]. Thus, the selection of excipients for paediatric use is a complex task that must account for developmental differences in metabolism and organ function. As children's bodies are still developing, they may be more sensitive to certain excipients. Thus, formulation scientists must ensure that all ingredients in a paediatric medicine are safe for the intended age group and used within acceptable limits [21, 24]. In practice, this often means using only excipients with a history of safe use in children from paediatric formulations or even the food industry [25] and conducting additional toxicity studies if a novel excipient is required [21]. Regulatory agencies actively encourage the use of resources like the STEP database (Safety and Toxicity of Excipients for Paediatrics) to identify excipients generally regarded as safe for children and to identify those with potential risks [21]. For example the European Medicines Agency recommends avoiding excipients known to have carcinogenic or genotoxic potential in paediatric products. In such cases, juvenile animal studies may be recommended to assess excipient safety in very young age groups when existing data are insufficient [21].

Practical considerations also influence paediatric formulation development. Medicines for children should minimise treatment burden by reducing dosing frequency, facilitating reliable administration, and avoiding disruption of daily routines. An optimal paediatric formulation would minimise dosing frequency (to reduce the burden on the child and caregiver) and be easy to administer reliably (e.g. a once-daily, pleasant-tasting liquid) [4]. It should also have a robust shelf life and avoid complex handling or storage requirements that increase the risk of errors. Given the wide paediatric age range, from preterm neonates to adolescents, a single formulation rarely meets all needs; instead, multiple age-appropriate formulations may be required [4].

Historically, the lack of paediatric-friendly options has led to widespread off-label use of adult medicines in children [26], with caregivers or healthcare providers having to manipulate adult dosage forms [4]. Such practices can result in dosing inaccuracies, stability issues, and safety concerns. For example a multicentre European study reported that nearly half of paediatric prescriptions (46%) were off-label or unlicensed [26], and 67% of hospitalised children received at least one such medicine during their stay [27]. These practices raise concerns about efficacy, safety, and quality control, underscoring the urgent need for licensed paediatric medicines [27]. Moreover, the limited commercial profitability of paediatric markets reduces industry incentives to invest in child-specific drug products, despite legislative frameworks such as the EU Paediatric Regulation and the FDA Paediatric Research Equity Act designed to stimulate development [28].

In summary, paediatric drug formulation must balance accurate dosing, acceptable dosage form (swallowability and taste), excipient safety, and practical administration [9]. Each of these challenges must be addressed to achieve sufficient bioavailability and adherence in children. Innovative formulation technologies, including amorphous solid dispersions (ASDs), offer promising solutions to several of these hurdles by enhancing drug solubility, enabling flexible dosage forms, and using excipients with established safety profiles.

## 10.3 Role of ASDs in Enhancing Paediatric Drug Bioavailability

One key rationale for employing amorphous solid dispersions in paediatric formulations is their proven ability to enhance the oral bioavailability of poorly water-soluble drugs. Poor solubility is a common issue for many modern drugs, and it can be particularly problematic in children, where the absolute doses are smaller and the volume of administration is limited. If a child cannot swallow a large tablet, the dose may need to be given as a diluted suspension or solution. However, a poorly soluble drug may not dissolve sufficiently in a small volume of fluid, resulting in sub-therapeutic exposure. By dispersing the drug molecularly within a polymer matrix, ASDs increase the apparent solubility and dissolution rate of the drug [29], enabling therapeutic levels to be achieved within the small volumes typical of paediatric administration.

### *10.3.1 Why ASDs Map Well To Paediatric Physiology?*

Paediatric patients present several physiological differences that can further compromise oral absorption. Neonates and infants have higher gastric pH, which reduces the solubility of weakly basic drugs. They also exhibit lower bile salt levels and sometimes shorter intestinal transit times, which can limit the dissolution and absorption of lipophilic compounds [30, 31]. In these scenarios, ASDs are advantageous, because they present the drug in a high-energy, amorphous form that is less dependent on GI conditions for dissolution.

A hallmark feature of ASDs is their ability to generate and stabilise a supersaturated solution. Upon dissolution, the amorphous drug rapidly reaches high concentrations ("the spring"), whilst the polymer inhibits recrystallisation, prolonging supersaturation ("the parachute") [32–34]. This mechanism is particularly relevant in paediatrics, where limited GI fluid volume and rapid gastric emptying may otherwise prevent adequate dissolution. In some cases, ASDs also yield colloidal nanoparticles that remain readily absorbable with the aid of polymers and surfactants [33].

### *10.3.2 Mechanistic Levers Relevant To Children*

Two formulation levers are especially pertinent for paediatrics:

1. **Polymer selection and drug–polymer interactions**. Enteric and amphiphilic polymers (e.g. HPMCAS, copovidone, Soluplus®) can both promote supersaturation and retard precipitation via hydrogen bonding or hydrophobic interactions.

Selection based on thermodynamic/kinetic miscibility and quantitative tools (e.g. Hansen solubility parameters) improves the likelihood that supersaturation will be sustained within paediatric GI fluids [35].

2. **Drug load versus performance.** Lower drug loads (higher polymer fractions) often extend supersaturation and increase flux across membranes, a trade-off that is usually acceptable for paediatric doses, which are smaller and can accommodate more polymer without excessive mass/volume. Whilst high-drug-load ASDs are possible when needed, paediatric scenarios commonly benefit from polymer-rich matrices that maximise stability [36].

### *10.3.3 Translation into Paediatric-Relevant Dosage Forms*

ASDs can be downstream-processed into mini-tablets, sprinkle granules/sachets, ODTs, or small-volume reconstitutable powders aligning with paediatric swallowability and dose-flexibility needs. Process choices also influence downstream handling which are critical when filling low-mass unit doses for children [12].

### *10.3.4 Representative Examples with Paediatric Relevance*

There are numerous examples where ASDs have significantly boosted the bioavailability of a drug, and these benefits are being translated to paediatric use. A recent case study highlighted the antifungal drug terbinafine hydrochloride (TER), which is effective for treating scalp fungal infections (tinea capitis) common in children but has very low aqueous solubility (TER is classified as BCS Class II). A rational, child-centred screen identified Soluplus®, HPMCAS, and copovidone as suitable carriers for an ASD of terbinafine HCl (BCS II) [37]. Spray-dried Soluplus ASDs delivered the lowest dose number and maintained prolonged supersaturation, indicating feasibility to meet paediatric dose requirements within small GI volumes, an advantage over crystalline terbinafine.

Another example can be seen with the antiretroviral drug nevirapine (NVP), widely used in paediatric HIV treatment. Nevirapine is BCS Class II (poor solubility, good permeability) with an aqueous solubility on the order of ~0.1 mg/mL at neutral pH [38]. In its crystalline form, it can be challenging to formulate into a palatable liquid of reasonable volume for children. Researchers prepared an ASD of nevirapine with PVP (polyvinylpyrrolidone) using a solvent evaporation method and found that all tested ASD formulations significantly increased the aqueous solubility of nevirapine compared to the pure drug [38]. Lower drug loads ($\approx$10% w/w NVP) produced the strongest supersaturation and highest permeation, illustrating the benefit of polymer-rich matrices for paediatric doses. This type of improvement means that a paediatric dose of nevirapine could be delivered in a much smaller volume or at a higher concentration of a liquid, or perhaps in solid granules, thereby improving compliance and efficacy. This underscores a key formulation principle: increasing drug loading often reduces ASD stability and dissolution performance. Because paediatric doses are smaller, a higher polymer-to-drug ratio is usually feasible and can improve both physical stability and apparent solubility [38].

Another example for the successful application of ASD technology is posaconazole (PCL), an antifungal agent used in the prophylaxis and treatment of invasive fungal infections. The marketed delayed-release (DR) tablets are based on an HPMCAS amorphous solid dispersion manufactured by hot-melt extrusion [39]. This design enables pH-dependent release, with limited drug release in gastric acid and enhanced dissolution in the intestine, resulting in higher and more consistent systemic exposure compared with the earlier oral suspension. Recent paediatric studies have demonstrated that DR tablets can achieve therapeutic drug levels in children when guided by therapeutic drug monitoring (TDM), supporting their clinical translatability [40]. In practice, DR tablets are increasingly preferred in paediatric patients who can swallow solid dosage forms, as they overcome the variability associated with suspensions and reduce the overall treatment burden.

Overall, ASDs enhance paediatric drug bioavailability by achieving higher concentrations in the GI tract from smaller, more palatable dosage forms. This is crucial for children, who often require low-dose, high-potency medicines in formats such as mini-tablets or sprinkle granules rather than large, unpalatable suspensions. The polymer matrix not only improves solubility but can also help stabilize drug release, reducing sharp peaks and sustaining levels. In summary, ASDs represent a powerful tool for developing child-appropriate formulations of poorly soluble drugs.

## 10.4 Manufacturing Methods of ASDs in Paediatrics

Several techniques are available for preparing amorphous solid dispersions, each with specific advantages and limitations. Method selection depends on the drug's properties, the polymer matrix, and practical considerations for scale-up and dosage form. In paediatrics, the chosen process must also enable small, flexible formats: for example powders for sachets and extruded strands for mini-tablets. Amongst the available methods, spray drying (SD) and hot-melt extrusion (HME) are the most widely used, whilst solvent evaporation, freeze-drying, co-precipitation, electrospinning, KinetiSol®, and 3D printing are emerging alternatives. The following subsections review these approaches with a focus on their paediatric relevance.

### *10.4.1 Spray Drying (SD)*

Spray drying is a solvent-based technique extensively utilized for ASD production [12]. In this process, the drug and polymer are dissolved in a volatile solvent or solvent mixture to form a homogeneous feed solution. This solution is atomised into fine droplets via a nozzle or rotary atomiser and introduced into a drying chamber with a hot gas stream (usually air or nitrogen). The rapid evaporation of the solvent yields solid microparticles in which the drug is molecularly dispersed within the polymer matrix. The resulting ASD powder is collected, typically using a cyclone separator, to produce a free-flowing material suitable for downstream processing [12].

Spray drying is a well-established method for ASD preparation and is used in numerous commercial products [12]. It offers several advantages: (a) Rapid solidification that "traps" the drug in an amorphous state, even for drugs with a strong tendency to crystallize (the fast drying can outpace crystallization kinetics). (b) Uniform mixing at the molecular level, since the drug and excipient were in proper solution prior to drying. (c) It is a continuous process and relatively easy to scale by using larger spray dryers or parallel nozzles. For paediatric applications, spray drying is particularly advantageous due to its formulation flexibility. The resulting ASD powder can be incorporated into a variety of child-friendly dosage forms such as sachets, mini-tablets, orodispersible tablets, or reconstitutable suspensions. This adaptability is valuable when addressing diverse age-related needs in children.

A practical challenge with spray drying is the need for suitable solvents and the removal of those solvents. The solvent must dissolve both the drug and the polymer; common choices include mixtures such as dichloromethane/methanol, acetone, or ethanol/water, depending on solubility requirements. From a paediatric safety perspective, residual solvents should be minimal. Fortunately, modern spray-drying processes are highly efficient at removing solvents, and the final products typically meet the ICH guidelines for residual solvents [41, 42, 116]. For paediatric use, formulators often prefer relatively benign solvents, such as ethanol (which is acceptable in limited residual amounts for oral products), over solvents like dichloromethane (which would have very tight residual limits due to its toxicity). Another challenge is that powder flowability can be poor, as spray-dried particles are typically fine, amorphous, and electrostatically charged. This can complicate tableting, particularly for mini-tablets. To address these issues, formulation strategies such as granulation, addition of flow aids (e.g. microcrystalline cellulose), or co-processing with excipients like lactose have been successfully employed. For instance an indomethacin ASD was successfully compressed into 2 mm mini-tablets using a blend of MCC and lactose to improve flow and compressibility [12].

Spray drying is used in several paediatric commercial formulations, including ivacaftor granules (Kalydeco®) and posaconazole powder for oral suspension (Noxafil® PowderMix) [4]. These examples demonstrate the technology's ability to support low-dose, taste-masked, and age-appropriate oral products (ivacaftor granules in sachets and posaconazole dry powder kits for suspension). Additionally, variants such as spray freeze-drying, which involves freezing the sprayed solution before vacuum drying, offer potential advantages for heat-sensitive drugs, though they are less common due to cost factor and complexity [43].

## *10.4.2 Hot Melt Extrusion (HME)*

Hot melt extrusion [32, 44–48] is a widely used, solvent-free method [44] for producing ASDs, well suited for both new drug development and product enhancement [45]. In this continuous process, the drug, polymer, and excipients are fed as powders into a heated extruder, where rotating screws apply shear and heat, melting the polymer and dispersing the drug molecularly within the matrix. The extrudate solidifies upon cooling to form the amorphous dispersion. The extrudate can be milled into granules, compressed into tablets, filled into capsules, shaped into films, or calibrated into filaments for 3D printing [49].

HME offers key advantages, including solvent-free processing, excellent scalability, and the ability to combine mixing and formulation in a single step. It has been used in commercial paediatric ASD formulations such as ritonavir oral powder (Norvir®) and ivacaftor mini-tablets, typically employing polymers like HPMCAS and copovidone [4]. However, the method requires thermally stable drugs and thermoplastic polymers. Plasticisers (e.g. PEG, triacetin) can reduce processing temperatures and limit degradation. Short residence times also help minimise thermal stress. Figure 10.2 illustrates a paediatric-friendly application, where filaments produced via HME are used in fused deposition modelling (FDM) 3D printers to fabricate chewable tablets in appealing shapes (e.g. stars, animals), enabling personalised dosing and potentially enhancing adherence in children [49–54].

Whilst HME offers many benefits, it also has limitations. High temperatures and shear can degrade heat-sensitive drugs, and only thermoplastic polymers like PVP-VA, HPMC, and HPMCAS are suitable [55, 56]. Newer polymers like Soluplus® can simplify processing by acting as both matrix and plasticizer [57, 58]. The dense extrudate often requires milling, which must be carefully managed to avoid recrystallisation. Despite these challenges, HME accounts for more than 50% of marketed ASDs, and is especially useful in paediatrics for producing small, solvent-free dosage forms like mini-tablets, granules, and 3D-printed dosage forms [59].

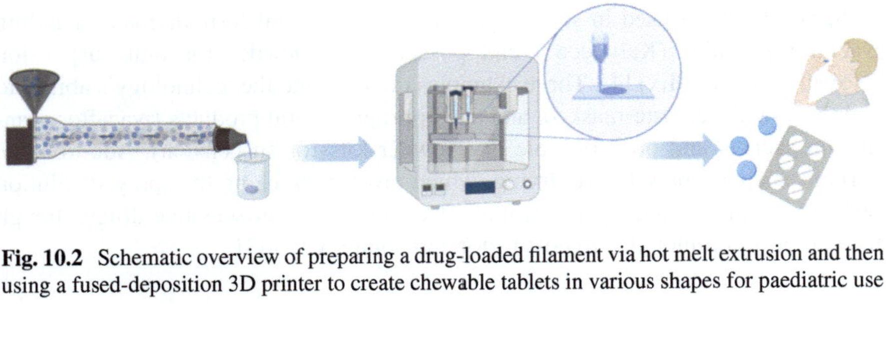

**Fig. 10.2** Schematic overview of preparing a drug-loaded filament via hot melt extrusion and then using a fused-deposition 3D printer to create chewable tablets in various shapes for paediatric use

### 10.4.3 Other Preparation Methods

Beyond spray drying and hot-melt extrusion, several alternative techniques are used for ASD production, primarily at the laboratory or small-scale manufacturing level:

(i) *Solvent Evaporation*: A simple lab-scale method where the drug and polymer are dissolved in a solvent, cast as a thin film, and dried under heat or vacuum [60]. The resulting solid is pulverised into ASD powder. Although widely used in early research [61], t is batch-based and not easily scalable. A gentler variant, freeze-drying, is suitable for heat-sensitive drugs but is costly and time-consuming, it has been successfully applied, for example, to prepare Spironolactone ASDs for oral dispersible films [62].

(ii) *Co-precipitation*: Antisolvent co-precipitation is used to produce ASDs at both small and large industrial scales. It typically yields larger particles than spray drying, reducing the need for densification [60] and improving compaction. It also allows high drug loading (up to ~70%) and is suitable for thermally sensitive drugs [60]. The drug and polymer are dissolved in a non-volatile solvent (e.g. DMA, DMF, DMSO) and added to a chilled, pH-controlled antisolvent (usually water), forming an amorphous complex. Key requirements include insolubility in the antisolvent, full miscibility of the solvents, and a glass transition temperature ($T_g$) above the precipitation temperature [60]. Limitations include the removal of residual solvents, pH sensitivity, and potential stability issues during drying [60]. This strategy has been applied to develop a mefenamic acid nanosuspension for paediatric use, addressing solubility limitations and enabling low-dose, better-accepted oral administration in children [63].

(iii) *Melt Quenching*: A simple method involving direct heating of the drug and polymer to form a molten mixture, which is then rapidly cooled to trap the drug in its amorphous state. Whilst suitable for lab-scale batches, it lacks the shear and mixing uniformity of extrusion, limiting its scalability [60]. A notable application of this method is the preparation of griseofulvin ASDs with saccharin, which not only improved dissolution but also helped mask the drug's bitterness, an important consideration for paediatric antifungal therapies requiring improved taste and oral acceptability [64].

(iv) *Electrospinning*: Electrospinning (Fig. 10.3) uses high voltage to create ultrafine fibres from a drug–polymer solution, forming mats with high surface area and excellent dissolution properties. These fibres can be processed into fast-dissolving webs or orodispersible films suitable for paediatric dosing [65–67]. However, electrospinning is still primarily a research tool; scaling it up and handling the fluffy, fibrous material in a manufacturing environment presents significant challenges. Co-axial electrospinning (where a two-fluid nozzle creates core-shell fibres) is an interesting variant, as it can be used for taste masking: for instance a core of drug-polymer ASD could be coated with a tasteless polymer shell in one step [68]. This has been investigated as a way to create taste-masked ASD fibres for paediatric use [68]. Overall, whilst promising, electrospinning has yet to see broad application in commercial ASD products.

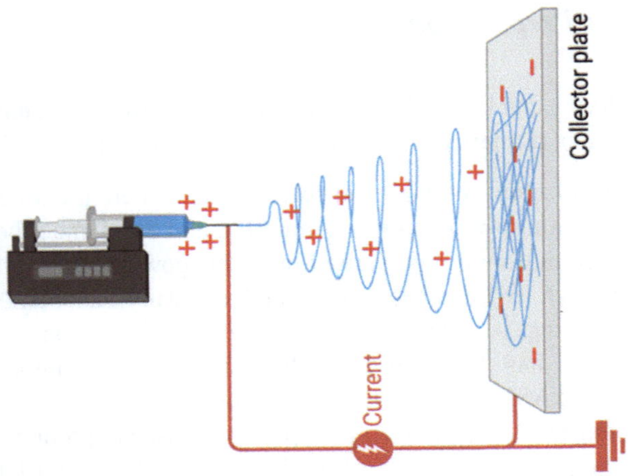


**Fig. 10.3** Schematic description of an electrospinning process. In electrospinning, a polymer solution (optionally drug-loaded) is ejected through a needle under a high voltage. The electric field stretches the solution into a fine filament that solidifies into fibres. The collected non-woven mat can be converted into dosage units. This technique has been used for the production of fast-dissolving ASD films for potential paediatric use

(v) *KinetiSol®*: KinetiSol® is a solvent-free, high-shear fusion process that rapidly melts and mixes drug–polymer blends using frictional heat generated by a rotating rotor [34, 69, 70]. It uses a high-energy rotor in a small chamber to mix the drug and excipients through intense shear and friction-generated heat for a very short time (on the order of a few seconds) [34, 60]. Unlike HME, the process occurs within seconds, making it suitable for heat-sensitive drugs. It has shown success in improving the dissolution of compounds like abiraterone and ivacaftor [34, 69] and offers a promising, though still specialised, alternative for ASD development. Although not present in current commercial formulations, KinetiSol® allows broader compatibility with varied drug physicochemical profiles and polymer systems [4].

(vi) *Three-dimensional (3D) printing*: 3D printing enables the creation of personalised dosage forms by layering drug–polymer materials into defined shapes [60]. In ASD applications, this often involves producing drug-loaded filaments via HME and using FDM printers to fabricate tablets or other shapes [49, 53, 54, 71–73]. Powder-based methods, like binder jetting or selective laser sintering, can also induce in situ amorphization during printing [2]. The potential of 3D printing lies in its ability to create personalised dosage forms [54], tailoring the dose, shape, and even release profile for an individual patient. This technology is particularly promising for paediatrics, allowing dose, shape, and release tailoring—ideal for children who require flexible, chewable, or fast-disintegrating forms. Although still experimental, 3D printing offers potential for on-demand ASD manufacturing in clinical and compounding settings [2].

## 10.5 Formulation Considerations for Paediatrics

### *10.5.1 Polymer Function and Selection*

Selecting appropriate polymers and excipients is critical in ASD development, as these components influence both stability and performance. In paediatrics, additional factors—safety, acceptability, and dose suitability, must guide selection [18, 24, 74]. Typically, the carrier is a high-molecular-weight polymer or a polymer blend, often accompanied by a small-molecule stabiliser. The polymer serves to

(i) Maintain the drug in an amorphous state.
(ii) Prevent crystallisation via intermolecular interactions.
(iii) Modulate dissolution through wetting and viscosity control.

Commonly employed polymers [75–77] in ASDs include polyvinylpyrrolidone (PVP; povidone), copovidone (PVP-VA, e.g. Kollidon VA64 or Plasdone S-630), hypromellose (HPMC) and its derivatives like HPMC-AS (HPMCAS, hypromellose acetate succinate), hydroxypropyl cellulose (HPC), polyethylene glycol (PEG) and related copolymers (like PEG-PVA copolymers), polymethacrylates (e.g. Eudragit® grades), and newer polymers like Soluplus® (polyvinyl caprolactam–polyvinyl acetate–PEG graft copolymer). Many ASDs utilise polymer-to-drug ratios in the range of 1:1 to 4:1 (polymer: drug), although the optimal ratio varies depending on the required stability and dissolution performance. Generally, higher polymer fractions improve stability (up to the saturation limit of the drug in polymer), as discussed earlier.

For paediatric use, the polymer must be inert and non-toxic, as a potentially large amount of polymer might be ingested relative to the drug. Fortunately, the polymers mentioned above are all generally regarded as safe (GRAS) for oral use. Many of them are even used as food additives or in other consumer products. For example PVP (povidone) has been used as a binder in tablets and as a suspending agent in liquids for decades (including in paediatric medicines) and is considered safe, because it is not appreciably absorbed systemically. HPMC is simply a cellulose derivative that is not digested; it acts like a dietary fibre and passes through the GI tract, and it has an excellent safety record in all ages. HPMCAS, although slightly more complex due to succinoyl and acetyl groups, is also not absorbed; it has been used in oral formulations and notably is the polymer in the ivacaftor ASD granules for infants (Kalydeco®) [69], demonstrating safety in even the youngest patients. Copovidone (PVP-VA) is a 60:40 vinyl pyrrolidone:vinyl acetate copolymer; it too is used in many pharmaceuticals and is considered safe for children (e.g. it appears in some paediatric chewable tablets as a binder). Soluplus® is a newer polymer explicitly designed for solid dispersion applications. It has been used in oral products (primarily in adults so far), and toxicology studies indicate it is well tolerated; whilst not yet common in paediatric medications, there is no known contraindication in principle. In the earlier-mentioned paediatric ASD development study, Soluplus was chosen, because it had an established safety profile in children and showed good solubilisation capacity for the drug [37]. This illustrates the decision process: amongst the viable polymers, pick the one that balances performance with paediatric acceptability.

### 10.5.2 Auxiliary Excipients and Safety Concerns

Additional excipients may be added to facilitate manufacturing or improve product performance. Plasticisers aid hot-melt extrusion, surfactants enhance dissolution, and antioxidants or moisture scavengers can help stabilise the formulation. In paediatrics, excipient safety must be carefully assessed on a case-by-case basis [21, 78–80]. For instance polysorbate 80 is commonly used even in neonates [74] but can be toxic at high doses [81]. It is generally considered safe in low oral quantities, but caution is advised. Sodium lauryl sulphate (SLS), whilst effective in enhancing wetting, is usually avoided due to its bitter taste and potential for gastrointestinal irritation [21]. Where possible, formulators may choose polymers with built-in amphiphilic properties, such as Soluplus® or copovidone, eliminating the need for additional surfactants.

### 10.5.3 Taste Masking and Palatability

Palatability is especially important in paediatric formulations delivered as orally disintegrating tablets, chewables, or granules [17, 46, 47, 68, 82, 83]. In such cases, taste masking becomes essential. Eudragit E, for example, can be used to delay drug release in the oral cavity, reducing bitterness [84, 85]. However, it is not commonly used in paediatrics due to concerns over long-term exposure and limited safety data. More often, simple strategies such as sweeteners and flavouring agents are employed. In Posaconazole PowderMix, a fruit-flavoured vehicle is used to reconstitute the ASD powder for improved taste [4]. In another example, beetroot powder was used in indomethacin ASD mini-tablets to create a child-friendly pink colour, enhancing acceptability without compromising safety [12].

### 10.5.4 Exposure Limits and Excipient Quality

Regulatory authorities require justification of excipient levels, especially in chronic paediatric medications. Whilst many polymers such as HPMC and PVP are tolerated even at gram-scale doses, others like PEG 4000 can cause osmotic diarrhoea at high levels [86]. It is also essential to use high-purity excipients when impurity-related degradation is a risk. For example the use of low-peroxide copovidone (e.g. S-630 Ultra) has been shown to reduce oxidative degradation of sensitive drugs such as quetiapine fumarate [87]. This was demonstrated with an ASD of quetiapine fumarate, where the use of Ultra grade copovidone significantly reduced the oxidative degradation of the drug [87]. Residual solvents from manufacturing processes like spray drying must be eliminated, particularly for paediatric use [116], to meet ICH Q3C guidelines. Secondary drying steps or vacuum ovens may be needed to ensure minimal solvent content.

### *10.5.5 Formulation Principles for Polymer Selection*

An ideal polymer for paediatric ASD development should combine several characteristics: good miscibility with the drug, a high glass transition temperature or strong crystallisation inhibition, low hygroscopicity, rapid dissolution in gastrointestinal fluids, and a bland profile without pharmacological activity or unpleasant taste [88].

Polymers like HPMC, HPMCAS, copovidone, and HPC often meet these criteria and are found in marketed ASD products for children [4]. Preformulation tools, such as solubility parameter modelling, Flory–Huggins interaction calculations [88], and spectroscopic techniques like FTIR or solid-state NMR [89], are used to predict compatibility and confirm stabilising interactions such as hydrogen bonding. Rational polymer selection relies on both predictive tools and experimental data to ensure safety in children and optimal ASD performance. Each drug–polymer pair has a specific amorphous solubility limit, above which phase separation and recrystallisation may occur. To ensure long-term stability, drug loading is typically kept below this limit, which is often determined through modelling or lab testing. Most marketed ASDs contain 5–50% drug, rarely exceeding 50% due to stability risks [90].

### *10.5.6 Final Dosage Form*

Beyond the ASD itself, the final dosage form must also rely on excipients with established paediatric safety profiles. Fillers like lactose and mannitol, lubricants such as magnesium stearate, and sweeteners like sucralose or xylitol are commonly used and well tolerated. Resources such as the EMA's Article 5(3) report and the FDA Inactive Ingredient Database provide formulation teams with valuable guidance on age-appropriate excipient use.

If a required excipient presents a potential concern, risks may be mitigated by choosing a purer grade or adjusting processing methods. For instance a secondary drying step may remove residual solvents, or a high-purity polymer may minimise degradation risks [87]. In all cases, careful selection and control of formulation components ensure that paediatric ASDs remain safe, effective, and acceptable.

## 10.6 Stability of ASDs in Paediatrics

One of the key challenges in the development of ASDs is ensuring physical and chemical stability throughout the product's shelf life [91]. The amorphous form of a drug is thermodynamically metastable, characterised by higher free energy compared to the crystalline state, making it prone to recrystallisation during manufacturing, storage, or even after administration [2]. For ASDs to be effective, the drug must remain amorphous at least until dissolution occurs in the gastrointestinal tract. This is particularly critical in paediatrics, where flexible dosing formats and reconstitution practices add further complexity.

### *10.6.1 Paediatric-Specific Stability Practices*

To preserve ASD stability in paediatrics, unit-dose formats, such as sachets, granules, or powders, are preferred. These formats limit moisture exposure and reduce the risk of degradation or crystallisation during storage and use. Products like ritonavir oral powder (Norvir®) and tacrolimus granules are dispensed in single-use packets to prevent moisture exposure after opening [4].

In contrast, multi-dose bottles may be prone to cumulative moisture ingress. If such formats are unavoidable, they must include desiccants, and in-use stability testing is required to assess performance across multiple openings [118, 119]. Packaging and labelling should guide caregivers to store tightly closed containers below specified temperature and humidity thresholds.

When ASDs are formulated into chewable or orodispersible tablets, which are commonly used in children, additional precautions are needed. These formats typically have higher porosity to promote disintegration, which can also increase moisture sensitivity. Volatile flavouring agents (e.g. strawberry aroma) may also be prone to evaporation or odour absorption if not properly sealed, further highlighting the need for protective, child-appropriate packaging.

### *10.6.2 Formulation Strategies to Enhance Stability*

Polymer selection plays a central role in stabilising the amorphous form. Ideal polymers form strong interactions with the drug, such as hydrogen bonding, and help maintain a high glass transition temperature ($T_g$) to reduce molecular mobility. In the development of indomethacin ASD mini-tablets, for instance, PVP (K30) was used alongside insoluble fillers like lactose and MCC to improve compressibility and reduce moisture uptake during tableting [12]. Sometimes, combining polymers can be advantageous. One polymer may enhance drug-polymer interaction, whilst another may provide additional moisture protection or increase $T_g$. Co-amorphous systems, in which a small molecule (e.g. an amino acid) replaces the polymer, are under investigation but are less commonly used in paediatrics due to the need for additional toxicological safety assessments [35].

Packaging for paediatric ASDs must minimise exposure to humidity, heat, and oxygen. Common solutions include foil blisters, sachets, and alu-alu strips, which offer excellent moisture barriers. If a multi-dose bottle is used, it typically includes a desiccant and clear caregiver instructions. Shelf life is often limited to 2 years, and in-use instructions (e.g. "use within 1 hour of mixing") are standard for reconstitutable formats like posaconazole and tacrolimus [4]. In-use stability testing is required when products will be used over time. For example sprinkle capsules or powders in a bottle may be tested over a 30-day period for moisture uptake, physical changes, or crystallisation. If stability cannot be ensured, developers may choose unit-dose packaging to eliminate the need for extended storage after opening. The single-dose Norvir® powder packets are an excellent example: once opened, the entire dose is used, avoiding any storage-related risks [4].

ASD formulations must withstand mechanical stresses such as shipping or accidental dropping. In some cases, tableting pressure can induce drug crystallisation. In the indomethacin mini-tablet study, compression-induced crystallisation was observed but successfully mitigated by using magnesium stearate as a lubricant and optimising filler ratios [12]. Special care is also needed to avoid high storage temperatures (e.g. in vehicles), and such warnings should be clearly communicated to caregivers.

In conclusion, ensuring the stability of amorphous solid dispersions in paediatric formulations requires a multifaceted approach, encompassing careful selection of polymers, protective packaging, and thoughtful product design. From a molecular perspective, polymers play a critical role in maintaining stability by increasing the glass transition temperature, forming specific interactions with the drug (e.g. hydrogen bonding), and creating a viscous environment that suppresses molecular mobility and inhibits crystallization [89]. These stabilising mechanisms are especially important in paediatrics, where formulations must remain robust under real-world conditions yet be flexible enough for child-friendly administration. When supported by appropriate in-use studies and guided by regulatory expectations, paediatric ASDs can achieve both chemical and physical stability, ensuring therapeutic efficacy and safety throughout the product's intended shelf life.

## 10.7 Regulatory Considerations for Paediatric ASDs

The development of paediatric medicines is governed by dedicated legislation, in both the United States and the European Union [92], that ensures formulations are age-appropriate, evidence-based, and not simply scaled-down adult products. Given the growing adoption of ASDs in paediatrics, aligning these technologies with regulatory expectations is essential across all stages, from design to approval.

In the United States, the Paediatric Research Equity Act (PREA) and the Best Pharmaceuticals for Children Act (BPCA) provide the regulatory backbone for paediatric development. PREA (first enacted in 2003, Pub. L. 108-155) mandates that most new drug applications (NDAs) include paediatric study plans, unless waived or deferred. The BPCA (2002, Pub. L. 107-109) incentivises voluntary paediatric studies with a 6-month extension of market exclusivity. The FDA recommends integrating paediatric formulation work early, ideally in parallel with adult development, to avoid delays [121]. Where standard tablets are unsuitable for children, alternatives, such as oral solutions, reconstitutable suspensions, or mini-tablets, are expected. These alternatives often benefit from ASD technology, which enables flexible, bioavailable formats like granules or sprinkles that meet FDA exposure requirements across paediatric age groups.

In the European Union, the Paediatric Regulation (EC No 1901/2006) requires a Paediatric Investigation Plan (PIP) to be submitted early in development. The plan outlines the clinical strategy and dosage forms for each age group. EMA's guidance [117] provides specific direction on formulation development, emphasising palatability, age-appropriate excipients, and administration feasibility. The Paediatric Committee (PDCO) reviews every proposed formulation and its excipients. Novel or unusually high levels of excipients (e.g. plasticisers or surfactants in ASDs) may trigger requests for additional toxicological data. However, many ASD-enabling polymers, such as PVP, HPMC, and HPMCAS, are well established and generally accepted when accompanied by a sound justification [21, 93]. Successful completion of a PIP can result in valuable rewards, such as a 6-month patent extension or, for orphan drugs, a 2-year market exclusivity extension [94].

Both FDA and EMA require thorough benefit–risk evaluations for each excipient in a paediatric product. Ingredients with known age-related toxicity, such as propylene glycol or benzyl alcohol, are restricted or subject to strict dose limits. For ASD formulations, this means only using well-qualified polymers or presenting supporting data when using newer or less-characterised materials [4]. Choosing polymers like HPMCAS, PVP/VA, or HPC, which have a history of paediatric use, helps streamline review. Toxicology concerns around other formulation aids (e.g. solvents or surfactants used in spray drying) must be addressed through analytical testing and appropriate residual control.

Product acceptability and palatability have become explicit parts of regulatory review. Regulators increasingly require evidence that paediatric patients can accept and use medicines appropriately. This might include, for example, swallowability research for a novel mini-tablet in toddlers or a palatability trial where children are given a placebo version of the medicine to ensure that the taste is acceptable.

Paediatric ASDs must also meet stringent quality standards, including solid-state characterisation, content uniformity, and stability. When formulations differ from adult products, bridging studies, such as bioequivalence or relative bioavailability trials, may be needed to ensure similar drug exposure. Quality-by-Design (QbD) approaches are encouraged, with attention to attributes like low-dose precision, dispersibility, and taste masking.

Incentives such as market exclusivity extensions have encouraged the development of ASD-based paediatric formulations, including mini-tablets, chewables, and reconstitutable granules [92]. Regulators also support patient-centric design, encouraging developers to involve caregivers in formulation decisions to ensure usability. This aligns well with the flexibility of ASDs, which can be adapted to child-friendly formats.

In summary, regulatory success for paediatric ASDs requires early alignment with regional frameworks (e.g. PREA, BPCA, and EU PIPs), justification of excipients, and evidence of acceptability and quality. When these elements are addressed proactively, ASD-based medicines can safely and effectively serve paediatric populations.

## 10.8 Clinical Applications and Case Studies

Amorphous solid dispersions have already proven instrumental in expanding therapeutic options for children. A growing number of ASD-based paediatric formulations, ranging from life-saving antivirals to innovative antifungal therapies, demonstrate how this technology addresses longstanding challenges in solubility, bioavailability, and child-friendly dosing. Below are key clinical examples, summarised in Table 10.2.

(i) *Cystic Fibrosis Transmembrane Regulator (CFTR) modulators*: The treatment of cystic fibrosis in infants and toddlers has been transformed by ASD-enabled formulations of CFTR modulators. Ivacaftor (Kalydeco®), originally approved as tablets for adults, is now available as granules for children as young as 1 month. The formulation, based on a spray-dried ASD with HPMCAS, improves solubility and allows for low-dose administration in a small volume of soft food or liquid [4]. Similar ASD-based sprinkle granules have been developed for combination therapies such as lumacaftor/ivacaftor (Orkambi®) and elexacaftor/tezacaftor/ivacaftor (Trikafta®), facilitating dosing in non-tablet-swallowing children [4]. These formats deliver comparable systemic exposure to adult tablets, making treatment accessible to even the youngest CF patients.

(ii) *HIV protease inhibitors (Ritonavir®)*: a poorly soluble and bitter [83] HIV protease inhibitor, was reformulated as an ASD powder (Norvir®) [4] for paediatric use. The powder is based on a copovidone matrix prepared via hot-melt extrusion and packaged in unit-dose sachets [4]. Mixed with strongly flavoured foods such as chocolate pudding, the formulation enables precise weight-based dosing for infants and avoids the ethanol content and crystallisation issues associated with older liquid forms. Since its approval in 2017, the powder has become an essential component of paediatric HIV regimens [4].

(iii) *Posaconazole powder for suspension (Noxafil®)*: The development of a reconstitutable ASD-based formulation of posaconazole (Noxafil® PowderMix)

**Table 10.2** Summary of key paediatric ASD case studies and products (Abbreviations: *AUC* = area under the concentration-time curve; PK = pharmacokinetics; NG-tube = nasogastric tube; ↑ = increase)

| API | Polymer/Carrier | Manufacturing method | Dosage form | Approval status | Key outcomes | Reference |
|---|---|---|---|---|---|---|
| Ivacaftor (Kalydeco®) | HPMCAS-LF (1:2 w/w) | Spray drying | Granules in 50 mg & 75 mg sachets | FDA 2018, EMA 2019 (<6 years) | 15× solubility ↑; paediatric $AUC_0$–12 h within ±10% of adults; low-volume (<5 mL) dosing in soft food. | [4, 97] |
| Lumacaftor/ Ivacaftor (Orkambi®) | HPMCAS or HPMC | Spray drying | Sprinkle granules (combination sachets) | FDA 2019 (< 6 years) | Comparable exposure to tablets; mixed into ≤5 mL soft food; eliminated large suspensions. | [4] |
| Elexacaftor/ Tezacaftor/ Ivacaftor (Trikafta®) | HPMCAS or HPMC | Spray drying | Sprinkle granules | FDA 2022 (<6 years) | Achieved adult-equivalent PK; small-volume administration; high carer acceptance | [98] |
| Ritonavir (Norvir®) | Copovidone (PVP-VA 64; 1:1) | Hot-melt extrusion | 100 mg sachets of oral powder | FDA 2017 (neonates–adolescents) | 20× solubility ↑; $AUC_{0-12}$h ≈ 1200 ng·h/mL in neonates; low-volume (≤2 mL) reconstitution; taste-masked with flavoured vehicle; use within 2 h. | [4, 83] |
| Posaconazole (Noxafil® PowderMix) | HPMCAS-MF (1:2.5 w/w) | Spray drying | Dual-chamber kit (ASD + 15 mL diluent) | FDA 2021 (≥ 2 years) | Reconstituted to 400 μg/mL in 5 min vs. 45 μg/mL; paediatric $AUC_{0-24}$h within ±8% of tablets; consume within 60 min to avoid crystallisation. | [4] |

| | | | | | | |
|---|---|---|---|---|---|---|
| Tacrolimus (Prograf®) | HPMC E5 LV (1:3 w/w) | Solvent granulation | 0.2 mg &1 mg unit-dose granules | EMA 2018 (infants, children) | ≥ 95% dissolution in 10 min; trough levels in therapeutic window; immediate dispersion in ≤5 mL water; removed need for NG-tube or capsule opening. | [4] |
| Itraconazole (SUBA®) | Soluplus® | Hot-melt extrusion | Adult capsules; paediatric sprinkles in development | Approved (adults); paediatric TBD | 4–5× bioavailability ↑ vs. cyclodextrin solution; paves way for sprinkle formulations in children. | [4, 95] |
| Lopinavir/Ritonavir (Kaletra®, AbbVie for adults and Cipla Ltd. for paediatric) | HPMCAS PEF 4000 | Hot-melt extrusion | Oral pellets in capsules containing solid dispersion | EMA 2007 | Paediatric dosing is based on age and weight; available in liquid form for infants and younger children, with a tablet formulation for older children. | [100, 120] |

addressed the limitations of the previous cyclodextrin suspension. Using spray-dried posaconazole with HPMCAS-MF, the powder is mixed with a fruit-flavoured diluent to form a palatable liquid (~16 mL) for oral intake [4]. Dosing instructions include immediate consumption within 1 h to avoid precipitation. The formulation improved both patient acceptability and absorption consistency, making it a more reliable option for paediatric antifungal prophylaxis [4].

(iv) *Tacrolimus granules for oral suspension (Prograf®)*: For paediatric transplant patients requiring immunosuppression, ASD-based granules of tacrolimus (Prograf®) offer precise, easy-to-administer dosing. Each sachet contains either 0.2 or 1 mg of drug in an HPMC-based ASD matrix prepared by solvent evaporation [4]. The granules are mixed with water and administered via oral syringe. This product eliminates the need for capsule opening or compounding and has been well accepted since its EU approval in 2018, offering improved dosing accuracy and patient compliance.

(v) Itraconazole (SUBA®) Antifungal Therapy: Itraconazole is another antifungal that historically had a problematic cyclodextrin solution (Sporanox® oral liquid) for children, it was hard to use due to the large volume and unpleasant taste (plus it required an acidic beverage to aid absorption) [95]. The SUBA®-Itraconazole platform, originally developed for adults (Tolsura®), uses an ASD with Soluplus® via HME to enhance itraconazole bioavailability by 4–5× over the original solution [4]. Although approved for adult use, efforts are ongoing to adapt this formulation into paediatric-friendly forms such as sprinkles or mini-tablets [4]. Clinical studies are ongoing to adapt this for younger patients [95]. This case illustrates how ASD technology enables oral delivery of poorly soluble antifungals, reducing reliance on IV therapy even in systemic infections.

These case studies highlight how ASDs have enabled the development of age-appropriate medicines that overcome key challenges in paediatric therapy. Beyond improving bioavailability, ASDs contribute to better compliance, more consistent efficacy, and safer administration, especially for drugs previously limited by poor solubility or taste. The success of formulations like ritonavir powder and ivacaftor granules opens doors for similar innovations, such as ASD-based integrase inhibitors for HIV [96], or mini-tablets for other BCS Class II drugs like certain anticonvulsants or NSAIDs.

## 10.9 Future Perspectives and Research Directions

The use of amorphous solid dispersions in paediatric pharmaceutics is expected to expand in the coming years, driven by technological advances and a deeper understanding of both formulation science and paediatric needs. Here are several key future perspectives and research directions in this field:

(i) *Advanced manufacturing techniques*: Future development will likely involve greater adoption of innovative manufacturing approaches, such as 3D printing and continuous manufacturing. Three-dimensional printing, in particular, holds promise for creating personalised paediatric medications [50–52, 73, 100]. Whilst technical and regulatory hurdles remain, such as dose uniformity and printer calibration, on-demand, patient-specific production could transform paediatric care. Continuous spray drying and integrated hot-melt extrusion (HME)-tablet production lines may further increase manufacturing efficiency, particularly for niche conditions or small patient populations.

(ii) *New polymers and materials*: Most current ASDs rely on a limited number of pharmaceutically accepted polymers [89]. Ongoing research seeks to expand this list with polymers offering better crystallisation inhibition, taste masking, or biodegradability. Naturally derived biopolymers, such as plant gums, alginates, or proteins, may support regulatory acceptance in children. Modifying existing polymers (e.g. lower-viscosity grades of HPMC) or engineering excipients at the molecular level may enhance performance for specific drug classes. Future materials may offer tailored functionality, such as improved mucoadhesion or ultra-high glass transition temperature ($T_g$) for hot climates [101].

(iii) *Combined technologies*: ASDs may increasingly be integrated with other drug delivery technologies. Co-amorphous systems, involving a second small molecule (e.g. an amino acid or another drug), can enhance solubility and sometimes mask taste [102]. These systems reduce polymer burden and may be especially useful in paediatrics. Novel hybrid technologies, such as in situ self-assembling nanoparticles (ISNPs) [103], may also expand ASD utility beyond immediate-release products.

(iv) *Taste-masking innovations*: ASDs can increase solubility but may intensify bitterness. Future formulations will integrate taste-masking techniques, such as saliva-resistant polymers, nano-coatings, or coaxial electrospinning for core-shell [46, 68, 82, 92, 104]. Electronic tongue [102] and trained sensory panels [105] are already used in formulation screening, and their role is expected to grow. Ensuring acceptable taste is essential for paediatric adherence, particularly in long-term therapy.

(v) *Improving drug loading in ASDs*: Achieving higher drug loads in ASDs can reduce excipient intake and dosage volume, important for paediatrics. Strategies include polymer blending, crystallisation inhibitors, and porous materials like mesoporous silica. These "hybrid ASDs" provide spatial confinement to prevent crystallisation and may allow >50% drug content whilst maintaining stability [4, 36]. Higher drug loads could reduce the number of tablets or granules a child must take.

(vi) *Enhanced pharmacokinetic modelling and in silico tools*: Computational modelling is improving, enabling simulation of ASD behaviour under paediatric gastrointestinal (GI) conditions. PBPK models tailored to children [16, 106, 107] can help predict dissolution and absorption for different age groups, reducing the need for empirical testing [108]. Coupled with in vitro–in vivo correlation studies, these tools will streamline the development process [109]. A mechanistic model can, for instance, predict that an ASD mini-tablet will dissolve to a given extent in a toddler's stomach, characterized by higher gastric pH and lower fluid volume, and estimate the resulting plasma concentration–time profile. If the projected exposure is insufficient, the formulation can be iteratively optimized in silico (e.g. adjusting polymer type or level, modifying drug load or particle size) and re-simulated before any paediatric testing. Such model-informed, hypothesis-driven refinement can reduce experimental burden and focus on subsequent studies, and it is increasingly recognized by regulators [110] as supportive evidence during development.

(vii) *Greater focus on age-specific formulation needs*: The diversity of the paediatric population necessitates age-appropriate ASD designs. Different subgroups may require different dosage forms: liquids for infants, granules for toddlers, and chewable or orally disintegrating tablets for older children [30, 93]. Solutions include tube-compatible dispersions for neonates and microparticles for aspiratable suspensions. Novel administration aids (e.g. reusable dispensers or sprinklers) may improve caregiver usability.

(viii) *Safety and regulatory science advancements*: New excipient safety databases (e.g. STEP) support more confident use of excipients in children [21]. Regulatory frameworks are evolving, with potential guidelines on 3D printing and modelling-based data packages. Involving carers and patients in the design process is increasingly valued, and social acceptance studies [111] may become standard to ensure formulation acceptability.

(ix) *Application to global health*: ASDs can address unmet needs in low- and middle-income countries for diseases like HIV [71, 112], tuberculosis (TB) [66, 67, 113], and malaria. Future work may focus on developing ASD formulations that are stable in tropical climates, which are characterised by high heat and humidity, and are cost-effective. This might involve using more readily available excipients (perhaps locally sourced starch derivatives or inexpensive polymers) or simplifying the manufacturing process so that it could be done in regional centres. Organisations such as the WHO have emphasised the need for 'child-size medicines' [114], that is, dosage forms of essential medicines that are optimally tailored to children. ASDs could help create, for example, a TB drug combo that dissolves well and can be given as sprinkles or dispersible tablets, replacing the current bitter crushed tablets. Solid dispersions that can be dispersed in water to form palatable drinks are also of interest. A specific example, namely, Praziquantel [104, 115], used for schistosomiasis (a parasitic disease) in African children, is extremely bitter and only comes in large tablets. Researchers are developing orally disintegrating tablets or films of praziquantel for children with ASD, making it more child friendly [35, 72]. This is a case where a low-cost, outdated drug is being reformulated using cutting-edge techniques for a global health impact. The hope is that by ensuring these innovations are cost-effective, NGOs and governments can adopt ASD-based paediatric formulations for essential medicines, closing the treatment gap for children globally. Initiatives like "Medicines Patent Pool" and others may play a role in sharing technology for producing such formulations in resource-limited settings.

In conclusion, the future of ASDs in paediatrics is promising and evolving. As research addresses current challenges, stability, palatability, manufacturing efficiency, and explores personalised and global solutions, ASDs are expected to move from specialised to standard tools in paediatric formulation. Continued interdisciplinary collaboration will be key to realising these advances and ensuring better, safer therapies for children worldwide.

## 10.10 Conclusion

Amorphous solid dispersions (ASDs) have emerged as a transformative strategy for enhancing the solubility and bioavailability of poorly water-soluble drugs. In recent years, their application has expanded meaningfully into paediatric pharmaceutics, where formulation challenges are often compounded by age-specific physiological, behavioural, and developmental factors. Throughout this chapter, we have explored how ASDs offer effective solutions, from improving dissolution in small gastrointestinal fluid volumes to enabling child-friendly dosage forms such as granules, mini-tablets, and reconstitutable powders.

Key considerations for paediatric use include selecting excipients with established safety in children, adapting manufacturing techniques like spray drying and hot-melt extrusion to produce age-appropriate formats, and implementing robust stability strategies to preserve the amorphous state. Clinical successes, such as ASD-based formulations for cystic fibrosis, HIV, fungal infections, and immunosuppression, clearly demonstrate their practical value in paediatric care, not only enhancing drug performance but also improving adherence through more acceptable dosage forms.

Looking ahead, the field is poised for continued innovation. Regulatory frameworks in both the USA and EU now support and incentivise the development of paediatric-specific formulations, providing a fertile environment for advancements such as personalised 3D-printed ASDs, novel biopolymers, and modelling-guided design. These technologies offer the potential to further refine dose accuracy, palatability, and therapeutic outcomes, especially in vulnerable populations such as neonates and children with chronic diseases.

In summary, ASDs represent a significant advancement in paediatric drug development. They exemplify how pharmaceutical innovation can bridge the gap between drug efficacy and real-world usability in children. With continued interdisciplinary collaboration and a growing emphasis on patient-centric design, ASDs are well positioned to become a foundational technology in the creation of safe, effective, and age-appropriate medicines, helping ensure that no child is left behind in accessing high-quality therapeutics.

## References


1. Chiou WL, Riegelman S. Pharmaceutical applications of solid dispersion systems. J Pharm Sci. 1971;60(9):1281–302. https://doi.org/10.1002/jps.2600600902.
2. Tambe S, Jain D, Meruva SK, Rongala G, Juluri A, Nihalani G, et al. Recent advances in amorphous solid dispersions: preformulation, formulation strategies, technological advancements and characterization. Pharmaceutics. 2022;14(10):2203. https://doi.org/10.3390/pharmaceutics14102203.
3. Pandi P, Bulusu R, Kommineni N, Khan W, Singh M. Amorphous solid dispersions: an update for preparation, characterization, mechanism on bioavailability, stability, regulatory considerations and marketed products. Int J Pharm. 2020;586:119560. https://doi.org/10.1016/j.ijpharm.2020.119560.
4. Moseson DE, Tran TB, Karunakaran B, Ambardekar R, Hiew TN. Trends in amorphous solid dispersion drug products approved by the U.S. Food and Drug Administration between 2012 and 2023. Int J Pharm X. 2024;7:100259. https://doi.org/10.1016/j.ijpx.2024.100259.
5. Ernest TB, Elder DP, Martini LG, Roberts M, Ford JL. Developing paediatric medicines: identifying the needs and recognizing the challenges. J Pharm Pharmacol. 2007;59(8):1043–55. https://doi.org/10.1211/jpp.59.8.0001.
6. Salunke S, O'Brien F, Cheng Thiam Tan D, Harris D, Math MC, Ariën T, et al. Oral drug delivery strategies for development of poorly water soluble drugs in paediatric patient population. Adv Drug Deliv Rev. 2022;190:114507. https://doi.org/10.1016/j.addr.2022.114507.
7. Chachlioutaki K, Gioumouxouzis C, Karavasili C, Fatouros DG. Small patients, big challenges: navigating pediatric drug manipulations to prevent medication errors – a comprehensive review. Expert Opin Drug Deliv. 2023;20(11):1489–509. https://doi.org/10.1080/17425247.2023.2273838.
8. Bobillot M, Delannoy V, Trouillard A, Kinowski JM, Sanchez-Ballester NM, Soulairol I. Potentially harmful excipients: state of the art for oral liquid forms used in neonatology and pediatrics units. Pharmaceutics. 2024;16(1) https://doi.org/10.3390/pharmaceutics16010119.
9. Batchelor HK, Marriott JF. Formulations for children: problems and solutions. Br J Clin Pharmacol. 2015;79(3):405–18. https://doi.org/10.1111/bcp.12268.
10. Cornilă A, Iurian S, Tomuță I, Porfire A. Orally dispersible dosage forms for paediatric use: current knowledge and development of nanostructure-based formulations. Pharmaceutics. 2022;14(8):1621. https://doi.org/10.3390/pharmaceutics14081621.
11. Comoglu T, Dilek Ozyilmaz E. Orally disintegrating tablets and orally disintegrating mini tablets – novel dosage forms for pediatric use. Pharm Dev Technol. 2019;24(7):902–14. https://doi.org/10.1080/10837450.2019.1615090.
12. Autzen Virtanen A, Myślińska M, Healy AM, Power E, Madi A, Sivén M. The challenge of downstream processing of spray dried amorphous solid dispersions into minitablets designed for the paediatric population – a sustainable product development approach. Eur J Pharm Sci. 2024;196:106752. https://doi.org/10.1016/j.ejps.2024.106752.
13. Zhang D, Rumondor ACF, Zhu W, Colace T, Marota M, Mora J, et al. The development of minitablets for a pediatric dosage form for a combination therapy. J Pharm Sci. 2020;109(12):3590–7. https://doi.org/10.1016/j.xphs.2020.08.021.
14. Harris D, Hermans E, Klein S, Wagner-Hattler L, Walsh J. Age-appropriate solid oral formulations for pediatric applications with a focus on multiparticulates and minitablets: summary of September 2019 EuPFI workshop. Eur J Pharm Biopharm. 2020;153:222–5. https://doi.org/10.1016/j.ejpb.2020.06.012.
15. Lou H, Liu M, Wang L, Mishra SR, Qu W, Johnson J, et al. Development of a mini-tablet of co-grinded prednisone–Neusilin complex for pediatric use. AAPS PharmSciTech. 2013;14(3):950–8. https://doi.org/10.1208/s12249-013-9981-x.
16. Abdel-Rahman SM, Amidon GL, Kaul A, Lukacova V, Vinks AA, Knipp GT. Summary of the National Institute of Child Health and Human Development–Best Pharmaceuticals for Children Act Pediatric Formulation Initiatives Workshop–Pediatric Biopharmaceutics

Classification System Working Group. Clin Ther. 2012;34(11):S11–24. https://doi.org/10.1016/j.clinthera.2012.09.014.
17. Deng Y, Shen L, Yang Y, Shen J. Development of nanoparticle-based orodispersible palatable pediatric formulations. Int J Pharm. 2021;596:120206. https://doi.org/10.1016/j.ijpharm.2021.120206.
18. Kapoor DU, Garg R, Gaur M, Patel MB, Minglani VV, Prajapati BG, et al. Pediatric drug delivery challenges: enhancing compliance through age-appropriate formulations and safety measures. J Drug Deliv Sci Technol. 2024;96:105720. https://doi.org/10.1016/j.jddst.2024.105720.
19. Ranmal SR, Walsh J, Tuleu C. Poor-tasting pediatric medicines: part 1. A scoping review of their impact on patient acceptability, medication adherence, and treatment outcomes. Front Drug Deliv. 2025;5
20. Valeur KS, Hertel SA, Lundstrøm KE, Holst H. The cumulative daily tolerance levels of potentially toxic excipients ethanol and propylene glycol are commonly exceeded in neonates and infants. Basic Clin Pharmacol Toxicol. 2018;122(5):523–30. https://doi.org/10.1111/bcpt.12950.
21. Rouaz K, Chiclana-Rodríguez B, Nardi-Ricart A, Suñé-Pou M, Mercadé-Frutos D, Suñé-Negre JM, et al. Excipients in the paediatric population: a review. Pharmaceutics. 2021;13(3):387. https://doi.org/10.3390/pharmaceutics13030387.
22. Shehab N, Lewis CL, Streetman DD, Donn SM. Exposure to the pharmaceutical excipients benzyl alcohol and propylene glycol among critically ill neonates. Pediatr Crit Care Med. 2009;10(2). Available from: https://journals.lww.com/pccmjournal/fulltext/2009/03000/exposure_to_the_pharmaceutical_excipients_benzyl.20.aspx
23. Gershanik J, Boecler B, Ensley H, McCloskey S, George W. The gasping syndrome and benzyl alcohol poisoning. N Engl J Med. 1982;307(22):1384–8. https://doi.org/10.1056/NEJM198211253072206.
24. Van Riet-Nales DA, Schobben AFAM, Vromans H, Egberts TCG, Rademaker CMA. Safe and effective pharmacotherapy in infants and preschool children: importance of formulation aspects. Arch Dis Child. 2016;101(7):662–9. https://doi.org/10.1136/archdischild-2015-308227.
25. Batchelor H, Kaukonen AM, Klein S, Davit B, Ju R, Ternik R, et al. Food effects in paediatric medicines development for products co-administered with food. Int J Pharm. 2018;536(2):530–5. https://doi.org/10.1016/j.ijpharm.2017.05.011.
26. Bellis JR, Kirkham JJ, Nunn AJ, Pirmohamed M. Adverse drug reactions and off-label and unlicensed medicines in children: a prospective cohort study of unplanned admissions to a paediatric hospital: adverse drug reactions and off-label and unlicensed medicines in children. Br J Clin Pharmacol. 2014;77(3):545–53. https://doi.org/10.1111/bcp.12222.
27. Conroy S. Survey of unlicensed and off label drug use in paediatric wards in European countries. BMJ. 2000;320(7227):79–82. https://doi.org/10.1136/bmj.320.7227.79.
28. Christiansen H, De Bruin ML, Hallgreen CE. Mandatory requirements for pediatric drug development in the EU and the US for novel drugs – a comparative study. Front Med. 2022;9:1009432. https://doi.org/10.3389/fmed.2022.1009432.
29. Augustijns P, Wuyts B, Hens B, Annaert P, Butler J, Brouwers J. A review of drug solubility in human intestinal fluids: implications for the prediction of oral absorption. Eur J Pharm Sci. 2014;57:322–32. https://doi.org/10.1016/j.ejps.2013.08.027.
30. Kaneria NS, Tuleu C, Ernest T. Opportunities for enteral drug delivery for neonates, infants, and toddlers: a critical exploration. Expert Opin Drug Deliv. 2022;19(5):475–519. https://doi.org/10.1080/17425247.2022.2064449.
31. Mooij MG, De Koning BA, Huijsman ML, De Wildt SN. Ontogeny of oral drug absorption processes in children. Expert Opin Drug Metab Toxicol. 2012;8(10):1293–303. https://doi.org/10.1517/17425255.2012.698261.
32. Sarode AL, Sandhu H, Shah N, Malick W, Zia H. Hot melt extrusion (HME) for amorphous solid dispersions: predictive tools for processing and impact of drug–polymer interactions

on supersaturation. Eur J Pharm Sci. 2013;48(3):371–84. https://doi.org/10.1016/j.ejps.2012.12.012.

33. Schittny A, Huwyler J, Puchkov M. Mechanisms of increased bioavailability through amorphous solid dispersions: a review. Drug Deliv. 2020;27(1):110–27. https://doi.org/10.1080/10717544.2019.1704940.
34. Ellenberger DJ, Miller DA, Williams RO. Expanding the application and formulation space of amorphous solid dispersions with KinetiSol®: a review. AAPS PharmSciTech. 2018;19(5):1933–56. https://doi.org/10.1208/s12249-018-1007-2.
35. Polyzois H, Nguyen HT, De Alvarenga R, Junior B, Taylor LS. Amorphous solid dispersion formation for enhanced release performance of racemic and enantiopure praziquantel. Mol Pharm. 2024;21(10):5285–96. https://doi.org/10.1021/acs.molpharmaceut.4c00711.
36. Mudie DM, Buchanan S, Stewart AM, Smith A, Shepard KB, Biswas N, et al. A novel architecture for achieving high drug loading in amorphous spray dried dispersion tablets. Int J Pharm X. 2020;2:100042. https://doi.org/10.1016/j.ijpx.2020.100042.
37. Boza IAF, Da Silva SL, Guedes NB, Bazzo GC, Stulzer HK. Pediatric formulation optimization using a rational design: exploring amorphous solid dispersion technology with terbinafine hydrochloride as a case study. AAPS PharmSciTech. 2025;26(1):40. https://doi.org/10.1208/s12249-024-03012-4.
38. Santos KAD, Danda LJDA, Oliveira TCD, Soares-Sobrinho JL, Soares MFDLR. The drug loading impact on dissolution and diffusion: a case-study with amorphous solid dispersions of nevirapine. Res. Soc Dev. 2022;11(14):e168111436117. https://doi.org/10.33448/rsd-v11i14.36117.
39. Mudie DM, Stewart AM, Biswas N, Brodeur TJ, Shepard KB, Smith A, et al. Novel high-drug-loaded amorphous dispersion tablets of posaconazole; in vivo and in vitro assessment. Mol Pharm. 2020;17(12):4463–72. https://doi.org/10.1021/acs.molpharmaceut.0c00471.
40. Wu J, Chen C, Luo C, Ning B, Liu Y, Li Z, et al. Therapeutic drug monitoring of posaconazole delayed-release tablets and injections in pediatric patients. Antimicrob Agents Chemother. 2024;68(12):e01112–24. https://doi.org/10.1128/aac.01112-24.
41. Shepard KB, Dower AM, Ekdahl AM, Morgen MM, Baumann JM, Vodak DT. Solvent-assisted secondary drying of spray-dried polymers. Pharm Res. 2020;37(8):156. https://doi.org/10.1007/s11095-020-02890-0.
42. Hsieh DS, Yue H, Nicholson SJ, Roberts D, Schild R, Gamble JF, et al. The secondary drying and the fate of organic solvents for spray dried dispersion drug product. Pharm Res. 2015;32(5):1804–16. https://doi.org/10.1007/s11095-014-1577-y.
43. Asai R, Kondo K, Sawada M, Ohta S, Niwa T. Pharmaceutical characterization of xerogel pill with various gelling agents as a multiple-unit dosage form for pediatric and geriatric patients. J Drug Deliv Sci Technol. 2023;79:104069. https://doi.org/10.1016/j.jddst.2022.104069.
44. Agrawal AM, Dudhedia MS, Zimny E. Hot melt extrusion: development of an amorphous solid dispersion for an insoluble drug from mini-scale to clinical scale. AAPS PharmSciTech. 2016;17(1):133–47. https://doi.org/10.1208/s12249-015-0425-7.
45. Patil H, Tiwari RV, Repka MA. Hot-melt extrusion: from theory to application in pharmaceutical formulation. AAPS PharmSciTech. 2016;17(1):20–42. https://doi.org/10.1208/s12249-015-0360-7.
46. Li J, Li C, Zhang H, Gao X, Wang T, Wang Z, et al. Preparation of azithromycin amorphous solid dispersion by hot-melt extrusion: an advantageous technology with taste masking and solubilization effects. Polymers. 2022;14(3):495. https://doi.org/10.3390/polym14030495.
47. Zhao M, You D, Yin J, Sun W, Yin T, Gou J, et al. Quaternary enteric solid dispersion prepared by hot-melt extrusion to mask the bitter taste and enhance drug stability. Int J Pharm. 2021;597:120279. https://doi.org/10.1016/j.ijpharm.2021.120279.
48. Tian Y, Jacobs E, Jones DS, McCoy CP, Wu H, Andrews GP. The design and development of high drug loading amorphous solid dispersion for hot-melt extrusion platform. Int J Pharm. 2020;586:119545. https://doi.org/10.1016/j.ijpharm.2020.119545.

49. Suárez-González J, Magariños-Triviño M, Díaz-Torres E, Cáceres-Pérez AR, Santoveña-Estévez A, Fariña JB. Individualized orodispersible pediatric dosage forms obtained by molding and semi-solid extrusion by 3D printing: a comparative study for hydrochlorothiazide. J Drug Deliv Sci Technol. 2021;66:102884. https://doi.org/10.1016/j.jddst.2021.102884.
50. Tabriz AG, Mithu MS, Antonijevic MD, Vilain L, Derrar Y, Grau C, et al. 3D printing of LEGO® like designs with tailored release profiles for treatment of sleep disorder. Int J Pharm. 2023;632:122574. https://doi.org/10.1016/j.ijpharm.2022.122574.
51. Scoutaris N, Ross SA, Douroumis D. 3D printed "Starmix" drug loaded dosage forms for paediatric applications. Pharm Res. 2018;35(2):34. https://doi.org/10.1007/s11095-017-2284-2.
52. Tabriz AG, Fullbrook DH, Vilain L, Derrar Y, Nandi U, Grau C, et al. Personalised tasted masked chewable 3D printed fruit-chews for paediatric patients. Pharmaceutics. 2021;13(8) https://doi.org/10.3390/pharmaceutics13081301.
53. Eduardo DT, Ana SE, José BF. A micro-extrusion 3D printing platform for fabrication of orodispersible printlets for pediatric use. Int J Pharm. 2021;605:120854. https://doi.org/10.1016/j.ijpharm.2021.120854.
54. Quodbach J, Bogdahn M, Breitkreutz J, Chamberlain R, Eggenreich K, Elia AG, et al. Quality of FDM 3D printed medicines for Pediatrics: considerations for formulation development, filament extrusion, printing process and printer design. Ther Innov Regul Sci. 2022;56(6):910–28. https://doi.org/10.1007/s43441-021-00354-0.
55. Nair AR, Lakshman YD, Anand VSK, Sree KSN, Bhat K, Dengale SJ. Overview of extensively employed polymeric carriers in solid dispersion technology. AAPS PharmSciTech. 2020;21(8):309. https://doi.org/10.1208/s12249-020-01849-z.
56. Zupan N, Yous I, Danede F, Verin J, Kouach M, Foulon C, et al. Impact of hot-melt extrusion on Glibenclamide's physical and chemical states and dissolution behavior: case studies with three polymer blend matrices. Pharmaceutics. 2024;16(8) https://doi.org/10.3390/pharmaceutics16081071.
57. Tousif Ayyub K, Moravkar K, Maniruzzaman M, Amin P. Effect of melt extrudability and melt binding efficiency of polyvinyl caprolactam polyvinyl acetate polyethylene glycol graft copolymer (Soluplus®) on release pattern of hydrophilic and high dose drugs. Mater Sci Eng C. 2019;99:563–74. https://doi.org/10.1016/j.msec.2019.01.126.
58. Lim H, Hoag SW. Plasticizer effects on physical–mechanical properties of solvent cast Soluplus® films. AAPS PharmSciTech. 2013;14(3):903–10. https://doi.org/10.1208/s12249-013-9971-z.
59. Huynh DTM, Hai HT, Hau NM, Lan HK, Vinh TP, Tran VD, et al. Preparations and characterizations of effervescent granules containing azithromycin solid dispersion for children and elder: solubility enhancement, taste-masking, and digestive acidic protection. Heliyon. 2023;9(6):e16592. https://doi.org/10.1016/j.heliyon.2023.e16592.
60. Bhujbal SV, Mitra B, Jain U, Gong Y, Agrawal A, Karki S, et al. Pharmaceutical amorphous solid dispersion: a review of manufacturing strategies. Acta Pharm Sin B. 2021;11(8):2505–36. https://doi.org/10.1016/j.apsb.2021.05.014.
61. Dharani S, Mohamed EM, Khuroo T, Rahman Z, Khan MA. Formulation characterization and pharmacokinetic evaluation of amorphous solid dispersions of dasatinib. Pharmaceutics. 2022;14(11):2450. https://doi.org/10.3390/pharmaceutics14112450.
62. Shamma R, Elkasabgy N. Design of freeze-dried Soluplus/polyvinyl alcohol-based film for the oral delivery of an insoluble drug for the pediatric use. Drug Deliv. 2016;23(2):489–99. https://doi.org/10.3109/10717544.2014.921944.
63. Perween N, Alshehri S, Easwari TS, Verma V, Faiyazuddin M, Alanazi A, et al. Investigating the feasibility of mefenamic acid nanosuspension for pediatric delivery: preparation, characterization, and role of excipients. Processes. 2021;9(4) https://doi.org/10.3390/pr9040574.
64. França MT, Martins Marcos T, Costa PFA, Bazzo GC, Nicolay Pereira R, Gerola AP, et al. Eutectic mixture and amorphous solid dispersion: two different supersaturating drug delivery system strategies to improve griseofulvin release using saccharin. Int J Pharm. 2022;615:121498. https://doi.org/10.1016/j.ijpharm.2022.121498.

65. Yu YQ, Yang X, Wu XF, Fan YB. Enhancing permeation of drug molecules across the skin via delivery in nanocarriers: novel strategies for effective transdermal applications. Front Bioeng Biotechnol. 2021;9:646554. https://doi.org/10.3389/fbioe.2021.646554.
66. Chachlioutaki K, Tzimtzimis EK, Tzetzis D, Chang MW, Ahmad Z, Karavasili C, et al. Electrospun orodispersible films of isoniazid for pediatric tuberculosis treatment. Pharmaceutics. 2020;12(5):470. https://doi.org/10.3390/pharmaceutics12050470.
67. Patil S, Celebioglu A, Uyar T. Orally fast-dissolving drug delivery systems for pediatrics: nanofibrous oral strips from isoniazid/cyclodextrin inclusion complexes. J Drug Deliv Sci Technol. 2023;85:104584. https://doi.org/10.1016/j.jddst.2023.104584.
68. Abdelhakim HE, Coupe A, Tuleu C, Edirisinghe M, Craig DQM. Utilising co-axial electrospinning as a taste-masking technology for paediatric drug delivery. Pharmaceutics. 2021;13(10):1665. https://doi.org/10.3390/pharmaceutics13101665.
69. Thompson SA, Davis DA, Miller DA, Kucera SU, Williams RO. Pre-processing a polymer blend into a polymer alloy by KinetiSol enables increased ivacaftor amorphous solid dispersion drug loading and dissolution. Biomedicines. 2023;11(5):1281. https://doi.org/10.3390/biomedicines11051281.
70. Gala U, Miller D, Williams RO. Improved dissolution and pharmacokinetics of abiraterone through KinetiSol® enabled amorphous solid dispersions. Pharmaceutics. 2020;12(4):357. https://doi.org/10.3390/pharmaceutics12040357.
71. Malebari AM, Kara A, Khayyat AN, Mohammad KA, Serrano DR. Development of advanced 3D-printed solid dosage pediatric formulations for HIV treatment. Pharmaceuticals. 2022;15(4):435. https://doi.org/10.3390/ph15040435.
72. Boniatti J, Januskaite P, Fonseca LBD, Viçosa AL, Amendoeira FC, Tuleu C, et al. Direct powder extrusion 3D printing of praziquantel to overcome neglected disease formulation challenges in paediatric populations. Pharmaceutics. 2021;13(8):1114. https://doi.org/10.3390/pharmaceutics13081114.
73. Patel H, Raje V, Maczko P, Patel K. Application of 3D printing technology for the development of dose adjustable geriatric and pediatric formulation of celecoxib. Int J Pharm. 2024;655:123941. https://doi.org/10.1016/j.ijpharm.2024.123941.
74. Kriegel C, Festag M, Kishore RSK, Roethlisberger D, Schmitt G. Pediatric safety of polysorbates in drug formulations. Children. 2019;7(1):1. https://doi.org/10.3390/children7010001.
75. Corrie L, Ajjarapu S, Banda S, Parvathaneni M, Bolla PK, Kommineni N. HPMCAS-based amorphous solid dispersions in clinic: a review on manufacturing techniques (hot melt extrusion and spray drying), marketed products and patents. Materials. 2023;16(20):6616. https://doi.org/10.3390/ma16206616.
76. D'Amico V, Arduino I, Vacca M, Iacobazzi RM, Altamura D, Lopalco A, et al. Colonic budesonide delivery by multistimuli alginate/Eudragit® FS 30D/inulin-based microspheres as a paediatric formulation. Carbohydr Polym. 2023;302:120422. https://doi.org/10.1016/j.carbpol.2022.120422.
77. Zhang W, Noland R, Chin S, Petkovic M, Zuniga R, Santarra B, et al. Impact of polymer type, ASD loading and polymer-drug ratio on ASD tablet disintegration and drug release. Int J Pharm. 2021;592:120087. https://doi.org/10.1016/j.ijpharm.2020.120087.
78. Kushwah V, Succhielli C, Saraf I, Paudel A. Amorphous solid dispersions: implication of method of preparation and physicochemical properties of API and excipients. Pharmaceutics. 2024;16(8):1035. https://doi.org/10.3390/pharmaceutics16081035.
79. Kokott M, Lura A, Breitkreutz J, Wiedey R. Evaluation of two novel co-processed excipients for direct compression of orodispersible tablets and mini-tablets. Eur J Pharm Biopharm. 2021;168:122–30. https://doi.org/10.1016/j.ejpb.2021.08.016.
80. Laquintana V, Asim MH, Lopedota A, Cutrignelli A, Lopalco A, Franco M, et al. Thiolated hydroxypropyl-β-cyclodextrin as mucoadhesive excipient for oral delivery of budesonide in liquid paediatric formulation. Int J Pharm. 2019;572:118820. https://doi.org/10.1016/j.ijpharm.2019.118820.

81. Preis M. Orally disintegrating films and mini-tablets – innovative dosage forms of choice for pediatric use. AAPS PharmSciTech. 2015;16(2):234–41. https://doi.org/10.1208/s12249-015-0313-1.
82. Trofimiuk M, Olechno K, Trofimiuk E, Czajkowska-Kośnik A, Ciosek-Skibińska P, Głowacz K, et al. Utilization of the Drug–polymer solid dispersion obtained by ball milling as a taste masking method in the development of orodispersible minitablets with hydrocortisone in pediatric doses. Pharmaceutics. 2024;16(8):1041. https://doi.org/10.3390/pharmaceutics16081041.
83. Morris JB, Tisi DA, Tan DCT, Worthington JH. Development and palatability assessment of Norvir® (ritonavir) 100 mg powder for Pediatric population. Int J Mol Sci. 2019;20(7):1718. https://doi.org/10.3390/ijms20071718.
84. Drašković M, Medarević D, Aleksić I, Parojčić J. In vitro and in vivo investigation of taste-masking effectiveness of Eudragit E PO as drug particle coating agent in orally disintegrating tablets. Drug Dev Ind Pharm. 2017;43(5):723–31. https://doi.org/10.1080/03639045.2016.1220572.
85. Lopalco A, Denora N, Laquintana V, Cutrignelli A, Franco M, Robota M, et al. Taste masking of propranolol hydrochloride by microbeads of EUDRAGIT® E PO obtained with prilling technique for paediatric oral administration. Int J Pharm. 2020;574:118922. https://doi.org/10.1016/j.ijpharm.2019.118922.
86. Hammer HF, Santa Ana CA, Schiller LR, Fordtran JS. Studies of osmotic diarrhea induced in normal subjects by ingestion of polyethylene glycol and lactulose. J Clin Invest. 1989;84(4):1056–62. https://doi.org/10.1172/JCI114267.
87. Butreddy A, Sarabu S, Bandari S, Batra A, Lawal K, Chen N, et al. Influence of Plasdone™ S630 ultra – an improved copovidone on the processability and oxidative degradation of quetiapine fumarate amorphous solid dispersions prepared via hot-melt extrusion technique. AAPS PharmSciTech. 2021;22 https://doi.org/10.1208/s12249-021-02069-9.
88. Zhang J, Guo M, Luo M, Cai T. Advances in the development of amorphous solid dispersions: the role of polymeric carriers. Asian J Pharm Sci. 2023;18(4):100834. https://doi.org/10.1016/j.ajps.2023.100834.
89. Shi Q, Chen H, Wang Y, Wang R, Xu J, Zhang C. Amorphous solid dispersions: role of the polymer and its importance in physical stability and in vitro performance. Pharmaceutics. 2022;14(8):1747. https://doi.org/10.3390/pharmaceutics14081747.
90. Iyer R, Petrovska Jovanovska V, Berginc K, Jaklič M, Fabiani F, Harlacher C, et al. Amorphous solid dispersions (ASDs): the influence of material properties, manufacturing processes and analytical technologies in drug product development. Pharmaceutics. 2021;13(10):1682. https://doi.org/10.3390/pharmaceutics13101682.
91. Liu X, Lu M, Guo Z, Huang L, Feng X, Wu C. Improving the chemical stability of amorphous solid dispersion with cocrystal technique by hot melt extrusion. Pharm Res. 2012;29(3):806–17. https://doi.org/10.1007/s11095-011-0605-4.
92. Thabet Y, Klingmann V, Breitkreutz J. Drug formulations: standards and novel strategies for drug Administration in Pediatrics. J Clin Pharmacol. 2018;58(S10) https://doi.org/10.1002/jcph.1138.
93. Slavkova M, Breitkreutz J. Orodispersible drug formulations for children and elderly. Eur J Pharm Sci. 2015;75:2–9. https://doi.org/10.1016/j.ejps.2015.02.015.
94. Penkov D, Tomasi P, Eichler I, Murphy D, Yao LP, Temeck J. Pediatric medicine development: an overview and comparison of regulatory processes in the European Union and United States. Ther Innov Regul Sci. 2017;51(3):360–71. https://doi.org/10.1177/2168479017696265.
95. Klingmann V, Linderskamp H, Meissner T, Mayatepek E, Moeltner A, Breitkreutz J, et al. Acceptability of Multiple Uncoated Minitablets in Infants and Toddlers: A Randomized Controlled Trial. J Pediatr. 2018 Oct;201:202–207.e1.
96. Satheesh D, et al. Effectiveness, safety, and tolerability of itraconazole oral solution in pediatric dermatophytosis. Indian J Paediatr Dermatol. 2020;21(4):259. https://doi.org/10.4103/ijpd.ijpd_34_20.

97. Breitkreutz DJ. Gedruckt mit der Genehmigung der Mathematisch-Naturwissenschaftlichen Fakultät der Heinrich-Heine-Universität Düsseldorf.
98. Kramer EL, Clancy JP. CFTR modulator therapies in pediatric cystic fibrosis: focus on ivacaftor. Expert Opin Orphan Drugs. 2016;4(10):1033–42. https://doi.org/10.1080/21678707.2016.1226800.
99. Truong NH, Benaboud S, Bouazza N, Barboura M, Bardin E, Miralles M, et al. Elexacaftor/tezacaftor/ivacaftor population pharmacokinetics in pediatric patients with cystic fibrosis. Clin Transl Sci. 2025;18(5):e70245. https://doi.org/10.1111/cts.70245.
100. Trasi NS, Bhujbal S, Zhou QT, Taylor LS. Amorphous solid dispersion formation via solvent granulation – a case study with ritonavir and lopinavir. Int J Pharm X. 2019;1:100035. https://doi.org/10.1016/j.ijpx.2019.100035.
101. Chacko IA, Ramachandran G, Sudheesh MS. Unmet technological demands in orodispersible films for age-appropriate paediatric drug delivery. Drug Deliv Transl Res. 2024;14(4):841–57. https://doi.org/10.1007/s13346-023-01451-3.
102. Lavan M, Wang X, McCain R, Jannasch A, Cooper B, Hostetler S, et al. Development of a pediatric mini-tablet formulation for expedited preclinical studies. AAPS PharmSciTech. 2021;22(1):40. https://doi.org/10.1208/s12249-020-01891-x.
103. Alopaeus JF, Göbel A, Breitkreutz J, Sande SA, Tho I. Investigation of hydroxypropyl-β-cyclodextrin inclusion complexation of two poorly soluble model drugs and their taste-sensation – effect of electrolytes, freeze-drying and incorporation into oral film formulations. J Drug Deliv Sci Technol. 2021;61:102245. https://doi.org/10.1016/j.jddst.2020.102245.
104. Pham K, Li D, Guo S, Penzak S, Dong X. Development and in vivo evaluation of child-friendly lopinavir/ritonavir pediatric granules utilizing novel in situ self-assembly nanoparticles. J Control Release. 2016;226:88–97. https://doi.org/10.1016/j.jconrel.2016.02.001.
105. Münster M, Schoch C, Schmidt C, Breitkreutz J. Multiparticulate system combining taste masking and immediate release properties for the aversive compound praziquantel. Eur J Pharm Sci. 2017;109:446–54. https://doi.org/10.1016/j.ejps.2017.08.034.
106. Golhen K, Buettcher M, Kost J, Huwyler J, Pfister M. Meeting challenges of Pediatric drug delivery: the potential of orally fast disintegrating tablets for infants and children. Pharmaceutics. 2023;15(4):1033. https://doi.org/10.3390/pharmaceutics15041033.
107. Batchelor H. Paediatric biopharmaceutics classification system: current status and future decisions. Int J Pharm. 2014;469(2):251–3. https://doi.org/10.1016/j.ijpharm.2014.02.046.
108. Batchelor HK, Fotaki N, Klein S. Paediatric oral biopharmaceutics: key considerations and current challenges. Adv Drug Deliv Rev. 2014;73:102–26. https://doi.org/10.1016/j.addr.2013.10.006.
109. Merchant HA, Liu F, Orlu Gul M, Basit AW. Age-mediated changes in the gastrointestinal tract. Int J Pharm. 2016;512(2):382–95. https://doi.org/10.1016/j.ijpharm.2016.04.024.
110. Niessen J, López Mármol Á, Ismail R, Schiele JT, Rau K, Wahl A, et al. Application of biorelevant in vitro assays for the assessment and optimization of ASD-based formulations for pediatric patients. Eur J Pharm Biopharm. 2023;185:13–27. https://doi.org/10.1016/j.ejpb.2023.02.008.
111. Mulberg AE, Conklin LS, Croft NM, Parikh A, Bousvaros A, DeZoeten E, et al. Pediatric extrapolation of adult efficacy to children is critical for efficient and successful drug development. Gastroenterology. 2022;163(1):77–83. https://doi.org/10.1053/j.gastro.2022.03.018.
112. Chaudhary S, Nair AB, Shah J, Gorain B, Jacob S, Shah H, et al. Enhanced solubility and bioavailability of dolutegravir by solid dispersion method: in vitro and in vivo evaluation – a potential approach for HIV therapy. AAPS PharmSciTech. 2021;22(3):127. https://doi.org/10.1208/s12249-021-01995-y.
113. Adeleke OA, Hayeshi RK, Davids H. Development and evaluation of a reconstitutable dry suspension containing isoniazid for flexible Pediatric dosing. Pharmaceutics. 2020;12(3):286. https://doi.org/10.3390/pharmaceutics12030286.

114. Nsabagasani X, Ogwal-Okeng J, Mbonye A, Ssengooba F, Nantanda R, Muyinda H, et al. The "child size medicines" concept: policy provisions in Uganda. J Pharm Policy Pract. 2015;8(1):2. https://doi.org/10.1186/s40545-015-0025-7.
115. Boniatti J. Development and characterization of praziquantel amorphous solid dispersions for paediatric use by hot melt extrusion and 3D printing; 2021
116. International Council for Harmonisation (ICH). ICH harmonised guideline: impurities – guideline for residual solvents Q3C(R9). Step 4 version. Geneva: ICH; 2024 [cited 2025 Sep 17]. Available from: https://database.ich.org/sites/default/files/ICH_Q3C-R9_Guideline_Step4_2024_0422.pdf
117. European Medicines Agency (EMA). Guideline on pharmaceutical development of medicines for paediatric use. EMA/CHMP/QWP/805880/2012 Rev. 2. London: EMA; 2013. Available from: https://www.ema.europa.eu/en/documents/scientific-guideline/guideline-pharmaceutical-development-medicines-paediatric-use_en.pdf
118. U.S. Food and Drug Administration. NORVIR (ritonavir) [package insert]. Silver Spring (MD): FDA; 2017 [cited 2025 Sep 17]. Available from: https://www.accessdata.fda.gov/drugsatfda_docs/label/2017/209512lbl.pdf
119. Committee for Proprietary Medicinal Products, Quality Working Party. Note for guidance on in-use stability testing of human medicinal products. CPMP/QWP/2934/99. London: European Medicines Agency; 2001 [cited 2025 Sep 17]. Available from: https://www.ema.europa.eu/en/documents/scientific-guideline/note-guidance-use-stability-testing-human-medicinal-products_en.pdf
120. U.S. Food and Drug Administration. Lopinavir and ritonavir oral pellets 40 mg/10 mg – prescribing information. Silver Spring (MD): FDA; 2015 [cited 2025 Sep 17]. Available from: https://www.accessdata.fda.gov/drugsatfda_docs/pepfar/205425PI.pdf
121. U.S. Food and Drug Administration. Guidance for industry: general clinical pharmacology considerations for Pediatric studies for drugs and biological products. Silver Spring (MD): FDA; 2014 [cited 2025 Sep 17]. Available from: https://www.fda.gov/media/85013/download

# Part V
# Biopharmaceutical Evaluation and Translation

# Chapter 11
# Assessing the In Vivo Performance of Orally Administered Amorphous Solid Dispersion Drug Products in Healthy Adults from In Vitro Data

**Christos Reppas and Maria Vertzoni**

**Abstract** During the last decade, significant progress has been made regarding the in vitro assessment of the luminal performance of orally administered amorphous solid dispersion drug products (ASDs). In this chapter, published model-independent and model-dependent methodologies for the assessment of in vivo performance of ASDs in adults from in vitro data are critically summarized. Better understanding of the luminal fate of ASDs, especially in the fed state, and harnessing of the uncertainty of global regulatory acceptance of noncompendial in vitro testing methods would further increase reliability of the in vitro assessment of luminal performance of orally administered ASDs.

**Keywords** Amorphous solid dispersion drug products · In vitro–in vivo relationships · Model-independent methods · Model-dependent methods

## 11.1 Introduction

For orally administered drug products, assessment of the in vivo performance from in vitro data is useful for understanding the impact of formulation, of dose, and of meal on drug absorption without performing costly clinical studies that are also associated with ethical issues as they are typically performed in healthy adults. Relevant information is crucial prior to performing pivotal clinical studies in patients, after postapproval changes in drug product formulation and production and it facilitates the development of generic drug products. It can also be used for setting/informing acceptable product specifications.

C. Reppas (✉) · M. Vertzoni
Department of Pharmacy, School of Health Sciences, National and Kapodistrian University of Athens, Zografou, Greece
e-mail: reppas@pharm.uoa.gr; vertzoni@pharm.uoa.gr



I. Nikolakakis, I. Partheniadis (eds.), *Amorphous Solid Dispersions*, AAPS Introductions in the Pharmaceutical Sciences 5,
https://doi.org/10.1007/978-3-032-24883-1_11

Depending on the number and the variability of processes that limit drug absorption, the development of in vivo relevant in vitro methodologies could be relatively easy (e.g., for extended-release products of BCS class I drugs) or relatively tedious (e.g., for high-risk products).

Especially during the last two decades, lipid-based formulations and amorphous solid dispersion drug products (ASDs) are the most frequently considered high-risk oral drug products [1]. ASDs are molecularly mixed dispersions of the drug with a polymer, and perhaps other excipients, for reducing crystallization tendency and enhancing dissolution rates [2].

Amorphous drug solubility is the concentration after the drug undergoes liquid–liquid or liquid–glass phase separation, forming a water-saturated-rich phase in metastable equilibrium with an aqueous phase containing molecularly dissolved drug. In presence of additive(s), the concentration of molecularly dissolved drug in solution, that is, the solute activity, may be decreased if the additive is distributed in the drug-rich phase (Fig. 11.1). Crystalline drug solubility is the concentration of the drug in an aqueous phase containing molecularly dissolved drug in equilibrium with a crystalline phase. Additive(s) do not affect the solute activity in equilibrium with a crystalline phase. Amorphous drug solubility is higher than crystalline solubility. Therefore, compared with crystalline drugs, amorphous drugs can temporarily achieve higher maximum concentrations of molecularly dissolved drug in an aqueous phase, that is, can temporarily supersaturate an aqueous phase (Fig. 11.1).

Nabilone-PVP (Cesamet®) was the first approved ASD in the USA [3] and by 2023, more than 50 drug products based on the amorphous formulation approach had been approved for oral administration [4]. Most of them are ASDs.

In this chapter, published model-independent and model-dependent methodologies for the assessment of in vivo performance of ASDs in adults from in vitro data are critically summarized.

Although an attempt was made to include as many relevant publications as possible, it is acknowledged that the resulting list cannot be considered comprehensive.

## 11.2 Model-Independent Methodologies for Assessing the In Vivo Performance of Amorphous Solid Dispersion Drug Products in Adults Based on In Vitro Data

### *11.2.1 Immediate-Release Amorphous Solid Dispersion Drug Products*

These methodologies are based on a model-independent relationship between in vitro data and data in humans.

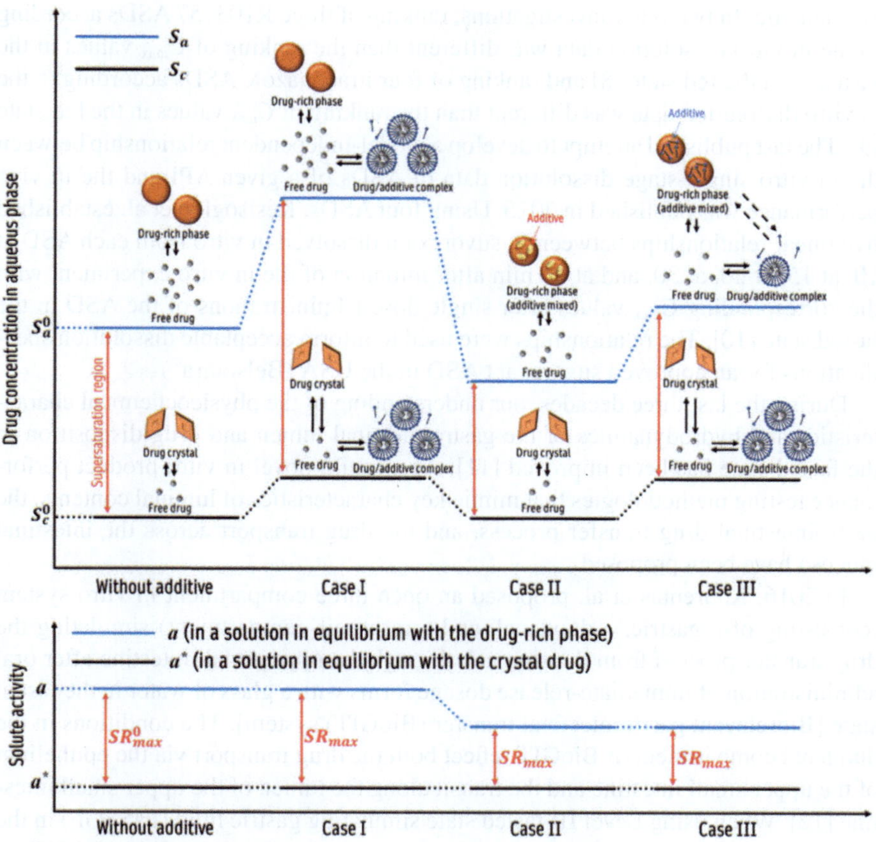


**Fig. 11.1** Schematic illustration of drug concentration changes at the crystalline and amorphous solubilities in the presence of additives. The solute activity (a) in a solution in equilibrium with crystalline and drug-rich phases in each system is represented at the bottom of the figure. Case I represents formation of a drug–additive complex in the aqueous phase. Case II represents distribution of a large amount of additive into the drug-rich phase. Case III represents a combination of both Case I and Case II mechanisms. *Sc* and *Sa* crystalline and amorphous solubilities, $SR_{max}$ maximum supersaturation ratio, *a* solute activity. Letters with superscript "0" are values in the absence of additives. (Reprinted from [5] with permission)

In early investigations, the usefulness of in vitro single-stage dissolution data of ASDs of a given active pharmaceutical ingredient (API) in simple aqueous media in ranking ASDs according to the maximum plasma concentration ($C_{max}$) measured values after single dose administrations to healthy adults was assessed. In vitro dissolution data of two indomethacin ASDs reflected the ranking of $C_{max}$ values in the fasted state [6] and in vitro dissolution data of two itraconazole ASDs reflected the ranking of $C_{max}$ values both in the fasted and in fed states [7]. It should be acknowledged, however, that in the former study [6], details on analytical method and the exact dosing conditions in the in vivo study are not provided. Also, in both studies [6, 7], in vitro data variability is not reported and filtration of in vitro samples was problematic. In two other investigations, ranking of three R103757 ASDs according to the in vitro dissolution data was different than the ranking of $C_{max}$ values in the fasted or in the fed state [8] and ranking of four itraconazole ASDs according to the in vitro dissolution data was different than the ranking of $C_{max}$ values in the fed state [9]. The last published attempt to develop a model-independent relationship between the in vitro single-stage dissolution data of ASDs of a given API and the in vivo performance was published in 2015. Using four ASDs, Kesisoglou et al. established five linear relationships between % suvorexant dissolved in vitro from each ASD at 10, at 15, at 20, at 30, and at 45 min after initiation of the in vitro experiment with the corresponding $C_{max}$ values after single dose administrations of the ASD in the fasted state [10]. The relationships were used to inform acceptable dissolution specifications for an approved suvorexant ASD in the USA (Belsomra®).

During the last three decades, our understanding of the physicochemical characteristics and hydrodynamics of the gastrointestinal lumen and drug disposition in the fasted state has been improved [11]. As a result, novel in vitro product performance testing methodologies that mimic key characteristics of luminal contents, the gastrointestinal drug transfer process, and the drug transport across the intestinal mucosa have been proposed.

In 2016, Kourentas et al. proposed an open three-compartment in vitro system (consisting of a gastric, a duodenal, and a reservoir compartment) simulating the drug transfer process from the stomach through the upper small intestine after oral administration of immediate-release dosage forms with a glass of water in the fasted state [Biorelevant gastrointestinal transfer (BioGIT) system]. The conditions in the duodenal compartment of BioGIT reflect both the drug transport via the epithelium of the upper small intestine and the transit along the lumen of the upper small intestine [12]. When using Level III fasted state simulating gastric fluid (FaSSGF) in the gastric compartment and Level II fasted state simulating intestinal fluid (FaSSIF) in the duodenal compartment, the apparent drug concentrations in the duodenal compartment of the BioGIT system simulate the apparent drug concentrations in the upper small intestine after administration of conventional or high-risk immediate-release drug products with a glass of water in the fasted state (Wilson et al. [13] and references therein). The BioGIT system has been shown to be a useful screening tool for assessing the impact of dose and/or formulation differences on early exposure, after administration of immediate-release products, including ASDs, with a glass of water in the fasted state [12], also, on an a priori basis [14]. The log-transformed mean plasma $AUC_{0\text{-}60min}$ ratios *vs.* the corresponding log-transformed mean $BioGIT_{0\text{-}50min}$ ratios are linearly related [$n = 16$ data points, slope 0.89(0.11), intercept 0.211(0.046), Pearson correlation coefficient 0.90] (Fig. 11.2). Importantly, potential limitations have also been identified [14].

In 2022, Holzem et al. investigated the usefulness of a two-stage in vitro methodology in formulation ranking of super-saturating immediate-release products in the fasted state [15]. The product is dispersed in Level I FaSSGF and the dispersion is transferred under constant flow rate conditions to the donor compartment a dissolution–permeation setup [PermeaLoop® setup [16]] containing Level II FaSSIF. Using three posaconazole formulations (including an ASD) the molecularly dispersed posaconazole permeated from the donor compartment into the

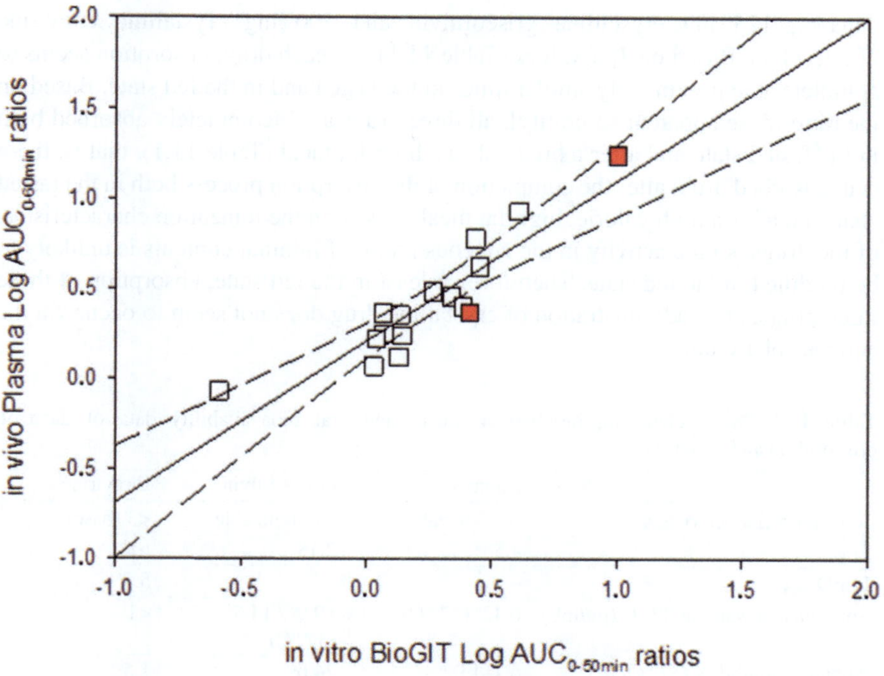


**Fig. 11.2** The best fitted line and the associated 95% confidence limits for the regression of log--transformed ratios of mean BioGIT $AUC_{0\text{-}50min}$ values vs. the log-transformed ratios of mean plasma $AUC_{0\text{-}60min}$ values. Ratios were estimated by using the lowest dose (when a total of two doses were investigated), the next lower dose (when more than two doses were investigated), or the formulation with the lowest BioGIT $AUC_{0\text{-}50min}$ (when a total of two formulations were investigated). Filled squares refer to ASDs. (Redrawn from [14])

acceptor compartment was linearly related with the extent of absorption in healthy adults (based on $AUC_{0\text{-}8h}$ values) [15]. Populating the relationship with additional products and/or defining potential limitations would increase the reliability of the approach, as the linear relationship was sensitive to the rate of transfer of the dispersion to the donor compartment of the dissolution–permeation setup.

It is interesting to note that Kourentas et al. focus to the apparent drug concentrations whereas Holzem et al. focus to the molecularly dispersed drug in the contents of upper small intestine [12, 15]. The difference relates to a long-standing topic in pharmaceutical literature. Although the importance of molecularly dissolved drug in the aqueous phase vs. total concentration in the aqueous phase on drug flux via dialysis membranes has been unequivocally shown [17–19], the situation in vivo is not clear; at least for lipophilic APIs. Based on data collected in diluted aspirates from the proximal small intestine of adults after administration of danazol in the triglyceride phase of an heterogenous liquid meal (500 mL, 750 kcal, 73% triglycerides, 14% proteins, 13% carbohydrates), coarse lipid particles directly contribute to danazol flux across Caco-2 cell lines [20]. A similar trend was also observed using undiluted aspirates and PAMPA membranes [21]. These data are in line with the positive food effect after single dose administrations of 200 mg crystalline danazol, 125 mg crystalline griseofulvin and 100 mg crystalline etravirine (Table 11.1). Based on $T_{max}$ values (Table 11.1), for each drug, absorption seems to complete at approximately similar times in the fasted and in the fed state. Based on the ratio [dose/apparent solubility], all three drugs are incompletely absorbed both in the fasted state and after a high-calorie, high-fat meal (Table 11.1), that is, there is unabsorbed drug, after the completion of the absorption process both in the fasted state and after a high-calorie, high-fat meal. Based on the ionization characteristics of the drugs, solute activity in the aqueous phase of luminal contents is unlikely to be modified in the fed state. Therefore, at least in the fed state, absorption of these three drugs, after administration of crystalline drug does not seem to occur via the aqueous phase only.

**Table 11.1** Physicochemical, biopharmaceutical, and oral bioavailability data of danazol, griseofulvin and etravirine

| | Danazol | Griseofulvin | Etravirine |
|---|---|---|---|
| Ionization characteristics | Nonionizable | Nonionizable | <3 (base)[h] |
| *LogP* | 4.53[a] | 2.18[e] | >5[h] |
| $LogD_{pH6.5}$ | – | – | 5.5[c] |
| Solubility in water at 37 °C (μg/mL) | 0.42[b] (37 °C) | 29.9[e] / 14.5[f] (37 °C) | <1[i] |
| Apparent solubility (37 °C) in aspirates from the duodenum in the fasted state (FaHIF) (μg/mL) | 6.7–13.2[c] | 24.6[c] | 1.5[c] |
| Apparent solubility (37 °C) in aspirates from the duodenum in the fed state (FeHIF) (μg/mL) | 40–95[c] | 60[c] | 4.1[c] |
| Dose (mg) | 200[d] | 125[g] | 100[h] |
| $\frac{Dose}{Solubility_{FaHIF}}$ (L) | 15–27.8 | 5.1 | 66.7 |
| $\frac{Dose}{Solubility_{FeHIF}}$ (L) | 2.1–5 | 2.1 | 24.4 |
| $T_{max}$ in the fasted state (h) | Range: 1–6[d] | Median: 6[g] | Range: 2–6[h] |
| $T_{max}$ in the fed state (h) | Range: 2–4[d] | Median: 5[g] | Range: 2–6[h] |
| High calorie, high-fat meal effect on bioavailability ($AUC_{0\text{-}tlast}$—fold increase) | 1.8 at the 200 mg dose level[d] | 1.5 at the 125 mg dose level[g] | 1.3 at the 100 mg dose level[h] |

[a]Bakatselou et al. [22]
[b]Pedersen et al. [23]
[c]Augustijns et al. [24]
[d]LLC S-AUS [25]
[e]Mithani et al. [26]
[f]Gramatté [27]
[g]Ahmed et al. [28]
[h]Schöller-Gyüre et al. [29]
[i]Weuts et al. [30]

### 11.2.2 Modified Release Amorphous Solid Dispersion Drug Products

Tsakiridou et al. investigated the usefulness of four in vitro methodologies in screening for product related differences in tacrolimus exposure after single oral administrations of three different ASDs with modified release characteristics in the fasted state; Advagraf® (a capsule containing amorphous tacrolimus dispersed in EC and HPMC), Envarsus® (a monolithic tablet containing amorphous tacrolimus) and a test tablet (containing a surfactant and amorphous tacrolimus dispersed in Eudragit-S100®) [31]. In vitro data collected with compendial apparatus IV (open loop mode) reflected the ranking of early exposure to healthy adults. A two-stage biphasic system (Inform platform, Pion Inc., MA, USA) was useful for predicting differences in early exposure between the non-disintegrating Envarsus® and the two disintegrating products. BioGIT data were useful for discussing differences in early exposure between the two disintegrating products. Prediction of product related differences in total exposure was successful only when the compendial apparatus IV (open loop mode) was used for comparing the two disintegrating products. In vitro performance testing with compendial apparatus IV was useful for qualitatively evaluating, on priori basis, differences in tacrolimus early and total exposure after administration of disintegrating products with modified release characteristics in the fasted state.

## 11.3 Model-Dependent Methodologies for Assessing the In Vivo Performance of Amorphous Solid Dispersion Drug Products in Adults Based on In Vitro Data

These methodologies are based on a model-dependent relationship between the simulated drug presence in the systemic circulation versus time after ASD administration and the observed drug presence in the systemic circulation in healthy adults versus time after ASD administration. The drug presence in the systemic circulation is simulated by using in vitro data (raw or estimated parameters) that are relevant to the drug absorption process and data (raw of estimated parameters) that are relevant to the drug disposition process.

Relevant methodologies started to appear in literature about two decades after the early publications of model-independent methodologies.

In 2016, Abuhelwa et al. established an in vitro–in vivo correlation (IVIVC) model, which provided insight into the observed variability of itraconazole pharmacokinetics after administration of two ASDs [32]. The model was based on fed/fasted pharmacokinetic data, single stage dissolution data collected under various pH conditions, and prior information about the typical and between-subject variability of luminal pH and transit times. Data were analyzed by using the population modelling approach as implemented in NONMEM.

Most of subsequent modelling procedures have been performed by applying physiologically based biopharmaceutics (PBB) modelling approaches that allowed for mechanistic understanding of the drug absorption process from the specific ASD.

Mitra et al. successfully simulated the presence of two Merck Sharp & Dohme (MSD) APIs in the systemic circulation after single dose administrations of ASDs in the fasted state using Gastroplus v. 8.5 [33]. The apparent solubility in the lumen of upper small intestine was estimated with a two-stage dissolution experiment (in Level I FaSSGF followed by Level II FaSSIF). No indication of crystallization was observed during the time frame of the in vitro experiment. With the first ASD (as an example of linear PK observed with dose escalation), the importance of apparent solubility in the upper small intestine for modelling the observed bioperformance of the product was shown. With the second ASD, the decreased fraction absorbed with dose escalation was successfully simulated, thus demonstrating the usefulness of the model to simulate the bioperformance in the scenario where saturation of absorption is observed [33].

Using the Simcyp® Simulator, version 15.0.86.0, Enami Riedmaier et al. successfully simulated venetoclax exposure after administration of an ASD in the fasted and in the fed state as well as the observed clinical drug–drug interactions [34]. Relevant simulations were based on the in vitro drug solubility and dissolution data and allowed for mechanistic understanding of venetoclax absorption process.

By using Simcyp version 17, release 1, in vitro solubility data, and data collected with in vitro transfer experiments, Litou et al. showed that, based on the calculated average fold error and the absolute average fold error, the scenario in which etravirine does not precipitate in the fed state is more representative of the in vivo data [35].

O'Dwyer et al. showed that precipitation parameters for itraconazole after administration of an ASD in the fasted state, which were estimated with a small-scale biphasic dissolution setup, were more useful than the default simulators parameters for precipitation (Simcyp V. 18, release 2) or the parameters estimated with a small-scale dissolution-permeation setup [36].

After modelling in vitro solubility and 2-stage dissolution data, and by using SimCyp v.18, release 2, Fiolka et al. built a PBB model for two ritonavir ASDs (100 mg Norvir® tablets and 100 mg Norvir® oral powder) in the fasted state (fold error predicted/observed 1.14, 0.95, 1.08 for $AUC$, $C_{max}$, and $T_{max}$, respectively) [37]. About the same period, O'Dwyer et al. built a PBB model for a ritonavir ASD (100 mg Norvir® tablets) in the fasted state with the same software version [38]. They used a precipitation rate constant derived from experimental kinetic solubility and crystalline solubility data, biphasic dissolution data, and a diffusion layer model scalar value estimated with the Erweka mini-paddle dissolution apparatus (fold error predicted/observed 1.01, 0.93, 0.68 for $AUC$, $C_{max}$, and $T_{max}$, respectively).

Arora et al. built a PBB model for a ritonavir ASD (100 mg Norvir® tablets) in the fasted state with an updated version of Simcyp (v.19); however, by using experimentally derived solubility data, pH shift experimental modelling to confirm the precipitation parameters, and compendial dissolution apparatus II data for dissolution modelling, for confirming the delayed release (disintegration) and for estimating the diffusion layer model scalar value (fold error predicted/observed 0.97, 0.82, 1.07 for *AUC*, $C_{max}$, and $T_{max}$, respectively) [39]. In the same study, Arora et al. captured the negative food effect of a ritonavir ASD (100 mg Norvir® tablets) by accounting for luminal fluid viscosity differences in the different prandial states (affecting drug diffusivity) and for the effect of drug free fraction on gut wall permeation rates [39].

Piscitelli et al. built a PBB model with Gastroplus 9.8 for an ASD of Encorafenib using in vitro solubility data [40]. The model successfully simulated the drug presence in the systemic circulation after administration in the fasted state, fed state, and under reduced gastric acid secretion conditions.

All PBB modelling approaches with ASDs summarized above have two main common characteristics:

The first is that they have focused on the complexity of drug supersaturation and drug distribution in the different species in gastrointestinal fluids. Absorption is assumed to be limited by the diffusion of drug species across the unstirred water layer with the effective permeability being a function of the unbound (molecularly dispersed) drug, the drug in the mixed micellar structures of the small intestine, and the corresponding diffusion coefficients. In relation to this, it should be mentioned that the contribution of amorphous drug nano-aggregates that may be formed when amorphous solubility in the unstirred water layer is exceeded to the effective permeability has also been considered (e.g., [41]). Nanoparticles, as bile acid micelles and potentially other colloidal species, could act as "shuttles" via diffusion across the unstirred layer attached to the intestinal epithelium sourcing unbound drug at the surface of the epithelium [42]. However, if amorphous nanoparticles are formed, and only the unbound drug can be transported via the intestinal epithelium, the effective permeability becomes epithelium limited. Experimental data relating to this argument are very limited. Recently, it was shown for the first time that after administration of Sporanox® ASD pellets to healthy adults in the fed state, itraconazole apparent concentrations in the micellar contents of the upper small intestine are well below the apparent crystalline solubility [43]. Although additional data are needed, this observation implies that, at least in the fed state, sink conditions are likely to prevail already in the bulk micellar contents. Based, also, on the apparent concentrations in the bulk colloidal contents (which were close to apparent crystalline solubility), itraconazole's effective permeability seems to be limited by the drug concentration in the colloidal bulk luminal contents [43].

The second common characteristic of PBB modelling approaches with ASDs employed today is that they have been useful for mechanistic understanding of the impact of formulation, dose, and/or food effects on the in vivo performance of ASDs, however, verification and validation activities have been very limited, if any at all. For regulatory decision-making, it has been advocated to establish that a model is credible for intended use, and that verification and validation activities are required for an *IVIVC m*odel to provide a surrogate for in vivo bioequivalence studies (i.e. biowaivers), specify in vitro dissolution acceptance parameters, and establish a link between drug product critical quality attributes and the in vivo performance of the drug product. As of December 2023, 3 out of 55 ASDs approved by the US FDA submitted an IVIVC as part of their regulatory package and none of these IVIVCs were acceptable by the FDA [44]. Very recently, Coutinho et al. published the first successfully developed compendial Level A IVIVC model of amorphous itraconazole solid dispersions [45]. The desire was to identify an in vitro dissolution method that would simulate human performance, as well as to determine formulation and manufacturing parameters that impacted in vivo itraconazole ASD tablet performance. The model was useful to assess the impact of grades of the same polymer, disintegrant level and dry granulation processing on the performance of spray-dried amorphous itraconazole solid dispersion tablets in humans. In addition, elements of the credibility of the developed IVIVC are discussed.

## 11.4 Conclusion

During the last decade, significant progress has been made regarding the in vitro assessment of the performance of orally administered ASDs. Based on published information to date, better understanding of the luminal fate of ASDs, especially in the fed state, and harnessing of the uncertainty of global regulatory acceptance of noncompendial in vitro testing methods would further increase reliability of the in vitro assessment of luminal performance of orally administered ASDs.

## References

1. Bennett-Lenane H, O'Shea JP, O'Driscoll CM, Griffin BT. A retrospective biopharmaceutical analysis of >800 approved Oral drug products: are drug properties of solid dispersions and lipid-based formulations distinctive? J Pharm Sci. 2020;109(11):3248–61.
2. Taylor LS, Zografi G. From unwanted annoyances to Oral delivery Saviors: the rollercoaster journey of amorphous drugs. Mol Pharm. 2025;22(9):5125–38.
3. Thakkar AL, Hirsch CA, Page JG. Solid dispersion approach for overcoming bioavailability problems due to polymorphism of nabilone, a cannabinoid derivative. J Pharm Pharmacol. 1977;29(12):783–4.
4. Moseson DE, Tran TB, Karunakaran B, Ambardekar R, Hiew TN. Trends in amorphous solid dispersion drug products approved by the U.S. Food and Drug Administration between 2012 and 2023. Int J Pharm: X. 2024;7:100259.
5. Ueda K, Moseson DE, Taylor LS. Amorphous solubility advantage: theoretical considerations, experimental methods, and contemporary relevance. J Pharm Sci. 2025;114(1):18–39.
6. Chowdary KPR, Suresh Babu KVV. Dissolution, bioavailability and ulcerogenic studies on solid dispersions of indomethacin in water soluble cellulose polymers. Drug Dev Ind Pharm. 1994;20(5):799–813.
7. Kapsi SG, Ayres JW. Processing factors in development of solid solution formulation of itraconazole for enhancement of drug dissolution and bioavailability. Int J Pharm. 2001;229(1–2):193–203.

8. Verreck G, Vandecruys R, De Conde V, Baert L, Peeters J, Brewster ME. The use of three different solid dispersion formulations--melt extrusion, film-coated beads, and a glass thermoplastic system--to improve the bioavailability of a novel microsomal triglyceride transfer protein inhibitor. J Pharm Sci. 2004;93(5):1217–28.
9. Six K, Daems T, de Hoon J, Van Hecken A, Depre M, Bouche MP, et al. Clinical study of solid dispersions of itraconazole prepared by hot-stage extrusion. Eur J Pharm Sci. 2005;24(2–3):179–86.
10. Kesisoglou F, Hermans A, Neu C, Yee KL, Palcza J, Miller J. Development of in vitro-in vivo correlation for amorphous solid dispersion immediate-release suvorexant tablets and application to clinically relevant dissolution specifications and in-process controls. J Pharm Sci. 2015;104(9):2913–22.
11. Vinarov Z, Abrahamsson B, Artursson P, Batchelor H, Berben P, Bernkop-Schnürch A, et al. Current challenges and future perspectives in oral absorption research: an opinion of the UNGAP network. Adv Drug Deliv Rev. 2021;171:289–331.
12. Kourentas A, Vertzoni M, Stavrinoudakis N, Symillidis A, Brouwers J, Augustijns P, et al. An in vitro biorelevant gastrointestinal transfer (BioGIT) system for forecasting concentrations in the fasted upper small intestine: design, implementation, and evaluation. Eur J Pharm Sci. 2016;82:106–14.
13. Wilson CG, Aarons L, Augustijns P, Brouwers J, Darwich AS, De Waal T, et al. Integration of advanced methods and models to study drug absorption and related processes: an UNGAP perspective. Eur J Pharm Sci. 2022;172:106100.
14. Kostantini C, Spilioti E, Bevernage J, Ceulemans J, Hansmann S, Hellemans K, et al. Screening for differences in early exposure in the fasted state with in vitro methodologies can be challenging: experience with the BioGIT system. J Pharm Sci. 2023;112(8):2240–8.
15. Holzem FL, Weck A, Schaffland JP, Stillhart C, Klein S, Bauer-Brandl A, et al. Biopredictive capability assessment of two dissolution/permeation assays, µFLUX™ and PermeaLoop™, using supersaturating formulations of Posaconazole. Eur J Pharm Sci. 2022;176:106260.
16. Sironi D, Rosenberg J, Bauer-Brandl A, Brandl M. PermeaLoop™, a novel in vitro tool for small-scale drug-dissolution/permeation studies. J Pharm Biomed Anal. 2018;156:247–51.
17. Elkhabaz A, Moseson DE, Brouwers J, Augustijns P, Taylor LS. Interplay of supersaturation and solubilization: lack of correlation between concentration-based supersaturation measurements and membrane transport rates in simulated and aspirated human fluids. Mol Pharm. 2019;16(12):5042–53.
18. Holzem FL, Petrig Schaffland J, Brandl M, Bauer-Brandl A, Stillhart C. Using molecularly dissolved drug concentrations in PBBMs improves the prediction of oral absorption from supersaturating formulations. Eur J Pharm Sci. 2024;194:106703.
19. Raina SA, Zhang GGZ, Alonzo DE, Wu J, Zhu D, Catron ND, et al. Impact of solubilizing additives on supersaturation and membrane transport of drugs. Pharm Res. 2015;32(10):3350–64.
20. Vertzoni M, Markopoulos C, Symillides M, Goumas C, Imanidis G, Reppas C. Luminal lipid phases after administration of a triglyceride solution of danazol in the fed state and their contribution to the flux of danazol across Caco-2 cell monolayers. Mol Pharm. 2012;9(5):1189–98.
21. Markopoulos C, Imanidis G, Vertzoni M, Symillides M, Parrott N, Reppas C. In vitro and ex vivo investigation of the impact of luminal lipid phases on passive permeability of lipophilic small molecules using PAMPA. Pharm Res. 2013;30(12):3145–53.
22. Bakatselou V, Oppenheim RC, Dressman JB. Solubilization and wetting effects of bile salts on the dissolution of steroids. Pharm Res. 1991;8(12):1461–9.
23. Pedersen BL, Müllertz A, Brøndsted H, Kristensen HG. A comparison of the solubility of danazol in human and simulated gastrointestinal fluids. Pharm Res. 2000;17(7):891–4.
24. Augustijns P, Wuyts B, Hens B, Annaert P, Butler J, Brouwers J. A review of drug solubility in human intestinal fluids: implications for the prediction of oral absorption. Eur J Pharm Sci. 2014;57:322–32.

25. LLC S-AUS. DANOCRINE®, Brand of DANAZOL CAPSULES, USP [Reference ID: 3061327]: FDA; 2011. Available from: https://www.accessdata.fda.gov/drugsatfda_docs/label/2011/017557s033s039s040s041s042lbl.pdf
26. Mithani SD, Bakatselou V, TenHoor CN, Dressman JB. Estimation of the increase in solubility of drugs as a function of bile salt concentration. Pharm Res. 1996;13(1):163–7.
27. Gramatté T. Griseofulvin absorption from different sites in the human small intestine. Biopharm Drug Dispos. 1994;15(9):747–59.
28. Ahmed IS, Aboul-Einien MH, Mohamed OH, Farid SF. Relative bioavailability of griseofulvin lyophilized dry emulsion tablet vs. immediate release tablet: a single-dose, randomized, open-label, six-period, crossover study in healthy adult volunteers in the fasted and fed states. Eur J Pharm Sci. 2008;35(3):219–25.
29. Schöller-Gyüre M, Boffito M, Pozniak AL, Leemans R, Kakuda TN, Woodfall B, et al. Effects of different meal compositions and fasted state on the oral bioavailability of etravirine. Pharmacotherapy. 2008;28(10):1215–22.
30. Weuts I, Van Dycke F, Voorspoels J, De Cort S, Stokbroekx S, Leemans R, et al. Physicochemical properties of the amorphous drug, cast films, and spray dried powders to predict formulation probability of success for solid dispersions: etravirine. J Pharm Sci. 2011;100(1):260–74.
31. Tsakiridou G, O'Dwyer PJ, Margaritis A, Box KJ, Vertzoni M, Kalantzi L, et al. On the usefulness of four in vitro methodologies in screening for product related differences in tacrolimus exposure after oral administration of amorphous solid dispersions with modified release characteristics in the fasted state. J Drug Deliv Sci Technol. 2022;69:102990.
32. Abuhelwa AY, Mudge S, Hayes D, Upton RN, Foster DJ. Population in vitro-in vivo correlation model linking gastrointestinal transit time, pH, and pharmacokinetics: itraconazole as a model drug. Pharm Res. 2016;33(7):1782–94.
33. Mitra A, Zhu W, Kesisoglou F. Physiologically based absorption Modeling for amorphous solid dispersion formulations. Mol Pharm. 2016;13(9):3206–15.
34. Emami Riedmaier A, Lindley DJ, Hall JA, Castleberry S, Slade RT, Stuart P, et al. Mechanistic physiologically based pharmacokinetic Modeling of the dissolution and food effect of a biopharmaceutics classification system IV compound-the venetoclax story. J Pharm Sci. 2018;107(1):495–502.
35. Litou C, Turner DB, Holmstock N, Ceulemans J, Box KJ, Kostewicz E, et al. Combining biorelevant in vitro and in silico tools to investigate the in vivo performance of the amorphous solid dispersion formulation of etravirine in the fed state. Eur J Pharm Sci. 2020;149:105297.
36. O'Dwyer PJ, Imanidis G, Box KJ, Reppas C. On the usefulness of two small-scale in vitro setups in the evaluation of luminal precipitation of lipophilic weak bases in early formulation development. Pharmaceutics [Internet]. 2020; 12(3):[272 p.].
37. Fiolka T, Van Den Abeele J, Augustijns P, Arora S, Dressman J. Biorelevant two-stage in vitro testing for rDCS classification and in PBPK modeling-case example ritonavir. J Pharm Sci. 2020;109(8):2512–26.
38. O'Dwyer PJ, Box KJ, Imanidis G, Vertzoni M, Reppas C. On the usefulness of four in vitro methods in assessing the intraluminal performance of poorly soluble, ionisable compounds in the fasted state. Eur J Pharm Sci. 2022;168:106034.
39. Arora S, Pansari A, Kilford P, Jamei M, Gardner I, Turner DB. Biopharmaceutic in vitro in vivo extrapolation (IVIV_E) informed physiologically-based pharmacokinetic model of ritonavir Norvir tablet absorption in humans under fasted and fed state conditions. Mol Pharm. 2020;17(7):2329–44.
40. Piscitelli J, Hens B, Tomaszewska I, Wollenberg L, Litwiler K, McAllister M, et al. Effect of food and a proton-pump inhibitor on the absorption of encorafenib: an in vivo-in vitro-in silico approach. Mol Pharm. 2023;20(5):2589–99.
41. Stewart AM, Grass ME. Practical approach to Modeling the impact of amorphous drug nanoparticles on the oral absorption of poorly soluble drugs. Mol Pharm. 2020;17(1):180–9.

42. Sugano K, Kataoka M, Mathews Cda C, Yamashita S. Prediction of food effect by bile micelles on oral drug absorption considering free fraction in intestinal fluid. Eur J Pharm Sci. 2010;40(2):118–24.
43. Dietrich S, Dimoula M, Argyropoulos T, Ceulemans J, Goumas K, Vertzoni M, et al. On the processes limiting oral drug absorption when amorphous solid dispersions are administered after a high-calorie, high-fat meal: Sporanox® pellets. Eur J Pharm Sci. 2024;199:106798.
44. Zhao G WK, Nguyen D, Lim HL, Rage B, Lu D. FDA perspective on dissolution testing for development of high-risk oral drug products containing amorphous solid dispersions (T1230–09–051). American Association of Pharmaceutical Scientist (AAPS) PharmSci 360 meeting—Salt Lake City—October 22, 2024.; 2024. Available from: https://posters.aaps.org/aaps/?menu=16&browseby=3&sortby=2&ce_id=2692&trend=19504#!*menu=16*browseby=3*sortby=2*ce_id=2692*trend=19504
45. Coutinho AL, Adhikari A, Krug S, Kane M, Hollenbeck RG, Hoag SW, et al. In vitro-in vivo correlation of amorphous solid dispersion enabled itraconazole tablets. Pharm Res. 2025;42(3):485–502.

# Chapter 12
# Regulatory Aspects and Commercialization of Drug Products Containing Amorphous Solid Dispersions

**Margarita Maria Gkantanidou, Ioannis Nikolakakis, Ioannis Partheniadis, and Maria Malamatari**

**Abstract** Amorphous solid dispersions (ASDs) have evolved to a commonly adopted approach for enhancing the bioavailability of poorly water-soluble drugs. This chapter discusses the regulatory framework relevant to the development of products containing ASDs, highlighting key ICH guidelines and their implications for stability, impurity control, specifications, pharmaceutical development, and bioequivalence studies. Differences in the Summary of Product Characteristics (SmPC) between ASD-based and conventional formulations are outlined, including aspects related to food effect, administration requirements, and packaging. A case study of Kalydeco® tablets and oral granules illustrates how products containing ASDs are developed, assessed, and managed throughout the product lifecycle. Overall, science-based and regulatory-informed pharmaceutical development plays a major role in the quality, safety, and efficacy of medicinal products containing ASDs.

**Keywords** ICH guidelines · Bioavailability · Quality by Design (QbD) · Regulatory assessment · Ivacaftor · Pharmaceutical development

---

M. M. Gkantanidou · I. Nikolakakis (✉)
Laboratory of Pharmaceutical Technology, School of Pharmacy, Faculty of Health Sciences, Aristotle University of Thessaloniki, Thessaloniki, Greece
e-mail: mgkanta@pharm.auth.gr; yannikos@pharm.auth.gr

I. Partheniadis
Department of Pharmacy, Faculty of Health and Medical Sciences, University of Copenhagen, Copenhagen, Denmark
e-mail: ioannis.partheniadis@sund.ku.dk

M. Malamatari
Institute of Pharmaceutical Science, Faculty of Life Sciences & Medicine, Franklin-Wilkins Building, King's College London, London, UK

Institute for Globally Distributed Open Research and Education (IGDORE), London, UK
e-mail: maria.malamatari@igdore.org



I. Nikolakakis, I. Partheniadis (eds.), *Amorphous Solid Dispersions*, AAPS Introductions in the Pharmaceutical Sciences 5,
https://doi.org/10.1007/978-3-032-24883-1_12

## 12.1 Introduction

Amorphous solid dispersions (ASDs) are homogeneous amorphous mixtures of a drug substance and water-soluble polymer [1]. While in the past, ASDs were avoided mainly due to concerns regarding their physical stability, currently, they are recognized as an enabling approach designed to ultimately enhance bioavailability. This shift from avoidance to adoption is reflected by the number of authorized drug products which contain ASDs [2]. Moreover, ASDs are used even from early drug discovery and preclinical development with a direct line of sight to clinical development to increase the probability of success during the clinical phase [3].

This chapter discusses regulatory aspects of drug products containing ASDs. It covers the key international guidelines to consider during the pharmaceutical development of such products. Differences in the summary of product characteristics (SmPC) between conventional and ASD-based dosage forms are highlighted, as such differences are underpinned by the effects of ASDs on biopharmaceutics, pharmacokinetics, stability, and conditions of use of the drug product. The example of an authorized ASD-based drug product (i.e., Kalydeco® tablets) is presented with the aim of shedding light on aspects pertaining to the regulatory assessment and commercialization of ASDs.

## 12.2 ICH Guidelines Related to ASDs

The International Council for Harmonization of Technical Requirements for Human Use (ICH) is a unique harmonization initiative, which brings together regulatory authorities and the pharmaceutical industry [4]. Its primary goal is to produce harmonized technical requirements to ensure the development and authorization of medicinal products, which are safe, effective, and of high quality. Since its foundation in 1990, ICH has produced nearly 80 guidelines categorized into quality (Q), safety (S), efficacy (E), and multidisciplinary (M) topics. The ICH guidelines are adopted by regulatory authorities around the world, including its founding regulatory members: EU, USA, and Japan.

The following section outlines the key ICH guidelines to consider during the pharmaceutical development of drug products containing ASDs. These guidelines also play a vital role during the assessment of initial marketing authorization applications (MAAs) and variations by regulatory agencies. However, it should be highlighted that guidelines also allow flexibility and the use of alternative approaches if sufficient scientific justification and evidence are provided.

### 12.2.1 ICH Q1A-Q1F Guidelines on Stability

The energy level difference between the amorphous and crystalline forms of a drug substance is linked to enhanced properties such as increased solubility and dissolution of the amorphous form. At the same time, the inherent thermodynamic instability of the amorphous form renders it susceptible to recrystallisation over time [5]. Thus, stability of ASDs remains a core challenge for formulation scientists. Apart from the physical stability of ASDs, their chemical stability is also of concern, as the molecules in the amorphous state have higher free energy and could be more prone to chemical degradation [6].

The ICH Q1A(R2) guideline provides recommendations on the stability data package to be submitted as part of the initial marketing authorization application for drug substances and drug products. The purpose of stability testing is to evaluate the influence of environmental factors (i.e., temperature, humidity, light) on the quality of the drug product. As per the general case, long-term stability testing (25 °C ± 2 °C/60% RH ± 5% RH or 30 °C ± 2 °C/65% RH ± 5% RH) on at least three primary batches should be submitted to the authorities (i.e., batches with the same formulation and container closure system as proposed for marketing produced by a manufacturing method representative of the commercial-scale method). Studies under accelerated conditions (40 °C ± 2 °C/75% RH ± 5% RH), and, where applicable, intermediate conditions (30 °C ± 2 °C/65% RH ± 5% RH) should also be submitted. Accelerated studies increase the rate of chemical degradation or physical change of a drug product and are used to evaluate the impact of short-term excursions on product quality (e.g., excursions during shipping). They could also be used to support long-term stability. While limited extrapolation of the real-time data beyond the observed range could be accepted upon justification (i.e., when the long-term and accelerated data demonstrate minimal or no change over time), it should be noted that risk-based predictive stability tools (e.g., accelerated stability assessment model), which are routinely used within pharmaceutical development, are not broadly used to support regulatory submissions [7].

Based on the results of the stability testing, a shelf life and storage conditions are established. When long-term stability data does not cover the proposed shelf life at the time of approval, a stability commitment should be given to continue the stability studies post approval to confirm the shelf life. The commitment should also include carrying out studies on production-scale batches, when such data were not available prior to approval. At least one production batch of the product manufactured in every strength and every primary packaging type should be included in the stability program (unless none is produced during that year). Out-of-specification results should be investigated and reported to the competent regulatory authorities.

Apart from the ICH Q1A(R2) guideline, the ICH Q1 guidelines on stability include the ICH Q1B guideline on photostability testing, the ICH Q1C guideline on stability testing of new dosage forms, the ICH Q1D guideline on the reduced designs for stability testing, the ICH Q1E on evaluation of stability data, and the ICH Q1F on the stability data package for registration in climatic zones III (hot and dry) and

IV (hot and humid) [8]. Currently, the ICH Q1 draft guideline has been published combining the various guidelines (ICH Q1A-F and ICH Q5C) into a single guideline on stability issues.

### 12.2.2 ICH Q3A-Q3D Guidelines on Impurities

ICH Q3 guidelines set the regulatory expectations for controlling the impurities in the drug substances and drug products.

Specifically, the ICH Q3C(R9) guideline focuses on the use of solvents in the manufacturing process, setting limits for residual solvents in drug products. Organic solvents (e.g., methanol, ethanol, isopropanol, acetone) are commonly used in the preparation of ASDs by spray drying [9]. The guideline classifies solvents based on risk assessment, and the limit proposed for each organic solvent should be considered. Secondary drying may often be required as a step of the manufacturing process to reduce residual solvent below the acceptable concentration limit.

ICH Q3D(R2) guideline refers to the control of elemental impurities in new drug products. Potential sources of elemental impurities include the synthesis of the drug substance and excipients, the manufacturing equipment, and the container closure arrangement. The guideline mentions hot-melt extrusion as an example of a process that may lead to the incorporation of elemental impurities derived from the manufacturing equipment.

### 12.2.3 ICH Q6A Guidelines on Specifications

The ICH Q6A guideline focuses on the establishment of a set of specifications for new drug substances and new drug products [10]. Specifications include the tests, procedures, and acceptance criteria, which play a major role in assuring the quality of the new drug product at release and at shelf life. The tests/criteria are divided into categories: the universal tests/criteria, which are applicable to all new drug products, and the specific tests/criteria, which depend on the dosage form. Universal tests for drug products include description, identification, assay, and impurities arising from degradation of the drug substance or from the manufacturing process. Specific tests for solid oral dosage forms, such as tablets and hard capsules, include dissolution, disintegration, hardness/friability, uniformity of dosage units, water content, and microbial limits. The specifications proposed should be justified based on relevant development data, compendial standards, data from batches of the drug product used in clinical studies, and a reasonable range of analytical and manufacturing variabilities. In case the acceptance criteria proposed are too broad, points are raised during assessment requesting the limits to be tightened in line with batch analysis data and compendial requirements, unless sufficient justification is provided.

Crystallinity and dissolution performance are two (often inherently linked) tests that are critical for drug products containing ASDs [11]. Given that dissolution is a prerequisite for absorption, dissolution testing is the only quality-control tool that could be linked with the in vivo performance of a drug. A perspective from the IQ Consortium Dissolution Working Group defines different types of dissolution methods used during development, registration, and commercialization: quality-control (QC) dissolution method, biorelevant dissolution method, and clinically relevant dissolution method. It should be noted that the dissolution testing included as part of the drug product specifications is usually sufficient to be a well-developed QC method, which would be able to detect variations during routine product manufacturing and changes during storage that could impact product performance.

Regarding crystallinity, it may be found in an ASD either due to incomplete amorphization or due to recrystallization (structure reorganization). It can be a complex critical quality attribute of ASDs, as it may be challenging to detect low levels of crystallinity, but also, it is not straightforward how to translate crystallinity to release performance and clinical impact [12].

### *12.2.4 ICH Q8 Guidelines on Pharmaceutical Development*

The ICH Q8(R2) guideline describes science- and risk-based approaches for pharmaceutical development recognizing that "*quality cannot be tested into products but should be built in by design*" [10]. Moreover, it highlights the link between enhanced process and product understanding, and regulatory flexibilities. In the annex of the guideline, the principles of the Quality by Design (QbD) are outlined. As per the guideline, QbD is defined as *"a systematic approach to development that begins with predefined objectives and emphasises product and process understanding and process control, based on sound science and quality risk management."* The elements of the QbD paradigm in pharmaceutical development are presented in Fig. 12.1. ICH Q8 guideline, ICH Q9 guideline on quality risk management, and ICH Q10 guideline on pharmaceutical quality system work together, forming a unique framework at different stages of the product lifecycle (e.g., product development, process development, commercial manufacturing, and continual verification and improvement). Implementation of ICH Q8, Q9, and Q10 provides a systematic, risk- and science-based approach to pharmaceutical manufacturing and development [10]. The following ICH guidelines are also building blocks of the ICH quality vision: ICH Q11 on development and manufacture of drug substances, ICH Q12 on lifecycle management, ICH Q13 on continuous manufacturing of drug substances and drug products, and ICH Q14 on analytical procedure development [13].

While important for all drug products, pharmaceutical development approaches and regulatory systems, implementation of the ICH quality vision is of great value for dosage forms containing ASDs, where enhanced product and process understanding are required to assure product quality. A recent review on modern approaches of QbD for ASDs is provided by Koo et al. [14].

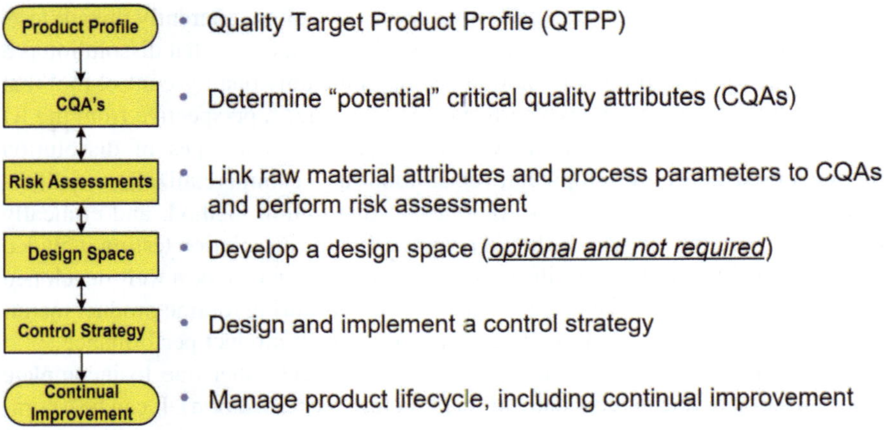


**Fig. 12.1** Elements of the QbD paradigm in pharmaceutical development in line with the ICH Q8(R2) guideline. (Adapted from ICH Q-IWG Basic Training (2010))

### *12.2.5 ICH M13A on Guideline of Immediate-Release Solid Oral Dosage Forms*

ICH M13A guideline is the first part of the ICH M13 guideline series (M13A-C) on the bioequivalence for immediate-release solid oral dosage form. It came into effect in January 2025 and it sets the requirements for the study design and data analysis to support bioequivalence of immediate-release solid oral dosage forms. ICH M13A introduces the definition of high-risk products where *"the drug substance characteristics in combination with the complexity of the formulation design or manufacturing process lead to an increased likelihood that in vivo performance will be impacted differently by varying gastrointestinal conditions between fasted and fed conditions*." The guideline lists solid dispersions of poorly soluble drugs as an example of high-risk products due to their complex formulation and/or manufacturing method. For high-risk products, bioequivalence studies should be conducted under both fasting and fed conditions, irrespective of the drug labelling relating to food intake, as long as there are no safety/ethical concerns for the study participants. Sufficient justification that there is no formulation-dependent impact on the absorption of the drug could waive the need for studies under both states. Such justification could be based on differences between the composition of formulations, the BCS classification of the drug substance, in vitro testing in biorelevant media, and modelling (e.g., physiologically based biopharmaceutics modelling) [15].

## 12.3 Summary of Product Characteristics of Medicinal Products Containing ASDs

The summary of product characteristics (SmPC) is a legally binding regulatory document, which describes the quality, safety, and efficacy of a medicinal product. It forms the basis of information for health-care professionals on how to use the product [16]. SmPC is a live document, with its information being updated throughout the lifecycle of the product. Although the structure of the SmPC template is harmonized across pharmaceutical products, frequently, there are differences in the content of the SmPC sections between conventional formulations and dosage forms based on ASDs. Such differences reflect the effects of ASDs on biopharmaceutics, pharmacokinetics, stability, and conditions of use. Below, a brief description of such differences is given, illustrated with examples from marketed authorized products.

Part 2 of the SmPC focuses on qualitative and quantitative composition of the medicinal product. Products based on ASDs typically include functional excipients (e.g., polymers and surfactants), which are critical for the product's stability (e.g., retaining the amorphous form upon storage), bioavailability (e.g., enhancing solubility), and manufacturability. Unlike conventional excipients, these components tend to have direct implications for product performance requiring sufficient regulatory justification. Part 3 of the SmPC contains information on the pharmaceutical form. An ASD is an intermediate, which is formulated into a final dosage form. The most common oral dosage form for ASDs is tablets, followed by granules/pellets, powders/granules for oral suspension, and capsules [2].

Under clinical particulars, part 4.2 of the SmPC informs on posology and method of administration for each indication and subpopulation. Several oral drug products based on ASDs contain statements that the tablets, capsules or granules should be swallowed whole, and should not be chewed, broken or crushed. For example, Kaletra® 200 mg/50 mg film-coated tablets contain a fixed combination of lopinavir and ritonavir, formulated as a solid dispersion by melt extrusion [17]. It is indicated for the treatment of HIV in adults, and the pediatric population above the age of 2 years (SmPC Kaletra 200 mg/50 mg FCT). A study in pediatric patients found that the exposure of lopinavir and ritonavir was reduced by 40% upon administration of crushed tablets compared to whole tablets, with implications for medication resistance and treatment failure [18]. Potential physicochemical mechanisms involved behind the impaired bioavailability of ASDs upon crushing/grinding of the dosage form include decreased particle size (resulting in rapid supersaturation, early nucleation, and crystal growth), induction of crystallization, and/or amorphous phase separation [19].

Products based on ASDs may contain different statements on food intake compared to conventional formulations, due to differences in food effect. Most approved ASDs exhibit positive food effect, while there are fewer products which exhibit negative food effect. There are also examples where ASDs have used a bio-enabling approach to formulate away food effect. For example, Noxafil® oral suspension should be administered immediately or after a meal, as absorption of posaconazole was significantly increased leading to a three-fold increase in systemic exposure under fed conditions compared to fasted conditions [20]. In contrast, Noxafil® gastro-resistant tablets, which contain posaconazole formulated as a polymeric solid dispersion, may be taken without regard to food intake [21]. The absence of food effect for the delayed-release solid dispersion was attributed to the sustained supersaturation in human intestine, leading to increased absorption in the fasted state [22]. Among other reasons, such difference in the food effect resulted in SmPC warnings, which highlight the non-interchangeability between the two formulations.

More rarely, differences could also be found in parts 4 and 4 of the SmPC which focus on interactions with other medicinal products. Such differences could be related to different pH-dependent drug–drug interactions, mainly observed upon coadministration with acid-reducing agents (ARAs), which elevate the gastric pH. For example, Phyrago® tablets (dasatinib ASD) were bioequivalent to Sprycel® tablets (crystalline dasatinib monohydrate). In contrast to Sprycel®, Phyrago's bioavailability was not affected upon coadministration with omeprazole or famotidine [23]. Such difference could be attributed to the reduced pH-dependent solubility of the ASD formulation compared to the crystalline form [24]. Thus, the clinical benefit of Phyrago is that there are no restrictions on concomitant use with ARAs, such as proton pump inhibitors and H2 receptor antagonists.

Under pharmacological properties, part 5 of the SmPC informs on the pharmacokinetic properties, including absorption, distribution, metabolism, and excretion (ADME), and pharmacokinetics in special populations (e.g., elderly, renal impairment). ASDs have been used as a formulation approach with "increased drug absorption which is mainly due to increased concentrations of molecularly dissolved API and facilitated diffusion through the unstirred layer by drug-rich particles" [25]. The increased absorption leads to enhanced bioavailability, including higher and less variable (within-subject and between-subject) exposure [24, 26]. A characteristic example of improved absorption is Tolsura® capsules, which contain a spray-dried ASD of itraconazole prepared by using the SUBA® proprietary technology [27]. Several multiple-dose relative bioavailability studies conducted under fed conditions showed that the steady-state $C_{max}$ and $AUC$ of itraconazole from Tolsura 130 mg capsules (2 × 65 mg) were 22% and 15% higher, when compared to Sporanox® 200 mg capsules (2 × 100 mg). The slightly higher overall exposure achieved by Tolsura® capsules versus Sporanox® capsules was deemed to be clinically insignificant, while the $C_{max}$ of itraconazole after administration of Tolsura® was less than the highest $C_{max}$ reported in the product information of Sporanox® capsules, providing assurance that the safety profile of Tolsura® is likely to be similar to that of Sporanox® capsules [28]. Tolsura® capsules were characterized as "super-bioavailable" itraconazole, as a dose of 130 mg once daily can be administered instead of 200 mg of Sporanox® capsules. Similar results were reached by a population pharmacokinetics (popPK) modelling, which included data from twelve Phase 1 single- and multiple-dose studies comparing the two products. Moreover, the popPK analysis indicated that Tolsura® capsules display approximately 30% reduction in exposure variability compared to Sporanox® capsules [29].

A randomized controlled trial comparing Tolsura® capsules to a conventional itraconazole formulation for treatment of endemic mycosis reported that Tolsura® capsules are well tolerated, with 50% less gastrointestinal adverse events compared to conventional itraconazole group [30]. Such observation could be attributed to the lower dose of itraconazole Tolsura® capsules. The study concluded that the adverse events and tolerability were similar between the two treatments, potentially due to the inadequate statistical power of the study to reliably detect true differences in the side effects between treatments.

Under pharmaceutical particulars, part 6.1 of the SmPC provides a list of excipients included in the drug product. Polymer selection is critical for the development of an ASD, as it is linked with the manufacturability of the dosage form, the stability of the amorphous drug substance in the solid state, as well as the in vitro (i.e., solubility, dissolution, supersaturation) and in vivo performance (i.e., bioavailability) of the drug. Polymers used in ASDs can be classified based on their chemical structure into polyvinylactam polymers (e.g., polyvinylpyrrolidone, copovidone), cellulosic polymers (hydroxypropyl methyl cellulose, hypromellose acetate succinate), acrylate and methacrylate (co-)polymers and other polymers (e.g., different types of Eudragit® S100) and other polymers (e.g., polyethylene glycol) [31]. Copovidone (PVPVA) is the most used polymer in the currently approved ASD-based products, followed by hypromellose acetate succinate (HPMCAS). Due to their different physicochemical properties (e.g., glass transition temperature), PVPVA and HPMCAS are more commonly used in ASDs produced by hot-melt extrusion and spray drying, respectively [2]. Apart from the polymeric carrier, other excipients added in the downstream processing of ASDs include diluents (e.g., microcrystalline cellulose), disintegrants (e.g., croscarmellose sodium), lubricants (e.g., magnesium stearate) and excipients for film coating [2, 32].

Parts 6.3 and 6.5 give information on the shelf life and the nature and contents of the container closure system (CCS). Such statements are granted upon assessment of the stability testing conducted in line with the ICH Q1 guidelines. Plastic bottles and blisters are the most widely used primary CCS for ASD-based tablets and capsules in the US and EU markets, respectively. High-density polyethylene (HDPE) bottles with polypropylene (PP) closures containing integrated desiccants (e.g., silica gel, molecular sieves) are commonly used, as they offer child-resistant packaging with the ability to protect the dosage form moisture and its implications for the stability of the ASD (i.e., amorphous–amorphous phase separation, drug crystallization). Several types of blisters have been used for tablets containing ASDs, including Alu/Alu blisters made of aluminum, and blisters made from thermoformable materials such as polyvinyl chloride (PVC), polyethylene (PE) and polychlorotrifluoroethylene (PCTFE). For example, in the case of Aquipta® tablets and Glecaprevir/Pibrentasvir tablets (also known as Maviret®), the primary CCS consists of aluminum foil and PVC/PE/PCTFE blisters [33, 34]. PCTFE is the most impermeable plastic film to the transmission of water vapor. Despite its higher cost compared to standard blister polymers (e.g., PVC), the use of PCTFE blisters for products containing ASDs is justifiable due to its superior ability to act as a moisture barrier [29]. It should be noted that products containing ASDs exhibit shelf life in the same range as conventional drug products (i.e., 2–6 years). This indicates that along with formulation development, selection of a suitable CCS is a critical step in product development as it contributes to the stability of the formulation upon storage [30].

## 12.4 Pharmaceutical Development and Regulatory Authorization of a Medicinal Product Throughout Its Lifecycle: The Example of Kalydeco® Tablets

In the following section, the drug development and authorization of Kalydeco® tablets is presented, including data published by Vertex Pharmaceuticals Incorporated (USA), the European public assessment report (EPAR), and the FDA application review files [35, 36]. Focus is placed on the quality and biopharmaceutics of the drug product.

Kalydeco® film-coated tablets contain the drug substance ivacaftor and are indicated for the treatment of cystic fibrosis (CF) in adults and pediatric patients >6 years of age, who have certain mutations in the cystic fibrosis transmembrane conductance regulator (CFTR) gene. Kalydeco was designated as an orphan medicinal product for the treatment of CF. Specifically, it was submitted in October 2011 as a full application under Art. 8 [3] of Directive 2001/83/EC in the EMA, and under section 505(b) of the Federal Food, Drug, and Cosmetic Act in the USA. Currently, it is approved in several countries including the USA, UK, Australia, Canada, and throughout the EU.

Ivacaftor was discovered after high-throughput screening of approximately 228,000 compounds from the Vertex Pharmaceuticals compound library. Following the stages of hit identification, hit-to-lead and lead optimization, ivacaftor was selected as the final compound for further preclinical and clinical development.

Based on in silico evaluation of the compound (i.e., Lipinski's rule of five, ADMET predictions), ivacaftor appeared to have the right profile to be orally absorbed (Fig. 12.2). However, measured physicochemical properties (i.e., melting point of 292 °C, Log$P > 5$, and poor aqueous solubility <0.05 mcg/ml) revealed that solubility is limited by its solid state and by solvation.

As ivacaftor fits primarily the definition of a "brick dust" molecule due to its very high melting point (i.e., >>200 °C), an enabling formulation approach is required to diminish the impact of the solid state on dissolution and solubility [38, 39]. Formation of an ASD was the formulation approach of choice, with the high glass transition temperature of ivacaftor ($T_g$: 175 °C) and its molecular weight (MW: 393 g/mol) being predictors of good long-term stability [40]. ASDs of ivacaftor prepared by spray drying exhibited a bioavailability in mice, rats, rabbits, and dogs, which was increased markedly compared to the bioavailability of the crystalline form in suspension. During the preformulation stage, the effect of drug loading on the solubility of ASDs of ivacaftor with HPMCAS was studied. The ASD with 50%

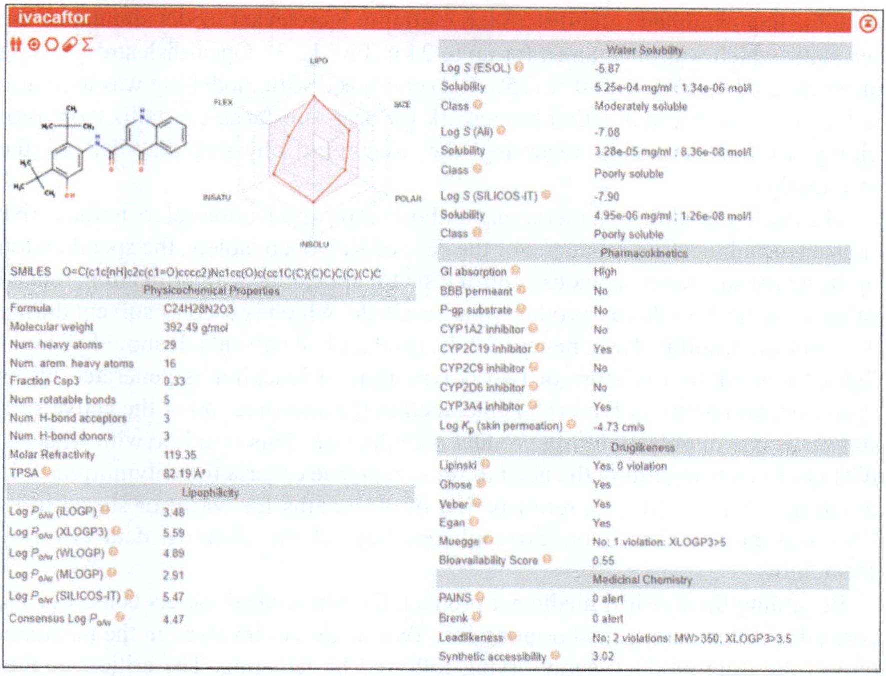


**Fig. 12.2** ADME parameters of ivacaftor computed by the SwissADME web tool [37]

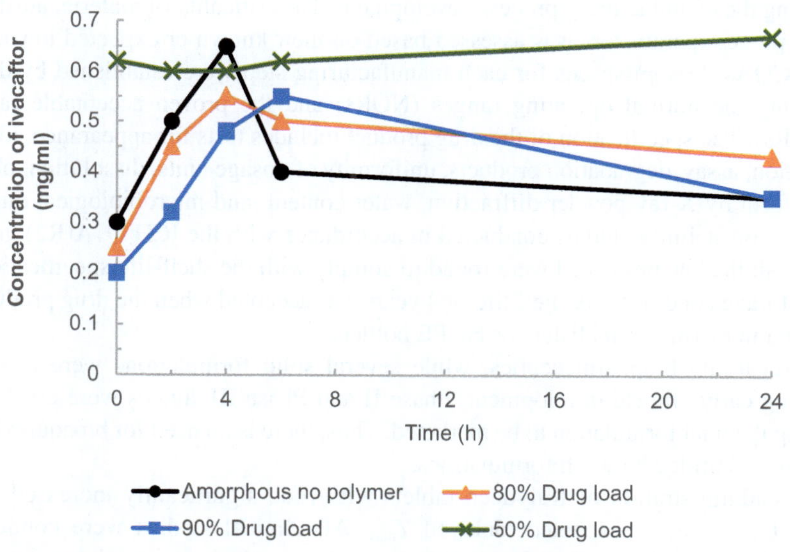


**Fig. 12.3** Apparent solubility of ivacaftor from solid dispersions with varying drug loading in biorelevant media simulating the fed state. (Adapted from [36])

drug loading exhibited solubility of 0.6 mg/ml in biorelevant media simulating the fed state, which was maintained for up to 24 h (Fig. 12.3). Open-dish stress testing at accelerated conditions (40 °C/75% RH) combined with modelling was used as a risk assessment and mitigation framework for drug substance crystallization, providing further assurance regarding the long-term physical stability of the formulation.

The quality module of a marketing authorization application refers to the active substance and the drug product. For the case of Kalydeco tablets, the specification of the active substance includes control tests for appearance, identification, related substances, acetamide (by-product of acetonitrile, which is used as solvent during the synthesis), sulfated ash, heavy metals, and residual solvents. It should be highlighted that while a mixture of two polymorphs of ivacaftor is generated by its synthesis, no test for polymorph is included in the specification of the active substance, as it is dissolved during product manufacture. This is in line with what the ICH Q6A states regarding the need to set acceptance criteria for polymorphism in the drug substance [41]. A retest period of 30 months for ivacaftor stored in its CSS was granted based on extrapolation beyond the stability data covering 18 months.

Regarding the finished medicinal product, the film-coated tablets consist of the core tablet, film coating, and printing ink. Two stages are involved in the manufacture of the drug product: spray drying followed by tableting. The critical quality attributes (CQAs) for the tablets included identity, assay, dissolution, and a blue color (FDA Chemistry Review). A Quality by Design (QbD) approach was employed during the manufacturing process development. The criticality of material attributes and process parameters was assessed based on their known or expected impact on the CQAs. Design spaces for each manufacturing step were established by determining the normal operating ranges (NORs) and the proven acceptable ranges (PARs). The specification of the drug product includes tests for appearance, identification, assay, degradation products, uniformity of dosage units, dissolution, physical form by X-ray powder diffraction, water content, and microbiological quality. Based on stability studies conducted in accordance with the ICH Q1A(R2) guideline, all the batches tested were found to comply with the shelf-life specification at all storage conditions. A shelf life of 4 years was accepted when the drug product is stored in thermoform blisters or HDPE bottles.

Regarding biopharmaceutics, while several solid formulations were assessed during early clinical development, Phase II and Phase III studies were conducted using the final formulation to be marketed. Thus, there is no need for bioequivalence studies to bridge between formulations.

Coadministration of Kalydeco tablets with food significantly increased *AUC* and $C_{max}$, and, as expected, delayed $T_{max}$. All clinical studies were conducted advising patients to take the medicinal product with food containing fat (i.e., high-fat, high-calorie meal). To avoid the impact that different types of meals may have on drug exposure, the product information specifies that Kalydeco tablets should be taken with fat-containing foods. Due to the absence of bioequivalence data between whole and crushed tablets, the tablets should be swallowed whole [42].

In 2015, the marketing authorization of Kalydeco was extended by adding a new pharmaceutical form to support an extension of the target population. Specifically, the new formulation consisted of oral granules and was indicated for the treatment of pediatric patients aged 2 years and older. The Quality Target Product Profile (QTPP) of the finished product included characteristics such as bioavailability, safety and efficacy, palatability, and an immediate age-appropriate pediatric formulation, which could be mixed with food, and with a shelflife of 2 years at room temperature [35]. Compared to tablets, the oral granules also contained mannitol as filler and sucralose as sweetener. The manufacturing process of the granules involves spray drying to produce the ASD followed by secondary drying, blending, compression, and filling into sachets. An extensive QbD approach was also employed for the granules. Product specification includes similar tests as for the tablets, while the acceptance criteria were justified based on pilot-scale batches of the oral granules. As several strengths of oral granules were produced differing only by the amount of granules filled into the sachet, a bracketing stability test design was used during formal stability studies where only the highest and lowest strengths were tested at all time points. In line with the ICH Q1D, the use of a reduced design, such as bracketing, when it is scientifically sound to assume that the stability of the intermediate samples is represented by the extremes, permits more efficient use of resources. Stability data for 1 year under long-term conditions (25 °C/60% RH) and 6 months under accelerated conditions (40 °C/75%) were provided with the variation application. In line with the ICH Q1E guideline and the CHMP guideline on the declaration of storage conditions, a 2-year shelf life and storage below 30 °C were granted initially. An in-use stability study where the granules were mixed with soft food (e.g., apple sauce, fruit puree) was also conducted to support the relevant labelling statement. Once further stability data were obtained showing that the product remains within specification for 4 years, the shelf life of the product was extended to 4 years upon approval of a variation [35].

In line with the guideline on pharmaceutical development of medicines for pediatric use, oral granules given with liquids (i.e., forming a liquid preparation) or semisolid food can be considered an age-appropriate formulation from birth or for infants above 6 months of age, respectively [43]. Currently, there are five strengths of Kalydeco granules (i.e., 13.4 mg, 25 mg, 50 mg, and 75 mg granules in sachet), which due to their dosing flexibility have extended the target population down to infants aged at least 1 month. The development of oral granules is not only an example of ASD-based pediatric medicinal products but also an example of a multifunctional product developed in line with the sustainability framework for pharmaceutical manufacturing, as formulation design allows for the product to be used by a wider range of patients [44]. Moreover, other authorized products containing ivacaftor and other CFTR modulator therapies (e.g., Okrambi® tablets and oral granules, Symkevi ®tablets) show that developing ASD-based tablets and granules is used as formulation platform for fixed combination medicinal products.

## 12.5 Conclusions

ASDs have evolved to an enabling formulation strategy for poorly water-soluble drugs, with several marketed medicinal products containing ASDs. Their successful development relies on understanding solid-state stability, polymer selection, supersaturation behavior, and control of critical quality attributes, supported by science- and risk-based regulatory frameworks such as the ICH guidelines.

As highlighted by the differences between the SmPCs of conventional dosage forms and ASD-based products, regulatory considerations for ASDs are primarily related to their quality (e.g., stability, specifications, packaging) and their biopharmaceutical performance (e.g., food effect).

The development history of Kalydeco® demonstrates how pharmaceutical development, which focuses on product and process understanding, supports regulatory approval and lifecycle management, including pediatric extensions. Overall, science-based and regulatory-informed pharmaceutical development plays a major role in the quality, safety, and efficacy of medicinal products containing ASDs.

## References


1. Newman A, Knipp G, Zografi G. Assessing the performance of amorphous solid dispersions. J Pharm Sci. 2012;101(4):1355–77.
2. Moseson DE, Tran TB, Karunakaran B, Ambardekar R, Hiew TN. Trends in amorphous solid dispersion drug products approved by the U.S. Food and Drug Administration between 2012 and 2023. Int J Pharm X. 2024;7:100259.
3. Anane-Adjei AB, Jacobs E, Nash SC, Askin S, Soundararajan R, Kyobula M, et al. Amorphous solid dispersions: utilization and challenges in preclinical drug development within AstraZeneca. Int J Pharm. 2022;614:121387.
4. ICH. Mission: Harmonisation for better health 2025. Available from: https://www.ich.org/page/mission
5. Hancock BC, Zografi G. Characteristics and significance of the amorphous state in pharmaceutical systems. J Pharm Sci. 1997;86(1):1–12.
6. Yu D, Li J, Wang H, Pan H, Li T, Bu T, et al. Role of polymers in the physical and chemical stability of amorphous solid dispersion: a case study of carbamazepine. Eur J Pharm Sci. 2022;169:106086.
7. McMahon M, Williams H, Debie E, Fu M, Bujalski R, Qiu F, et al. Utilization of risk-based predictive stability within regulatory submissions; industry's experience. AAPS Open. 2020;6(1):1.
8. ICH. Quality guidelines 2025. Available from: https://www.ich.org/page/quality-guidelines
9. Dedroog S, Adriaensens P, Van den Mooter G. Gaining insight into the role of the solvent during spray drying of amorphous solid dispersions by studying evaporation kinetics. Mol Pharm. 2022;19(5):1604–18.
10. Abraham J. International conference on harmonisation of technical requirements for registration of pharmaceuticals for human use. Handbook of transnational economic governance regimes: Brill Nijhoff; 2010. p. 1041–53.
11. Purohit HS, Zhang GGZ, Gao Y. Detecting crystallinity in amorphous solid dispersions using dissolution testing: considerations on properties of drug substance, drug product, and selection of dissolution media. J Pharm Sci. 2023;112(1):290–303.

12. Moseson DE, Taylor LS. Crystallinity: a complex critical quality attribute of amorphous solid dispersions. Mol Pharm. 2023;20(10):4802–25.
13. Simões A, Veiga F, Vitorino C. Question-based review for pharmaceutical development: an enhanced quality approach. Eur J Pharm Biopharm. 2024;195:114174.
14. Koo J, Jeon H, Cheong J, Joo Y, Han S, Kim H, et al. Modern approaches to quality by design for amorphous solid dispersion product development: a narrative review. J Pharm Investig. 2026;
15. Khan MS, Khan SM, Martins FS, Boddu R, Saini AK, Kollipara S. New era in bioequivalence global harmonization through ICH M13 initiative: critical review on new concepts, alternative approaches for high-risk products. Xenobiotica. 2025;55(6):464–77.
16. Committee S. Directive 2001/82/EC of the European Parliament and of the Council of 6 November 2001 on the Community code relating to veterinary medicinal products. Off J L. 2004;311(28/11):1–66.
17. Breitenbach J. Melt extrusion can bring new benefits to HIV therapy. Am J Drug Deliv. 2006;4(2):61–4.
18. Best BM, Capparelli EV, Diep H, Rossi SS, Farrell MJ, Williams E, et al. Pharmacokinetics of lopinavir/ritonavir crushed versus whole tablets in children. J Acquir Immune Defic Syndr. 2011;58(4)
19. Pappas PG, Spec A, Miceli MH, Proia L, Arauz AB, Hayes J, et al. 144. MSG-15: Pharmacokinetic (PK), Adverse Events (AEs), and Tolerability Data from an Open Label Randomized Clinical Trial (RCT) Comparing Oral Suba-itraconazole (SUBA-ITC) to Conventional Itraconazole (C-ITC) for Treatment of Endemic Mycosis (EM). Open Forum Infect Dis. 2020;7(Supplement_1):S202–S3.
20. (EMC) EMC. Noxafil 40 mg/ml oral suspension 2025. Available from: https://www.medicines.org.uk/emc/product/176/smpc
21. (EMC) EMC. Noxafil 100 mg Gastro-resistant Tablets 2025. Available from: https://www.medicines.org.uk/emc/product/5388/smpc
22. Hens B, Corsetti M, Brouwers J, Augustijns P. Gastrointestinal and systemic monitoring of posaconazole in humans after fasted and fed state administration of a solid dispersion. J Pharm Sci. 2016;105(9):2904–12.
23. Center for Drug Evaluation and Research. Application number: 216099Orig1s000, Clinical pharmacology review(s) 2022. Available from: https://www.accessdata.fda.gov/drugsatfda_docs/nda/2024/216099Orig1s000ClinPharmR.pdf
24. Lennernäs H, Brisander M, Liljebris C, Jesson G, Andersson P. Enhanced bioavailability and reduced variability of dasatinib and sorafenib with a novel amorphous solid dispersion technology platform. Clin Pharmacol Drug Dev. 2024;13(9):985–99.
25. Schittny A, Huwyler J, Puchkov M. Mechanisms of increased bioavailability through amorphous solid dispersions: a review. Drug Deliv. 2020;27(1):110–27.
26. Sugihara M, Takeuchi S, Sugita M, Higaki K, Kataoka M, Yamashita S. Analysis of intra- and Intersubject variability in oral drug absorption in human bioequivalence studies of 113 generic products. Mol Pharm. 2015;12(12):4405–13.
27. Tolsura. Itraconazole 65 mg capsules 2025. Available from: https://tolsura.com/
28. Center for Drug Evaluation and Research. Application number: 208901Orig1s000, Multidiscipline review 2018. Available from: https://www.accessdata.fda.gov/drugsatfda_docs/nda/2018/208901Orig1s000MultidisciplineR.pdf
29. Allen KW. A critical assessment of factors effecting blister pack formation in the pharmaceutical industry. Loughborough University; 2011.
30. Antosik-Rogóż A, Szafraniec-Szczęsny J, Knapik-Kowalczuk J, Kurek M, Gawlak K, Paluch M, et al. How does long-term storage influence the physical stability and dissolution of bicalutamide from solid dispersions and minitablets? Processes [Internet]. 2022; 10(5):[1002 p.].
31. Zhang J, Guo M, Luo M, Cai T. Advances in the development of amorphous solid dispersions: the role of polymeric carriers. Asian J Pharm Sci. 2023;18(4):100834.

32. Démuth B, Nagy ZK, Balogh A, Vigh T, Marosi G, Verreck G, et al. Downstream processing of polymer-based amorphous solid dispersions to generate tablet formulations. Int J Pharm. 2015;486(1–2):268–86.
33. (EMC) EMC. AQUIPTA 60 mg tablets 2025. Available from: https://www.medicines.org.uk/emc/product/15049/smpc#gref
34. (EMC) EMC. Glecaprevir/Pibrentasvir 100 mg/40 mg film-coated tablets 2025. Available from: https://www.medicines.org.uk/emc/product/763/smpc
35. European Medicines Agency. Kalydeco: ivacaftor 2025. Available from: https://www.ema.europa.eu/en/medicines/human/EPAR/kalydeco.
36. Hadida S, Van Goor F, Dinehart K, Looker AR, Mueller P, Grootenhuis PDJ. Chapter twenty-four—case history: Kalydeco® (VX-770, ivacaftor), a CFTR potentiator for the treatment of patients with cystic fibrosis and the G551D-CFTR mutation. In: Desai MC, editor. Annual reports in medicinal chemistry. 49: Academic; 2014. p. 383–98.
37. Daina A, Michielin O, Zoete V. SwissADME: a free web tool to evaluate pharmacokinetics, drug-likeness and medicinal chemistry friendliness of small molecules. Sci Rep. 2017;7(1):42717.
38. Wassvik CM, Holmén AG, Draheim R, Artursson P, Bergström CA. Molecular characteristics for solid-state limited solubility. J Med Chem. 2008;51(10):3035–9.
39. Quodbach J, Preis E, Karkossa F, Winck J, Finke JH, Steiner D. Novel strategies for the formulation of poorly water-soluble drug substances by different physical modification strategies with a focus on peroral applications. Pharmaceuticals [Internet]. 2025; 18(8):[1089 p.].
40. Mahlin D, Bergström CAS. Early drug development predictions of glass-forming ability and physical stability of drugs. Eur J Pharm Sci. 2013;49(2):323–32.
41. Elder D. ICH Q6A Specifications: test procedures and acceptance criteria for new drug substances and new drug products: chemical substances. ICH quality guidelines: an implementation guide. 2017. p. 433–66.
42. (EMC) EMC. Kalydeco 150 mg Film-coated Tablets 2025. Available from: https://www.medicines.org.uk/emc/product/3040/smpc.
43. European Medicines Agency. Guideline on pharmaceutical development of medicines for paediatric use. EMA/CHMP/QWP/805880/2012 Rev 2. 2013.
44. Autzen Virtanen A, Myślińska M, Healy AM, Power E, Madi A, Sivén M. The challenge of downstream processing of spray dried amorphous solid dispersions into minitablets designed for the paediatric population—a sustainable product development approach. Eur J Pharm Sci. 2024;196:106752.

# Chapter 13
# Future Perspectives: Artificial Intelligence, Continuous Manufacturing, and Personalized ASDs

**Ioannis Partheniadis and Ioannis Nikolakakis**

**Abstract** The development of Amorphous Solid Dispersions (ASDs) is undergoing a paradigm shift from empirical "trial-and-error" methods to a systematic, data-driven framework. This transition is anchored in Quality by Design (QbD) principles, utilizing risk assessments and Design of Experiments (DoE) to ensure product stability and performance. Artificial Intelligence (AI) and Machine Learning (ML) are now central to this evolution, providing high-precision predictions for drug–polymer miscibility, solubility enhancement, and physical stability. Furthermore, generative AI models are facilitating the de novo design of functional excipients tailored to specific molecular targets. Manufacturing strategies are simultaneously evolving through the adoption of Continuous Manufacturing (CM), primarily via integrated Hot Melt Extrusion (HME). The incorporation of Process Analytical Technology (PAT), such as NIR and Raman spectroscopy, enables real-time monitoring and closed-loop control to prevent recrystallization. On the frontier of patient-centric care, 3D printing allows for point-of-care production of personalized dosage forms with complex geometries that tune drug release kinetics. While challenges regarding regulatory alignment and long-term physical stability remain, the convergence of material science, digital twin technology, and advanced engineering is positioned to unlock the therapeutic potential of previously "unformulatable" poorly water-soluble compounds by 2030.

**Keywords** Machine learning · In silico modeling · Process Analytical Technology (PAT) · Molecular descriptors · Point of care

I. Partheniadis (✉)
Department of Pharmacy, Faculty of Health and Medical Sciences, University of Copenhagen, Copenhagen, Denmark
e-mail: ioannis.partheniadis@sund.ku.dk

I. Nikolakakis
Laboratory of Pharmaceutical Technology, School of Pharmacy, Faculty of Health Sciences, Aristotle University of Thessaloniki, Thessaloniki, Greece
e-mail: yannikos@pharm.auth.gr

 
I. Nikolakakis, I. Partheniadis (eds.), *Amorphous Solid Dispersions*, AAPS Introductions in the Pharmaceutical Sciences 5,
https://doi.org/10.1007/978-3-032-24883-1_13

## 13.1 Introduction

The development of amorphous solid dispersions (ASDs) has transitioned from historical, empirical "trial-and-error" approaches to a systematic, data-driven paradigm [1]. This shift is primarily driven by Quality by Design (QbD), which emphasizes the proactive integration of quality during the design phase rather than a reliance on retrospective end-product testing [2, 3]. QbD frameworks utilize risk assessments and Design of Experiments (DoE) to establish multidimensional design spaces. These spaces allow for the prediction of product stability and performance, thereby reducing the dependency on extensive experimental cycles and minimizing the risk of late-stage failures [4, 5].

Artificial Intelligence (AI) and Machine Learning (ML) are further transforming ASD formulation by predicting critical quality attributes with high precision. ML algorithms are increasingly deployed to forecast drug–polymer miscibility, solubility enhancement, and long-term physical stability, effectively streamlining the formulation selection process [6–8]. Furthermore, advanced computational techniques, such as Generative Adversarial Networks (GANs), enable generative approaches to identify optimal drug–excipient combinations and expedite the discovery of novel stabilization strategies [9].

The manufacturing of ASDs is similarly evolving through a transition from traditional batch processing to continuous manufacturing (CM) [10–12]. This shift is notably facilitated by integrated hot melt extrusion (HME) techniques, which offer a solvent-free, highly automated processing route. Within these continuous lines, Process Analytical Technology (PAT) (e.g., via the utilization of near-infrared (NIR) and Raman spectroscopy) enables real-time monitoring of the solid state [13–16]. The integration of PAT with automated algorithms facilitates closed-loop control, allowing for the dynamic adjustment of critical process parameters to ensure consistent product quality [10, 11].

Personalized medicine is increasingly viable through the integration of 3D printing technologies, such as fused deposition modeling (FDM), into ASD production [1, 17, 18]. These techniques enable the point-of-care manufacture of individualized dosage forms with customized strengths, release kinetics, and complex architectures tailored to specific patient requirements [1]. By leveraging the synergy between AI and 3D printing, developers can rapidly optimize therapies for pediatric, geriatric, and other niche patient populations.

Despite these advances, substantial hurdles remain regarding clinical translation, the "black box" nature of complex AI models, and global regulatory misalignment. Digital twin technologies, that is, computational replicas of manufacturing processes or patient physiological profiles, are being explored to facilitate real-time predictive monitoring and overcome these barriers [19]. Ultimately, the convergence of material science, advanced computation, and continuous engineering is positioned to address the challenges of poorly water-soluble drugs, expanding the boundaries of pharmaceutical innovation to render previously "unformulatable" compounds accessible.

## 13.2 Artificial Intelligence and Machine Learning in ASD Design

Artificial intelligence (AI) has transitioned from an emerging concept into a foundational tool for molecular modeling and the design of ASDs. Traditionally, the optimization of ASDs to enhance the bioavailability of poorly water-soluble drugs relied on iterative, empirical experimental approaches. Currently, machine learning (ML) frameworks provide theoretical guidance and predictive capabilities that streamline formulation design and reduce the experimental burden associated with traditional development cycles [6].

### *13.2.1 Predicting Miscibility*

Deep learning is transforming the prediction of miscibility between novel active pharmaceutical ingredients (APIs) and polymers by estimating the Flory–Huggins interaction parameter ($\chi$) [20–22]. Historically, determining $\chi$ to understand phase equilibria and thermodynamics required complex and resource-intensive techniques, such as differential scanning calorimetry (DSC) or vapor pressure osmometry [23].

Advanced neural network architectures can now encode the compositional, structural, and physicochemical features of components into high-dimensional latent spaces to predict temperature-dependent $\chi$ values [24]. This data-driven approach facilitates the rapid calculation of the Gibbs free energy of mixing ($\Delta G_{mix}$) and the identification of phase separation boundaries (binodal and spinodal curves) for specific drug–polymer pairs without exhaustive thermal analysis.

### *13.2.2 Solubility Enhancement Modeling*

To effectively enhance the bioavailability of poorly soluble compounds, ASDs must facilitate rapid initial dissolution followed by sustained supersaturation (a phenomenon termed as "spring and parachute" behavior). By training ML algorithms on high-throughput datasets encompassing molecular descriptors, polymer properties, and environmental variables, researchers can predict these complex dissolution kinetics [25]. Computational platforms utilizing random forests or gradient boosting machines can categorize and predict dissolution profiles, distinguishing between transient supersaturation and stable "parachute" states. This enables formulation scientists to anticipate the performance of drug–polymer combinations in biorelevant media, reducing the reliance on extensive in vitro dissolution testing [25].

### 13.2.3 Generative AI for Excipient Design

Beyond predictive modeling, generative AI offers pathways for the de novo design of excipients tailored to the functional groups of BCS Class II or IV drugs. Deep generative models, including recurrent neural networks (RNNs), Transformers, and variational autoencoders (VAEs), capture the underlying probability distributions of chemical space to propose novel molecular graphs [26]. By coupling these networks with reinforcement learning (RL), molecular generation can be guided by objective functions targeting specific properties, such as glass transition temperature ($T_g$) or hydrogen-bonding potential [27, 28]. This allows for the computational assembly of custom polymeric carriers that provide optimal thermodynamic stabilization for specific target APIs.

## 13.3 Transitioning to Continuous Manufacturing

The shift from traditional batch processing to continuous manufacturing (CM) represents a significant advancement in the production of ASDs. Unlike batch manufacturing, CM enables the integration of the entire production cycle into a single and uninterrupted flow, eliminating the logistical delays associated with intermediate hold times [29]. Hot melt extrusion (HME) is particularly suited for this paradigm as it is inherently a continuous and solvent-free process [30–32]. In an integrated HME setup, the extruder is coupled directly with downstream unit operations, such as milling, continuous blending, and tableting [29, 33]. Linking these operations into a single sequence reduces material handling and eliminates intermediate storage, which accelerates process development and minimizes the risks of environmental degradation or human error between manufacturing stages.

To ensure rigorous quality control within a continuous framework, Process Analytical Technology (PAT) is integrated directly into the manufacturing line. PAT utilizes sensors in the line and chemometric models to monitor critical quality attributes in real time, bypassing the need for destructive sampling off the line [34, 35]. Spectroscopy using Raman and Near-Infrared (NIR) light sources is frequently implemented within HME lines to monitor the physical state of the formulation and verify that the API remains in the desired amorphous form [34, 36, 37]. Given that the amorphous state is thermodynamically metastable, continuous monitoring is essential to prevent unintended recrystallization.

The integration of PAT enables an automated and closed loop control framework. If spectroscopic sensors detect crystalline signatures, which indicates incomplete amorphization or localized recrystallization, the system can immediately adjust critical process parameters. For instance, the control software can automatically modulate the extruder barrel temperature or the rotation speed of the twin screws to optimize thermal and shear forces. This adaptability in real time ensures that the final product consistently meets the predefined quality specifications for amorphization and stability.

## 13.4 Point-of-Care Manufacturing and Personalized ASDs

The transition from conventional to precision medicine aims to adapt treatments and dosages to each individual based on specific biomarkers, genomics, and metabolic profiles [38, 39]. Traditionally, pharmaceutical laboratories and industrial facilities have lacked the adaptability required to manufacture-tailored solid oral dosage forms [40]. Additive manufacturing, particularly 3D printing, disrupts this paradigm by enabling manufacturing at the point of care [41, 42]. In the near future, algorithms interpreting genotypic and clinical information could direct a 3D printer to create a specific and tailored dosage on demand. Because 3D printers have a small footprint and are cost efficient for small-scale production, they are suited for a decentralized network [43, 44]. This allows local hospital pharmacies or clinical sites to print and dispense personalized treatments directly to the patient.

### *13.4.1 Fused Deposition Modeling and Amorphous Solid Dispersions*

Among 3D printing technologies, fused deposition modeling (FDM) is the most investigated for pharmaceutical solid oral forms due to its simplicity and low cost [45, 46]. The most prominent method to utilize this is Hot Melt Extrusion coupled with FDM, where an active pharmaceutical ingredient and polymer excipients are mixed and extruded into a drug-loaded filament. This extrudate then serves as the feedstock for the FDM printer, where it is heated and deposited layer by layer to construct the tablet [47].

Crucially, this technique facilitates the preparation of ASDs. Because the drug is subjected to two distinct thermal cycles (a first cycle during the extrusion process and another one when passing through the printer nozzle), the dual heating effectively promotes the amorphization of the drug [40, 48]. Formulating the drug in an amorphous state rather than a crystalline form significantly enhances the dissolution kinetics and solubility of poorly water-soluble compounds.

### 13.4.2 Complex Geometries for Controlled Drug Release

One of the primary advantages of FDM is the ability to print complex geometric structures without the need for traditional molds. Because drug release kinetics are tied to the ratio of surface area to volume, modifying the digital design of the tablet allows for precise control over the dissolution rate. To achieve faster release, filaments can be printed into highly porous or channeled shapes. Research highlights the creation of slotted grids, internal scaffolding, and architectures with precise gaps between plates to accelerate water penetration and drug release [49, 50]. Furthermore, internal geometry can be manipulated using the printer infill settings [51]. Patterns such as hexagonal or linear grids dictate the internal porosity and erosion mechanisms of the tablet [52]. This provides a customizable tool to tune the drug release rate according to the specific metabolic requirements of the patient.

## 13.5 Conclusions and Outlook to 2030

The trajectory of amorphous solid dispersions is no longer defined solely by basic solubility enhancement but by a sophisticated convergence of material science, predictive computation, and advanced engineering. As the pharmaceutical industry looks toward 2030, the transition from empirical experimentation to a digitalized and systematic paradigm is becoming the standard. Research gaps in this field highlight several critical areas that must be addressed to fully realize this potential. Despite significant advances in formulation strategies and manufacturing techniques such as hot melt extrusion and spray drying, challenges remain in ensuring long-term physical stability and maintaining supersaturation during dissolution. These factors are crucial for achieving consistent bioavailability and ensuring that therapeutic outcomes remain predictable across diverse patient populations.

A deeper mechanistic understanding of drug polymer miscibility, molecular mobility, and phase separation phenomena is essential to better predict performance and prevent recrystallization. While machine learning– and artificial intelligence–based predictive models show promise in accelerating formulation design, their integration with experimental validation requires further development to improve accuracy and generalizability across different chemical spaces. Furthermore, novel polymeric carriers and excipients tailored for specific drugs remain underexplored. The regulatory pathways for such new materials are still evolving, which currently poses hurdles for clinical translation and the adoption of innovative stabilization strategies.

By 2030, the wider adoption of advanced characterization methods will be necessary to elucidate the microstructural features that influence stability and dissolution behavior. Addressing these existing gaps through interdisciplinary collaboration will be essential to develop robust, scalable, and personalized formulations. This requires a unified approach combining computational modeling, material science, process engineering, and regulatory science. Ultimately, the integration of these fields into a seamless digital thread will ensure that the most challenging and previously "unformulatable" compounds are successfully transformed into accessible and effective treatments for the global market.

## References


1. Yang S, Hu X, Zhu J, Zheng B, Bi W, Wang X, et al. Aspects and implementation of pharmaceutical quality by design from conceptual frameworks to industrial applications. Pharmaceutics. 2025;17(5):623.
2. Patel J, Dhingani A, Garala K, Raval M, Sheth N. Quality by design approach for oral bioavailability enhancement of Irbesartan by self-nanoemulsifying tablets. Drug Deliv. 2014;21(6):412–35.
3. Rathore AS, Winkle H. Quality by design for biopharmaceuticals. Nat Biotechnol. 2009;27(1):26–34.
4. Food U, Administration D. Guidance for industry: process validation: general principles and practices. Rockville: Center for Biologics Evaluation and Research US Department of Health and Human Services; 2011.
5. Yu LX. Pharmaceutical quality by design: product and process development, understanding, and control. Pharm Res. 2008;25(4):781–91.
6. Jiang J, Lu A, Ma X, Ouyang D, Williams RO. The applications of machine learning to predict the forming of chemically stable amorphous solid dispersions prepared by hot-melt extrusion. Int J Pharm: X. 2023;5:100164.
7. Mendyk A, Szlęk J, Jachowicz R. ME_expert 2.0: a heuristic decision support system for microemulsions formulation development. In: Formulation tools for pharmaceutical development. Elsevier; 2013. p. 39–71.
8. Zhang Z-h, Pan W-s. Expert system for the development and formulation of push–pull osmotic pump tablets containing poorly water-soluble drugs. In: Formulation tools for pharmaceutical development. Elsevier; 2013. p. 73–108.
9. Chen W, Liu X, Zhang S, Chen S. Artificial intelligence for drug discovery: resources, methods, and applications. Mol Ther Nucleic Acids. 2023;31:691–702.
10. Pestieau A, Krier F, Thoorens G, Dupont A, Chavez PF, Ziemons E, et al. Towards a real time release approach for manufacturing tablets using NIR spectroscopy. J Pharm Biomed Anal. 2014;98:60–7.
11. Portier C, Vigh T, Di Pretoro G, Leys J, Klingeleers D, De Beer T, et al. Continuous twin screw granulation: impact of microcrystalline cellulose batch-to-batch variability during granulation and drying—a QbD approach. Int J Pharm: X. 2021;3:100077.
12. Thiry J, Lebrun P, Vinassa C, Adam M, Netchacovitch L, Ziemons E, et al. Continuous production of itraconazole-based solid dispersions by hot melt extrusion: preformulation, optimization and design space determination. Int J Pharm. 2016;515(1–2):114–24.
13. Huang J, Kaul G, Utz J, Hernandez P, Wong V, Bradley D, et al. A PAT approach to improve process understanding of high shear wet granulation through in-line particle measurement using FBRM C35. J Pharm Sci. 2010;99(7):3205–12.
14. Nishal S, Kumar V, Phogat P, Garg V, Dureja H. Selection of design of PAT. In: Introduction to quality by design (QbD) from theory to practice. Springer; 2024. p. 59–89.
15. Roggo Y, Chalus P, Maurer L, Lema-Martinez C, Edmond A, Jent N. A review of near infrared spectroscopy and chemometrics in pharmaceutical technologies. J Pharm Biomed Anal. 2007;44(3):683–700.
16. Wahlich J. Review: continuous manufacturing of small molecule solid oral dosage forms. Pharmaceutics. 2021;13(8):1311.

17. Cornilă A, Iurian S, Tomuţă I, Porfire A. Orally dispersible dosage forms for paediatric use: current knowledge and development of nanostructure-based formulations. Pharmaceutics. 2022;14(8):1621.
18. Ruzycki CA, Tavernini S, Martin AR, Finlay WH. Characterization of dry powder inhaler performance through experimental methods. Adv Drug Deliv Rev. 2022;189:114518.
19. Chen Y, Yang O, Sampat C, Bhalode P, Ramachandran R, Ierapetritou M. Digital twins in pharmaceutical and biopharmaceutical manufacturing: a literature review. Processes. 2020;8(9):1088.
20. Liao H-C, Lin Y-H, Peng C-H, Li Y-P. Directed message passing neural networks for accurate prediction of polymer–solvent interaction parameters. ACS Eng Au. 2025;5(5):530–9.
21. Obaidullah AJ, Mahdi WA. Machine learning-based analysis on pharmaceutical compounds interaction with polymer to estimate drug solubility in formulations. Sci Rep. 2025;15(1):23683.
22. Walden DM, Bundey Y, Jagarapu A, Antontsev V, Chakravarty K, Varshney J. Molecular simulation and statistical learning methods toward predicting drug-polymer amorphous solid dispersion miscibility, stability, and formulation design. Molecules. 2021;26(1):182.
23. Lee JK, Yao SX, Li G, Jun MBG, Lee PC. Measurement methods for solubility and diffusivity of gases and supercritical fluids in polymers and its applications. Polym Rev. 2017;57(4):695–747.
24. Aoki Y, Wu S, Tsurimoto T, Hayashi Y, Minami S, Tadamichi O, et al. Multitask machine learning to predict polymer–solvent miscibility using Flory–Huggins interaction parameters. Macromolecules. 2023;56(14):5446–56.
25. Ge K, Shen J, Chen H, Wang J, Cui Y, Shen C, et al. Advancing amorphous solid dispersions design: insights into dissolution kinetics via thermodynamic descriptor and machine learning. Mol Pharm. 2025;22(7):4259–68.
26. Loeffler HH, He J, Tibo A, Janet JP, Voronov A, Mervin LH, et al. Reinvent 4: modern AI–driven generative molecule design. J Chem. 2024;16(1):20.
27. Elbadawi M, Gaisford S, Basit AW. Advanced machine-learning techniques in drug discovery. Drug Discov Today. 2021;26(3):769–77.
28. Meyers J, Fabian B, Brown N. De novo molecular design and generative models. Drug Discov Today. 2021;26(11):2707–15.
29. Nambiar AG, Singh M, Mali AR, Serrano DR, Kumar R, Healy AM, et al. Continuous manufacturing and molecular modeling of pharmaceutical amorphous solid dispersions. AAPS PharmSciTech. 2022;23(7):249.
30. Mishra SM, Richter M, Mejia L, Sauer A. Downstream processing of itraconazole:HPMCAS amorphous solid dispersion: from hot-melt extrudate to tablet using a quality by design approach. Pharmaceutics. 2022;14(7):1429.
31. Simões MF, Pinto RMA, Simões S. Hot-melt extrusion in the pharmaceutical industry: toward filing a new drug application. Drug Discov Today. 2019;24(9):1749–68.
32. Stanković M, Frijlink HW, Hinrichs WL. Polymeric formulations for drug release prepared by hot melt extrusion: application and characterization. Drug Discov Today. 2015;20(7):812–23.
33. Kshirsagar MM, Chatale BC, Dyawanapelly S, Vora LK, Amin PD. Continuous processing strategies for amorphous solid dispersions of itraconazole: impact of polymer selection and manufacturing techniques. Pharmaceutics. 2025;17(9):1090.
34. Islam MT, Scoutaris N, Maniruzzaman M, Moradiya HG, Halsey SA, Bradley MS, et al. Implementation of transmission NIR as a PAT tool for monitoring drug transformation during HME processing. Eur J Pharm Biopharm. 2015;96:106–16.
35. Saerens L, Vervaet C, Remon JP, De Beer T. Process monitoring and visualization solutions for hot-melt extrusion: a review. J Pharm Pharmacol. 2014;66(2):180–203.
36. Fazekas B, Péterfi O, Galata DL, Nagy ZK, Hirsch E. Process analytical technology based quality assurance of API concentration and fiber diameter of electrospun amorphous solid dispersions. Eur J Pharm Biopharm. 2024;204:114529.
37. Hattori Y. Raman and near-infrared spectroscopy for in-line sensors. Anal Sci. 2022;38(12):1455–6.

38. Vaithinathan AG, Asokan V. Public health and precision medicine share a goal. J Evid Based Med. 2017;10(2):76–80.
39. Vicini P, Fields O, Lai E, Litwack ED, Martin AM, Morgan TM, et al. Precision medicine in the age of big data: the present and future role of large-scale unbiased sequencing in drug discovery and development. Clin Pharmacol Ther. 2016;99(2):198–207.
40. Cailleaux S, Sanchez-Ballester NM, Gueche YA, Bataille B, Soulairol I. Fused Deposition Modeling (FDM), the new asset for the production of tailored medicines. J Control Release. 2021;330:821–41.
41. Chan HN, Shu Y, Xiong B, Chen Y, Chen Y, Tian Q, et al. Simple, cost-effective 3D printed microfluidic components for disposable, point-of-care colorimetric analysis. Acs Sensors. 2016;1(3):227–34.
42. Lyousoufi M, Lafeber I, Kweekel D, de Winter BC, Swen JJ, Le Brun PP, et al. Development and bioequivalence of 3D-printed medication at the point-of-care: bridging the gap toward personalized medicine. Clin Pharmacol Ther. 2023;113(5):1125–31.
43. Fuenmayor E, O'Donnell C, Gately N, Doran P, Devine DM, Lyons JG, et al. Mass-customization of oral tablets via the combination of 3D printing and injection molding. Int J Pharm. 2019;569:118611.
44. Trenfield SJ, Awad A, Goyanes A, Gaisford S, Basit AW. 3D printing pharmaceuticals: drug development to frontline care. Trends Pharmacol Sci. 2018;39(5):440–51.
45. Norman J, Madurawe RD, Moore CM, Khan MA, Khairuzzaman A. A new chapter in pharmaceutical manufacturing: 3D-printed drug products. Adv Drug Deliv Rev. 2017;108:39–50.
46. Vithani K, Goyanes A, Jannin V, Basit AW, Gaisford S, Boyd BJ. An overview of 3D printing technologies for soft materials and potential opportunities for lipid-based drug delivery systems. Pharm Res. 2018;36(1):4.
47. Awad A, Trenfield SJ, Gaisford S, Basit AW. 3D printed medicines: a new branch of digital healthcare. Int J Pharm. 2018;548(1):586–96.
48. Thiry J, Krier F, Evrard B. A review of pharmaceutical extrusion: critical process parameters and scaling-up. Int J Pharm. 2015;479(1):227–40.
49. Goyanes A, Robles Martinez P, Buanz A, Basit AW, Gaisford S. Effect of geometry on drug release from 3D printed tablets. Int J Pharm. 2015;494(2):657–63.
50. Tagami T, Nagata N, Hayashi N, Ogawa E, Fukushige K, Sakai N, et al. Defined drug release from 3D-printed composite tablets consisting of drug-loaded polyvinylalcohol and a water-soluble or water-insoluble polymer filler. Int J Pharm. 2018;543(1–2):361–7.
51. Fuenmayor E, Forde M, Healy AV, Devine DM, Lyons JG, McConville C, et al. Comparison of fused-filament fabrication to direct compression and injection molding in the manufacture of oral tablets. Int J Pharm. 2019;558:328–40.
52. Palekar S, Nukala PK, Mishra SM, Kipping T, Patel K. Application of 3D printing technology and quality by design approach for development of age-appropriate pediatric formulation of baclofen. Int J Pharm. 2019;556:106–16.

# Index




I. Nikolakakis, I. Partheniadis (eds.), *Amorphous Solid Dispersions*, AAPS Introductions in the Pharmaceutical Sciences 5,
https://doi.org/10.1007/978-3-032-24883-1

MIX
Papier aus verantwortungsvollen Quellen
Paper from responsible sources
FSC® C105338

If you have any concerns about our products,
you can contact us on
**ProductSafety@springernature.com**

In case Publisher is established outside the EU,
the EU authorized representative is:
**Springer Nature Customer Service Center GmbH**
**Europaplatz 3, 69115 Heidelberg, Germany**

Printed by Libri Plureos GmbH
in Hamburg, Germany